Reagents for Organic Synthesis

Fieser and Fieser's

Reagents for Organic Synthesis

VOLUME TEN

Mary Fieser
  Harvard University

1807 **WJ** 1982

A Wiley-Interscience Publication
JOHN WILEY & SONS
New York · Chichester · Brisbane ·
Toronto · Singapore

Library of Congress Cataloging in Publication Data:

Library of Congress Catalog Card Number: 66-27894

ISBN 0-471-86636-9

Printed in the United States of America

10 9 8 7 6 5 4 3 2 1

PREFACE

This volume of reagents includes references to papers published during 1980 and the first half of 1981. I am indebted to many colleagues for advice and help. Alan E. Barton and Paul B. Hopkins have again read the manuscript. I am especially grateful for invaluable proofreading by Dr. James V. Heck, Dr. Ving Lee, Professor Bruce Lipshutz, Professor Dale Boger, Dr. Michael McMillan, Daniel Brenner, Jeffrey C. Hayes, Jay W. Ponder, John E. Munroe, William P. Roberts, Steven Freilich, Thomas Eckrich, Robert E. Wolf, Jr., Marco Pagnotta, Marifaith Hackett, William McWhorter, Judith Hartman, William L. Seibel, Andrew G. Myers, John C. Schmidhauser, Daniel F. Lieberman, Charles Manly, Donald Wolanin, J. Jeffrey Howbert, and, especially, Katharine Brighty, who read almost the entire galley proof. The proofreading was coordinated by Peter T. Lansbury, Jr. The picture of the group was taken by Alan Barton.

MARY FIESER

Cambridge, Mass.
April, 1982

CONTENTS

A

Acetyl cyanide, CH_3COCN. Mol. wt. 69.16, b.p. 90–92°.

Acyl cyanides can be prepared by reaction of acid chlorides with cyanotrimethyl-silane at 60–70°[1] or with copper cyanide.[2]

Monocyanohydrins of β-diketones.[3] In the presence of $TiCl_4$, acetyl cyanide reacts with enol silyl ethers of ketones at −78° to afford monocyanohydrins of β-diketones in excellent yield. The corresponding reaction with enol silyl ethers of aldehydes proceeds in only about 35% yield. A low temperature is essential for this reaction. A similar reaction is possible with allyltrimethylsilane.

Examples:

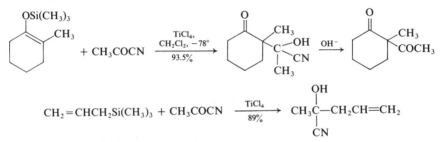

[1] K. Herrman and G. Simchen, *Synthesis*, 204 (1979).
[2] J. F. Normant and C. Piechucki, *Bull. Soc.*, 2402 (1972).
[3] G. A. Kraus and M. Shimagaki, *Tetrahedron Letters*, **22**, 1171 (1981).

Acetyl hypofluorite (1). Mol. wt. 78.04.

Preparation in $CH_3COOH–CFCl_3$:

$$NaX + CH_3COOH \longrightarrow HX + CH_3COONa \xrightarrow[50-80\%]{F_2} CH_3COOF + NaF$$

$(X = F, CH_3CO_2, CF_3CO_2)$ **1**

Fluorination.[1] The reagent reacts with *trans-* and *cis*-stilbene, preferentially by *syn*-addition. Only the *syn*-adducts are isolable in the reaction of **1** with ethyl *trans-* and *cis*-cinnamate (equation I).

$$(I)\ C_6H_5CH=CHC_6H_5 \xrightarrow{\ \mathbf{1}\ } \underset{F}{\overset{OAc}{C_6H_5CH-CHC_6H_5}}$$

	threo	erythro
trans	45%	7%
cis	11%	51%

[1] S. Rozen, O. Lerman, and M. Kol, *J.C.S. Chem. Comm.*, 443 (1981).

Acrolein, $CH_2=CHCHO$, **2**, 15; **3**, 5.

    Hetero-Diels-Alder reaction.[1] Acrolein undergoes a Diels-Alder reaction with furanoid and pyranoid exocyclic vinyl ethers at 25° (4–6 days). The spiroketal (**2**) obtained can be oxidized in CH_3OH by *m*-chloroperbenzoic acid to the ring

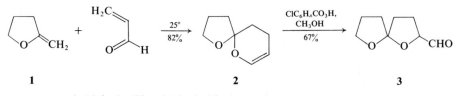

    1   **2**   **3**

contracted aldehyde. The aldehyde (**3**) obtained in this way was used to synthesize racemic chalcogran (**4**), an aggregation pheromone of a beetle. Some polyether antibiotics contain similar subunits.

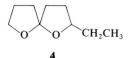

4

[1] R. E. Ireland and D. Habich, *Tetrahedron Letters*, **21**, 1389 (1980).

Alkyl diphenyl phosphonates, $(C_6H_5)_2\overset{\text{O}}{\overset{\|}{P}}CH_2R$.

    (E)- and (Z)-Alkenes. Both (Z)- and (E)-alkenes can be prepared stereoselectively by the Wittig-Horner reaction formulated in Scheme (I).[1] Thus it is possible to prepare the pure diastereoisomers of the alcohols **1** by chromatography and crystallization. The elimination of $(C_6H_5)_2\overset{\text{O}}{\overset{\|}{P}}ONa$ from either alcohol is stereospecific when both R groups are aliphatic.

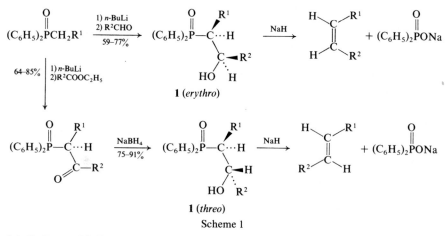

Scheme 1

[1] A. D. Buss and S. Warren, *J.C.S. Chem. Comm.*, 100 (1981).

Alkyllithium reagents, 9, 5–8.

Simplified preparation.[1] Alkyl- and aryllithium reagents can be prepared readily and in high yield by reaction of alkyl or aryl bromides with lithium containing 2% sodium by irradiation for 10–30 minutes in an ultrasound laboratory cleaner. No initiation is necessary and the solvent (THF) can be technical grade, although yields are higher in the absence of water. Reagents prepared in this way can be used for addition to carbonyl compounds (Barbier reaction[2]) in 60–90% isolated yields.

Determination of concentration. Two laboratories[3] recommend 1,3-diphenyl-2-propanone tosylhydrazone (m.p. 185–186° dec.) for determination of the concentration of RLi in THF solutions. The dianion of the tosylhydrazone has an intense orange color.

β-Amino alcohols.[4] Alkyllithium reagents undergo double addition to α-trimethylsilyloxy nitriles to form β-trimethylsilyloxy amines. These products are hydrolyzed by aqueous acetic acid to β-amino alcohols (equation I).

$$\text{(I)} \quad R^1\!-\!\underset{\underset{\displaystyle OSi(CH_3)_3}{|}}{\overset{\overset{\displaystyle R^2}{|}}{C}}\!-\!C\!\equiv\!N \quad \xrightarrow[\substack{2)\,H_2O \\ 75-90\%}]{1)\,2R^3Li} \quad R^1\!-\!\underset{\underset{\displaystyle (CH_3)_3SiO}{|}}{\overset{\overset{\displaystyle R^2}{|}}{C}}\!-\!\underset{\underset{\displaystyle R^3}{|}}{\overset{\overset{\displaystyle R^3}{|}}{C}}\!-\!NH_2 \quad \xrightarrow[75-90\%]{\substack{HOAc, \\ H_2O}} \quad R^1\!-\!\underset{\underset{\displaystyle HO}{|}}{\overset{\overset{\displaystyle R^2}{|}}{C}}\!-\!\underset{\underset{\displaystyle R^3}{|}}{\overset{\overset{\displaystyle R^3}{|}}{C}}\!-\!NH_2$$

Reaction with allene sulfoxides.[5] Allenic sulfoxides are desulfurized by reaction with methyllithium. An example is the transformation shown in equation (I). The propynylcarbinol **1** is treated with benzenesulfenyl chloride to form the allene **2**,

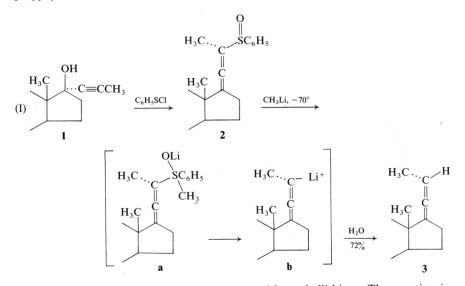

which undergoes cleavage to **3** on treatment with methyllithium. The reaction is stereospecific.

The same paper mentions a benzilic type rearrangement of a sulfoxide induced by methyllithium (equation II).

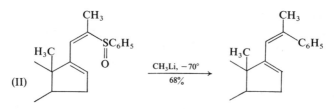

Reaction with p-*benzoquinones*.[6] Fischer and Henderson have found conditions that result in 1,2-addition of various RLi reagents to *p*-benzoquinone. The reaction when conducted in ether at $-78°$ results in 4-alkyl-4-hydroxycyclohexa-2,5-diene-1-ones. These products undergo further 1,2-addition with RLi in THF to give dialkyl-cyclohexa-2,5-diene-1,4-diols. Mixed dienediols can be obtained by use of two different RLi reagents.

    Example:

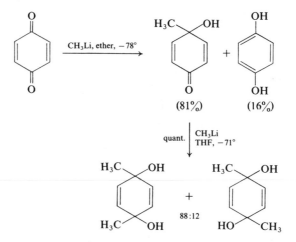

Reaction with isoprene epoxide.[7] Alkyllithium reagents undergo 1,4-addition to isoprene epoxide to give predominately (Z)-allylic alcohols, particularly in the presence of a base (equation I). The reaction was used to prepare α-santalol (**1**) from π-bromotricyclene.

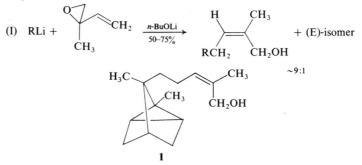

Phenyllithium reacts with the epoxide by 1,2- and 1,4-addition, with no stereoselectivity.

[1] J.-L. Luche and J.-C. Damiano, *Am. Soc.*, **102**, 7926 (1980).
[2] C. Blomberg and F. A. Hartog, *Synthesis*, 18 (1977).
[3] M. F. Lipton, C. M. Sorenson, A. C. Sadler, and R. H. Shapiro, *J. Organometal. Chem.*, **186**, 155 (1980).
[4] R. Amouroux and G. P. Axiotis, *Synthesis*, 270 (1981).
[5] G. Neef, U. Eder, and A. Seeger, *Tetrahedron Letters*, **21**, 903 (1980).
[6] A. Fischer and G. N. Henderson, *Tetrahedron Letters*, **21**, 701 (1980).
[7] M. Tamura and G. Suzukamo, *Tetrahedron Letters*, **22**, 577 (1981).

O-Allylhydroxylamine (1). Mol. wt. 73.10.

Preparation:

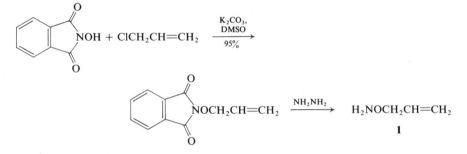

2,3-Disubstituted pyridines.[1] A typical reaction is the synthesis of 5,6,7,8-tetra-hydroquinoline (equation I). If the thermolysis is carried out in the absence of oxygen, the yield of the pyridine is 3%.

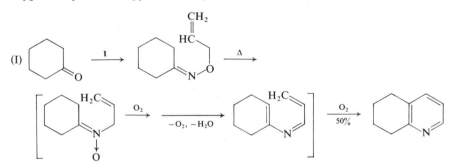

[1] H. Irie, I. Katayama, and Y. Mizuno, *Heterocycles*, **12**, 771 (1979).

$$O$$

Allyltetramethylphosphorodiamidate, $[(CH_3)_2N]_2POCH_2CH=CH_2$ **(1).** Mol. wt. 192.21, b.p. 57°/0.01 mm. The reagent is prepared by reaction of allyl alcohol and tetramethylphosphorodiamidic chloride in the presence of benzyltriethylammonium chloride in CH_2Cl_2 and aqueous NaOH (90% yield).

Adrenolactone (3).[1] A new synthesis of the lactone involves reaction of the tetrahydropyranyl derivative of 3β-hydroxy-17-ketoandrostene (2) with the dilithio derivative of 1.

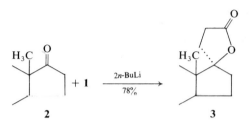

2   **3**

[1] G. Sturtz and J.-J. Yaouane, *Synthesis*, 289 (1980).

Allyltin difluoroiodide, CH_2=$CHCH_2SnIF_2$. The reagent is prepared *in situ* from allyl iodide and SnF_2 (dried).

Homoallylic alcohols.[1] The reagent reacts with aldehydes and ketones to form derivatives of homoallylic alcohols in satisfactory yield (equation I). 1,3-Dimethyl-2-imidazolidinone is the most satisfactory solvent.

(I) $$CH_2{=}CHCH_2I + R^1COR^2 \xrightarrow[70\text{–}95\%]{SnF_2} CH_2{=}CHCH_2\overset{\displaystyle R^1}{\underset{\displaystyle R^2}{\overset{|}{\underset{|}{C}}}}{-}OH$$

[1] T. Mukaiyama, T. Harada, and S. Shoda, *Chem. Letters*, 1507 (1980).

Allyltrimethylsilane, $(CH_3)_3SiCH_2CH{=}CH_2$ (1). Mol. wt. 114.27, b.p. 84–88°. Suppliers: Aldrich, Petrarch Systems.

Acetonylation.[1] In the presence of $TiCl_4$ or BF_3 etherate allyltrimethylsilane undergoes conjugate addition to α,β-enones. The δ,ε-enones formed are converted by ozonation or $KMnO_4$–KIO_4 to δ-keto aldehydes. Wacker oxidation can be used to obtain methyl ketones.

Examples:

$$(CH_3)_3SiCH_2CH{=}CH_2 + C_6H_5CH{=}CHCOCH_3 \xrightarrow[95\%]{\substack{TiCl_4, \\ CH_2Cl_2, -30°}}$$

$$\underset{\displaystyle CH_2CH{=}CH_2}{\overset{\displaystyle |}{C_6H_5CHCH_2COCH_3}} \xrightarrow[72\%]{O_3, CH_2Cl_2, -78°} \underset{\displaystyle CH_2CHO}{\overset{\displaystyle |}{C_6H_5CHCH_2COCH_3}}$$

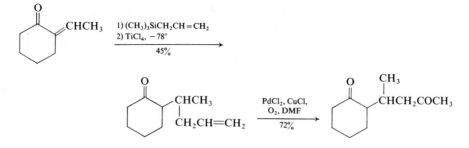

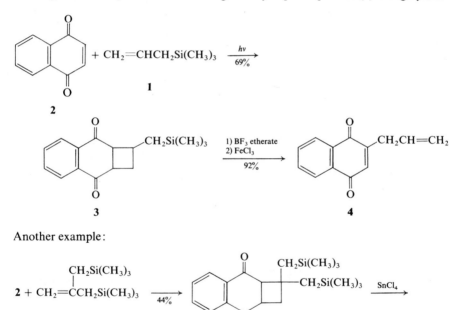

(2-Methyl-2-propenyl)trimethylsilane, $(CH_3)_3SiCH_2C(CH_3){=}CH_2$, can be used in this sequence as a route to 1,5-diketones.

Allylnaphthoquinones.[2] Allyltrimethylsilane (**1**) undergoes irradiation-assisted cycloaddition to 1,4-napthoquinone (**2**) to give a mixture of stereoisomers (**3**). Cleavage and subsequent oxidation of **3** gives allylnaphthoquinone (**4**) in high yield.

Another example:

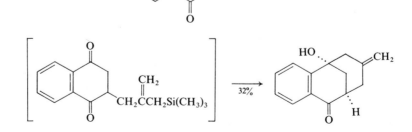

Trimethylsilyl ethers and esters.[3] The reaction of alcohols and allyltrimethyl-silane in acetonitrile with TsOH as catalyst (70–80°, 1–3 hours) results in trimethyl-silyl ethers in 85–95% yield with elimination of propene. The same reaction with carboxylic acids results in trimethylsilyl esters. Phenols do not undergo this reaction.

t-Butyldimethylsilyl ethers can be prepared in the same way from allyl-*t*-butyl-dimethylsilane (b.p. 65–66°/47 mm).

[1] A. Hosomi, H. Kobayashi, and H. Sakurai, *Tetrahedron Letters*, **21**, 955 (1980).
[2] M. Ochiai, M. Arimoto, and E. Fujita, *J.C.S. Chem. Comm.*, 459 (1981).
[3] T. Morita, Y. Okamoto, and H. Sakurai, *Tetrahedron Letters*, **21**, 835 (1980).

Allyltrimethylsilylzinc chloride, $[(CH_3)_3SiCH \cdots CH \cdots CH_2]^- ZnCl^+$ (**1**). This nucleo-phile is prepared by reaction of allyltrimethylsilane with *sec*-butyllithium and TMEDA in THF at $-70°$ and then with $ZnCl_2$ at $-40°$.

Spiro-γ-lactones.[1] This anion (**1**) has been used to prepare steroidal 17-spiro-γ-lactones (equation I) by reaction of a 17-keto steroid at the γ-position. The Sharpless oxidation step can be conducted in the presence of a Δ^4-3-keto group. When the

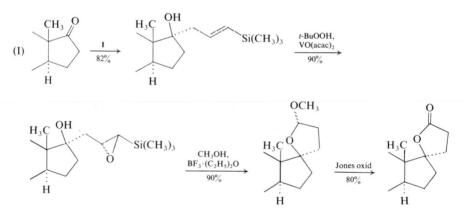

lithium anion of allyltrimethylsilane is used in the reaction some attack at the α-position also occurs.

[1] E. Ehlinger and P. Magnus, *Tetrahedron Letters*, **21**, 11 (1980).

Alumina, 1, 19–20; **2,** 17; **3,** 6; **4,** 8; **6,** 16–17; **7,** 5–7; **8,** 9–13; **9,** 8–11.

Cleavage of epoxides (**6,** 16–17; **8,** 10–12). Cleavage of epoxides catalyzed by Woelm neutral alumina is particularly useful in the case of medium-ring epoxides, which are generally rather unreactive and which are liable to undergo transannular reactions. For example, *cis*-cyclooctene oxide supported on neutral alumina is con-verted to *trans*-2-acetoxycyclooctanol in 78% yield by reaction with acetic acid at

25° for 24 hours.[1] Acetolysis under homogeneous conditions at 60° for 22 hours proceeds in 21% yield.[2]

Monoesterification of dicarboxylic acids.[3] Chemiabsorption of a dicarboxylic acid on alumina or silica can be used to effect selective esterification of one acid group with diazomethane. The method was demonstrated by conversion of terephthalic acid, C_6H_4-1,4-$(COOH)_2$, into the monomethyl ester in quantitative yield.

Michael reaction with exocyclic α,β-enones.[4] Conjugate addition of secondary amines to α,β-enones proceeds readily unless the enone is exocyclic. In such a case the reaction is very slow unless alumina is added as catalyst. Thus the reaction of **1** and diethylamine in the presence of alumina proceeds in 2 hours to give **2** quantitatively.

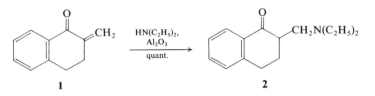

1   **2**

Diarylacetylenes.[5] Thermal decarbonylation of diarylcyclopropenones in refluxing *o*-dichlorobenzene to give diarylacetylenes proceeds in high yield in the presence of alumina (equation I).

(I)
$$Ar^1C{\equiv}CAr^2 + CO$$

(Al₂O₃, Δ, 80–95%)

[1] G. H. Posner, M. Hulce, and R. K. Rose, *Org. Syn.* submitted (1980).
[2] A. C. Cope, S. Moon, and P. E. Peterson, *Am. Soc.*, **81**, 1650 (1959).
[3] T. Chihara, *J.C.S. Chem. Comm.*, 1215 (1980).
[4] S. W. Pelletier, A. P. Venkov, J. Finer-Moore, and N. V. Mody, *Tetrahedron Letters*, **21**, 809 (1980).
[5] D. H. Wadsworth and B. A. Donatelli, *Synthesis*, 285 (1981).

Aluminum chloride, 1, 24–34; **2,** 21–23; **3,** 7–9; **4,** 10–15; **5,** 10–13; **6,** 17–19; **7,** 7–9; **8,** 13–15; **9,** 11–13.

[2 + 2] Cycloaddition. In the presence of either $C_2H_5AlCl_2$[1] or $AlCl_3$,[2] esters of 2,3-butadienoic acid (**1**) undergo [2 + 2] cycloadditions at the 3,4-double bond with acyclic or cyclic alkenes to give cyclobutylideneacetic esters. The reaction is considered to involve the vinyl cation $H_2C{=}\overset{+}{C}{-}CH{=}C(OR)O\bar{A}lCl_3$. A mixture of

(E)- and (Z)-isomers is formed, but the (E)-isomer is favored. The ester undergoes Diels-Alder reactions (thermal and catalyzed) at the 2,3-double bond.

Examples:

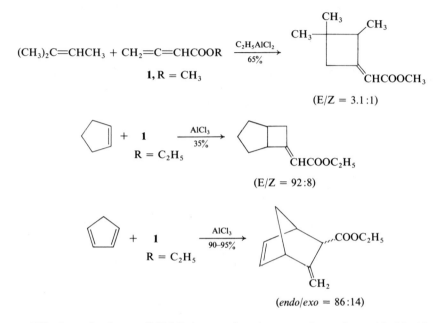

$(CH_3)_2C=CHCH_3 + CH_2=C=CHCOOR$ $\xrightarrow[65\%]{C_2H_5AlCl_2}$

1, R = CH_3

(E/Z = 3.1:1)

+ **1** $\xrightarrow[35\%]{AlCl_3}$

R = C_2H_5

(E/Z = 92:8)

+ **1** $\xrightarrow[90-95\%]{AlCl_3}$

R = C_2H_5

(endo/exo = 86:14)

Chiral α-amino ketones.[3] N-Methoxycarbonyl-protected α-amino acid chlorides undergo Friedel-Crafts reactions with high retention of chirality. Cbz-protected amino acids under the same conditions give intractable tars.

Examples:

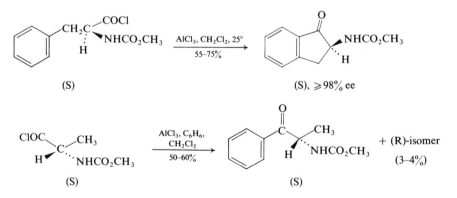

$\xrightarrow[55-75\%]{AlCl_3, CH_2Cl_2, 25°}$

(S)

(S), ≥98% ee

$\xrightarrow[50-60\%]{AlCl_3, C_6H_6, CH_2Cl_2}$

(S)

(S)

+ (R)-isomer

(3-4%)

Cyclic chiral α-amino ketones are reduced by BH_3, NaBH_4, or LiAlH_4 predominately to chiral *trans*-α-amino alcohols (80–90% yield).

[1] B. B. Snider and D. K. Spindell, *J. Org.*, **45**, 5017 (1980).
[2] H. M. R. Hoffmann, Z. M. Ismail, and A. Weber, *Tetrahedron Letters*, **22**, 1953 (1981).
[3] D. E. McClure, B. H. Arison, J. H. Jones, and J. J. Baldwin, *J. Org.*, **46**, 2431 (1981).

Aluminum chloride–Ethanethiol, 9, 13.

Dealkylation of esters.[1] The combination of an aluminium halide and a dialkyl sulfide is more potent than the aluminum halide–ethanethiol system. Dealkylation of higher esters is possible with the newer system. The order of activity of sulfides is tetrahydrothiophene > $(CH_3)_2S$ > $(C_2H_5)_2S$.

[1] M. Node, K. Nishide, M. Sai, K. Fuji, and E. Fujita, *J. Org.*, **46**, 1991 (1981).

(S)-(–)-2-Amino-1-methoxy-3-phenylpropane, H_2N ▶ $\overset{\displaystyle CH_2OCH_3}{\underset{\displaystyle CH_2C_6H_5}{\overset{\displaystyle |}{\underset{\displaystyle |}{C}}}}\cdots H$ **(1)**, α_D – 13.8°. The

methoxy amine is prepared from (S)-phenylalanine by $NaBH_4$ reduction followed by methylation.

Enantioselective alkylation of ketones. Chiral imines prepared from cyclic ketones and **1** on metalation and alkylation are converted to chiral 2-alkylcyclo-alkanones in 87–100% enantiomeric purity.[1] The high enantioselectivity is dependent on chelation of the lithium ion in the anion by the methoxyl group, which results in a rigid structure.

Example:

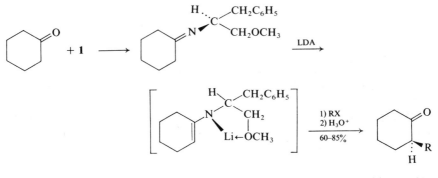

(S), 80–99% ee

On double alkylation of cyclohexanone, (2S,6S)-2,6-dimethylcyclohexanone is obtained in 85% ee.

Enantioselective alkylation of acylic ketones is also possible,[2] but in this case the enantioselectivity is dependent on the geometry of the chiral imine. Thus the product of thermodynamic alkylation (of the E-isomer) is the enantiomer of the product of kinetic alkylation of the (Z)-isomer. The optical yields of thermodynamic alkylation are generally higher than those of kinetic alkylation.

[1] A. I. Meyers, D. R. Williams, G. W. Erickson, S. White, and M. Druelinger, *Am. Soc.*, **103**, 3081 (1981).
[2] A. I. Meyers, D. R. Williams, S. White, and G. W. Erickson, *ibid.*, **103**, 3088 (1981).

Ammonia-borane, $BH_3:NH_3$. Mol. wt. 30.87. The borane is available from Callery Chem. Co. as a white, crystalline, air-stable compound; soluble in H_2O and CH_3OH; fairly soluble in THF, ether, $CHCl_3$, and $CH_3COOC_2H_5$.

Reduction of aldehydes and ketones.[1] Earlier work on amine–borane reagents was conducted mainly with tertiary amines and led to the conclusion that these borane complexes reduced carbonyl compounds very slowly, at least under neutral conditions, and that the yield of alcohols is low. Actually complexes of borane with primary amines, NH_3 or $(CH_3)_3CNH_2$, reduce carbonyl compounds rapidly and with utilization of the three hydride equivalents. $BH_3:NH_3$ is less subject to steric effects than traditional complex hydrides. A particular advantage is that $NH_3:BH_3$ and $(CH_3)_3CNH_2:BH_3$ reduce aldehyde groups much more rapidly than keto groups, but cyclohexanone can be reduced selectively in the presence of aliphatic and aromatic acyclic ketones.

[1] G. C. Andrews and T. C. Crawford, *Tetrahedron Letters*, **21**, 693 (1980); G. C. Andrews, *ibid.*, **21**, 697 (1980).

(2S,4S)-(Anilinomethyl)-1-ethyl-4-hydroxypyrrolidine, 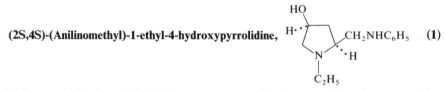 **(1)**

Mol. wt. 220.32, b.p. 150–170°/2 mm, α_D −58.1°. The pyrrolidine is prepared in several steps from (2S,4S)-N-benzyloxycarbonyl-4-hydroxyproline.

Asymmetric Michael addition.[1] This pyrrolidine is an efficient catalyst for asymmetric Michael addition of thiols to cyclohexenone (*cf.* **8**, 431). (S)-2-(Anilinomethyl)-1-ethylpyrrolidine, which lacks the 4-hydroxyl group of **1**, has little effect on enantioselection. The important role of a hydroxyl group in several asymmetric inductions has already been noted.

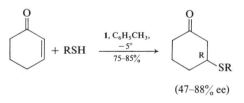

(47–88% ee)

[1] T. Mukaiyama, A. Ikegawa, and K. Suzuki, *Chem. Letters*, 165 (1981).

Arenechromium tricarbonyls, 6, 27–28; **7**, 71–72; **8**, 21–22; **9**, 21.

Regioselective lithiation.[1] The lithiation of 3-methoxybenzyl alcohol (**1**) occurs mainly at the 2-position because of coordination of the lithium substituent with both proximal oxygens. In contrast the chromium tricarbonyl complex (**4**) of **1** is

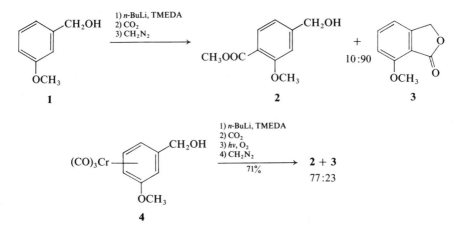

lithiated mainly *para* to the CH$_2$OH. The difference is even more striking in lithiation of the complex of 1,2,3,4-tetrahydro-7-methoxy-1-naphthol (**5**), in which the —CH$_2$OH group is locked into a rigid conformation.

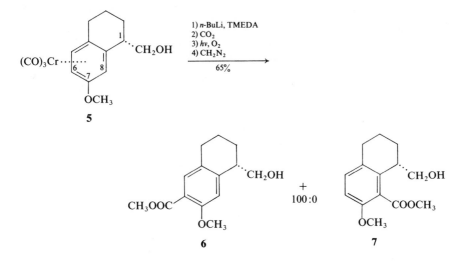

Ritter reaction.[2] This reaction involves the condensation of nitriles with an alcohol in a strongly acidic medium to form an amide.[3] Yields are generally low with primary and secondary alcohols, which form relatively unstable carbocations.

However, these cations are stabilized by an arenechromium tricarbonyl group in the α-position. Thus benzyl alcohol complexed with chromium tricarbonyl undergoes Ritter reactions rapidly and in high yield (equation I). On the other hand, the

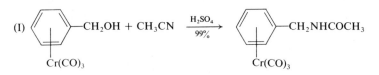

complexes of alcohols of the type $ArCR^1R^2OH$ are so stable that no reaction with nitriles occurs. Ritter reaction with complexes of secondary alcohols of the type $ArCHR^1OH$ proceed with practically 100% retention of configuration.

A preliminary experiment indicates the usefulness of similar complexation for Ritter reactions with a propargyl alcohol (equation II).

$$(II)\ HC{\equiv}CCH_2OH + CH_3CN\ \xrightarrow[\ 35\%\ ]{H_2SO_4}\ HC{\equiv}CCH_2NHCOCH_3$$
$$\underset{Co_2(CO)_6}{|}\qquad\qquad\qquad\qquad\underset{Co_2(CO)_6}{|}$$

[1] M. Uemura, N. Nishikawa, and Y. Hayashi, *Tetrahedron Letters*, **21**, 2069 (1980).
[2] S. Top and G. Jaouen, *J. Org.*, **46**, 78 (1981).
[3] L. I. Krimen and D. J. Cota, *Org. React.*, **17**, 213 (1969).

Azidomethyl phenyl sulfide, $C_6H_5SCH_2N_3$ **(1)**. Mol. wt. 165.22, b.p. 55–58°/0.23 mm, stable to temperatures of 105°. The reagent is prepared from thioanisole by chlorination (SO_2Cl_2) and azide displacement (NaN_3, NaI); 93% overall yield.

Amination of aromatic organometallics.[1] The reagent reacts with aromatic Grignard reagents or aryllithium reagents to form triazenes, which are converted to primary aromatic amines by hydrolysis with aqueous KOH or HCOOH (equation I). Grignard reagents react directly with **1**, but organolithium reagents require 1

$$(I)\ ArMgBr\ or\ ArLi\ \xrightarrow[40–90\%]{\textbf{1}}\ ArNHN{=}NCH_2SC_6H_5\ \xrightarrow[50–90\%]{H_2O}\ ArNH_2$$

equivalent of anhydrous $MgBr_2$.[2] This amination reaction cannot be extended to aliphatic substrates.

[1] B. M. Trost and W. H. Pearson, *Am. Soc.*, **103**, 2483 (1981).
[2] Prepared by reaction of Mg with $BrCH_2CH_2Br$ in ether.

Azidotrimethylsilane, 1, 1236; **3**, 316; **4**, 542; **5**, 719–720; **6**, 632; **9**, 21–22.

Isocyanates (**5**, 719).[1] The reaction of acid chlorides with azidotrimethylsilane is catalyzed by trace amounts of powdered KN_3 and 18-crown-6. For instance the acid chloride **1** only undergoes reaction with the silane under catalysis.

[1] J. D. Warren and J. B. Press, *Syn. Comm.*, **10**, 107 (1980).

B

Barium manganate, 8, 21; **9**, 23.

Dehydrogenation of 2-imidazolines.[1] Barium manganate is superior to MnO_2 (**2**, 258) for dehydrogenation of 2-imidazolines to imidazoles (three examples, 75–95% yield).

[1] J. L. Hughey, S. Knapp, and H. Schugar, *Synthesis*, 489 (1980).

Benzeneselenenyl halides, 5, 518–522; **6**, 459–460; **7**, 286–287; **9**, 25–32.

α,β-Unsaturated aldehydes. Direct selenenylation of aldehydes has been reported (**5**, 519), but a milder, general method utilizes the piperidine enamine followed by hydrolysis. The α,β-unsaturated aldehyde is then obtained in the usual manner. The overall yields range from 50 to 80%.[1]

Addition to 1-alkenes (**8**, 25). The critical steps in a recent synthesis of aldosterone (**4**) involve anti-Markovnikov addition of C_6H_5SeBr to **1**, oxidation to the allylic bromide, and acetate displacement to give **2**. The corresponding 21-monoacetate was converted to the triol **3** (OsO_4, N-methylmorpholine N-oxide). The final steps to **4** involved periodate cleavage and saponification.[2]

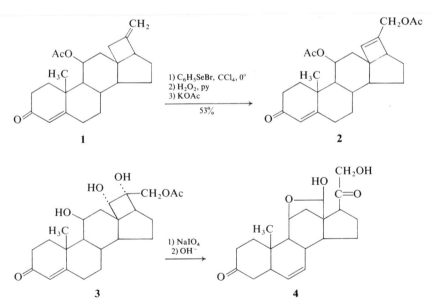

16

α-Phenylselenenylation of α,β-unsaturated esters.[3] (E)-α,β-Unsaturated esters are converted to α-phenylseleno-α,β-unsaturated esters in 20–65% yield by reaction with LDA followed by C_6H_5SeBr. The reaction is related to the reaction of α,β-enones with pyridine and C_6H_5SeCl (**9,** 28–29), and is also believed to involve conjugate addition of the base followed by selenenylation of the enolate.

The reaction is not applicable to (Z)-α,β-unsaturated esters.

1-Nitroalkenes. Anions of nitroalkanes react with benzeneselenenyl bromide in THF to form nitro(phenylseleno)alkanes, which usually are not isolated but rather are converted directly into nitroalkenes by oxidation with 35% aqueous H_2O_2 (equation I).[4]

(I) $C_6H_{13}CH_2CH_2NO_2$ $\xrightarrow{\text{\textit{n}-BuLi}}$

$$\left[C_6H_{13}CH_2CH=\overset{+}{N}\underset{OLi}{\overset{O^-}{\diagup}} \xrightarrow{C_6H_5SeBr} \underset{\underset{NO_2}{|}}{\overset{\overset{SeC_6H_5}{|}}{C_6H_{13}CH_2-CH}} \right] \xrightarrow[\substack{80\% \\ overall}]{H_2O_2} C_6H_{13}CH=CHNO_2$$

(61%)

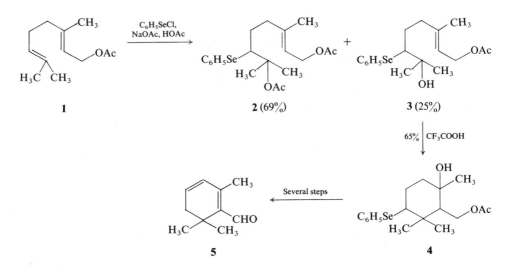

Olefinic cyclization.[5] Reaction of geranyl acetate (**1**) with C_6H_5SeCl does not result in the desired cyclized product **4,** but in **2** and **3.** The latter product when treated with various acids cyclizes to **4**. The phenylseleno group has been shown to be essential for this cyclization. The product was used for a synthesis of safranal (**5**).

   Phenylselenoetherification (**8**, 26–28). This cyclization has been described in detail.[6] The 16 examples reported indicate that the reaction is applicable to unsaturated primary, secondary, and tertiary alcohols as well as to phenols. The most important use is for synthesis of allylic ethers by *syn*-selenoxide elimination, which proceeds selectively away from the oxygen. The value of this methodology for synthesis of natural products is illustrated by a synthesis of a muscarine analog (**1**), outlined in equation (I).

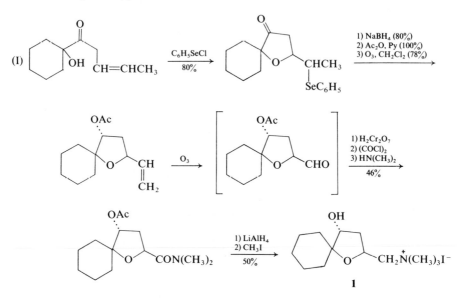

   Pyrrolidines; piperidines.[7] Reaction with benzeneselenenyl chloride converts Δ^4- and Δ^5-unsaturated ethyl urethanes into phenylseleno-substituted pyrrolidines and piperidines, respectively. Yields are considerably improved by the presence of silica gel, which presumably facilitates the ring closure subsequent to addition of the reagent to the double bond (equation I).

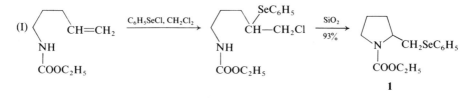

   α-Phenylseleno-α,β-unsaturated carbonyl compounds.[8] α-Phenylselenenylation of α,β-enones and α,β-enals can be effected with C_6H_5SeCl–pyridine (1:1). A related reaction is that of (E)-α,β-enoic acids with LDA and C_6H_5SeBr.[3] Both reactions are believed to involve conjugate addition of the base, selenenylation of the enolate, and elimination of the amine. Both reactions are hindered by substituents in the γ-position.

1,3-Enone transposition. A new method for this transposition depends on the regio- and stereoselectivity of addition of C_6H_5SeCl to allylic alcohols. An example is shown in equation (I).

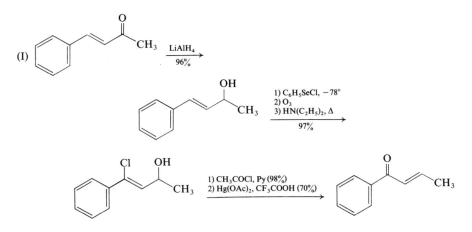

A more subtle transposition of a chiral cyclohexenone into the optical antipode is shown in equation (II).[9]

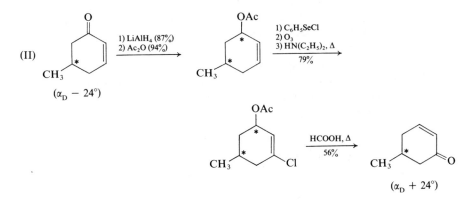

α-Chloro-α-phenylselenenyl ketones. Benzeneselenenyl chloride reacts with α-diazo ketones to give α-chloro-α-phenylselenenyl ketones in high yield. The products can be converted into either α-chloro- or α-phenylseleno-α,β-unsaturated ketones (equation I). The reaction is applicable to cyclic or acyclic α-diazo ketones.[10]

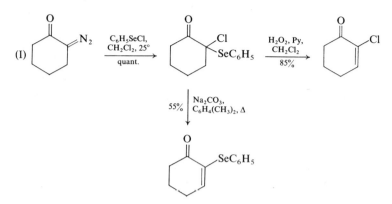

(I)

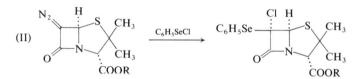

(II)

The reaction of C_6H_5SeCl with 6-diazopenicillins proceeds stereoselectively to give 6β-phenylselenenyl derivatives (equation II).[11]

β-Hydroxy and β-acetoamido selenides. β-Hydroxy selenides can be obtained in one step by reaction of an alkene with C_6H_5SeCl in aqueous CH_3CN at $25°$ (70–95% isolated yield). The reaction results in *trans*-adducts, with the C_6H_5Se group attached to the less substituted carbon atom. Vinyl acetates do not undergo addition but rather hydrolysis to the carbonyl compound. In the case of norbornene, the β-chloro selenide is isolable because this product requires higher temperatures for hydrolysis.[12]

The same reaction, but in the presence of triflic acid, results in β-acetamido selenides by a *trans*-addition. In the case of a terminal olefin, the phenylseleno group is introduced mainly at the terminal carbon atom.[13]

Example:

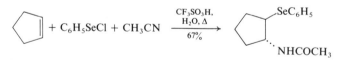

Glycosyloxyselenation.[14] This reaction can be used to prepare 2′-deoxydisacchar-ides. A typical reaction is shown in equation (I). The first step is reaction of benzeneselenenyl chloride with an enol ether (*cf.* **9**, 29–30). The adduct is then treated with pyridine and a second sugar. Finally the C_6H_5Se group is reduced by triphenyltin hydride.

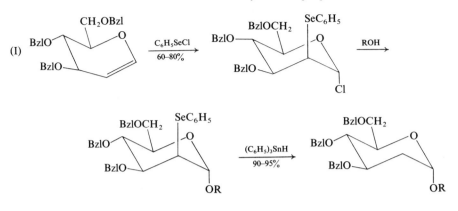

[1] D. R. Williams and K. Nishitani, *Tetrahedron Letters*, **21**, 4417 (1980).

[2] M. Miyano, *J. Org.*, **46**, 1846 (1981).

[3] T. A. Hase and P. Kukkola, *Syn. Comm.*, **10**, 451 (1980).

[4] T. Sakakibara, I. Takai, E. Ohara, and R. Sudoh, *J.C.S. Chem. Comm.*, 261 (1981).

[5] T. Kametani, K. Suzuki, H. Kurobe, and H. Nemoto, *Chem. Pharm. Bull. Japan*, **29**, 105 (1981).

[6] K. C. Nicolaou, R. L. Magolda, W. J. Sipio, W. E. Barnette, Z. Lysenko, and M. M. Joullie, *Am. Soc.*, **102**, 3784 (1980), for further examples *see* S. Uemura, A. Toshimitsu, T. Awai, and M. Okano, *Tetrahedron Letters*, **21**, 1533 (1980).

[7] D. L. J. Clive, A. Singh, C. K. Wong, W. A. Kiel, and S. M. Menchen, *J. Org.*, **45**, 2120 (1980).

[8] G. Zima and D. Liotta, *Syn. Comm.*, **9**, 697 (1979).

[9] D. Liotta and G. Zima, *J. Org.*, **45**, 2551 (1980).

[10] D. J. Buckley, S. Kulkowit, and A. McKervey, *J.C.S. Chem. Comm.*, 506 (1980).

[11] P. J. Giddings, D. I. John, and E. F. Thomas, *Tetrahedron Letters*, **21**, 395 (1980).

[12] A. Toshimitsu, T. Abai, H. Owada, S. Uemura, and M. Okano, *J.C.S. Chem. Comm.*, 412 (1980).

[13] A. Toshimitsu, T. Aoai, S. Uemura, and M. Okano, *J.C.S. Chem. Comm.*, 1041 (1980).

[14] G. Jaurand, J.-M. Beau, and P. Sinay, *J.C.S. Chem. Comm.*, 572 (1981).

Benzeneselenenyl hexafluorophosphate or -antimonate, $C_6H_5\overset{+}{Se}XF_6^-$ (X = P or Sb). These salts are prepared by reaction of C_6H_5SeCl with 1 equiv. of $AgSbF_6$ or $AgPF_6$.[1]

Cyclization of unsaturated β-keto esters. The salts react with certain unsaturated β-keto esters or 1,3-diketones to form cyclized products involving the carbonyl oxygen and/or the α-methylene group.[2] Products of the latter type are not obtainable with N-phenylselenophthalimide (**9**, 366–367).

Examples:

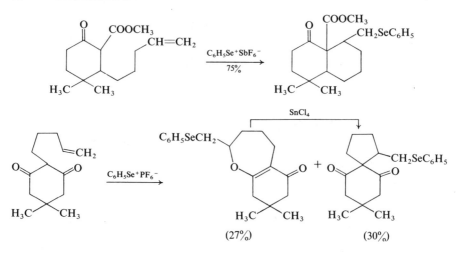

[1] G. H. Schmid and D. G. Garratt, *Tetrahedron Letters*, 3991 (1975).
[2] W. P. Jackson, S. V. Ley, and A. J. Whittle, *J.C.S. Chem. Comm.*, 1173 (1980).

Benzeneseleninic acid, 8, 28–29; **9,** 32.

Tetrazenes. 1,1-Disubstituted hydrazines (**1**) are oxidized by this reagent to tetrazenes (**2**) in 75–95% yield.[1]

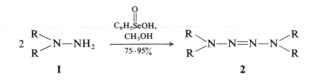

[1] T. G. Back, *J.C.S. Chem. Comm.*, 530 (1981).

Benzeneseleninic anhydride, 6, 240–241; **7,** 139; **8,** 29–32; **9,** 32–34.

Regeneration of carbonyl compounds from certain derivatives.[1] Ketones can be recovered in satisfactory yield from the phenylhydrazones, *p*-nitrophenyl-hydrazones, tosylhydrazones, oximes, and semicarbazones by reaction with 1 equivalent of $(C_6H_5SeO)_2O$. 2,4-Dinitrophenylhydrazones and N,N-dimethyl-hydrazones are inert under even rather vigorous conditions. The reagent can also be used to regenerate aldehydes from oximes or tosylhydrazones.

The reagent oxidizes hydrazo compounds to azo compounds and hydroxylamines to nitroso derivatives.

Angular hydroxylation of polycyclic ketones (**9,** 33–34). Oxidation of polycyclic ketones with $(C_6H_5SeO)_2O$ effects hydroxylation of an angular tertiary carbon adjacent to the carbonyl group. The more thermodynamically stable epimer is usually the major product.[2]

Examples:

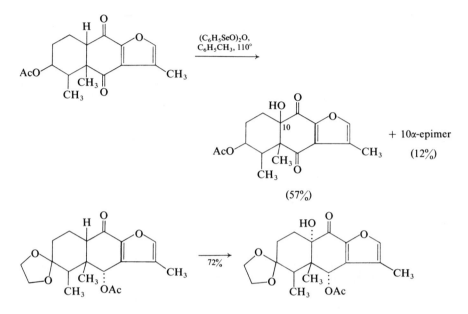

(57%)

+ 10α-epimer

(12%)

72%

Oxidation of phenols and hydroquinones.[3] Both 1- and 2-naphthol are oxidized by benzeneseleninic anhydride to 1,2-naphthoquinone in comparable yield (62–63%). This *ortho*-selectivity is general and is explained by a mechanism outlined in equation (I). As expected, the reagent oxidizes hydroquinones to the quinones in 84–92% yield.

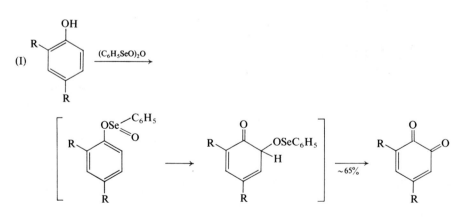

(I)

~65%

[1] D. H. R. Barton, D. J. Lester, and S. V. Ley, *J.C.S. Perkin I*, 1212 (1980).
[2] K. Yamakawa, T. Satoh, N. Ohba, R. Sakaguchi, S. Takita, and N. Tamura, *Tetrahedron*, **37**, 473 (1981).
[3] D. H. R. Barton, A. G. Brewster, S. V. Ley, C. M. Read, and M. N. Rosenfeld, *J.C.S. Perkin I*, 1473 (1981).

Benzenesulfenyl chloride; 5, 523–524; **6**, 30–32; **8**, 32–34; **9**, 35–38.

Tricyclenes. Benzenesulfenyl chloride reacts with 5-methylene-2-norbornene (**1**) to form the tricyclene derivative **2**. This reaction was used to synthesize tricyclo-*eka*-santalol (**3**)[1] and cyclosativene (**4**).[2]

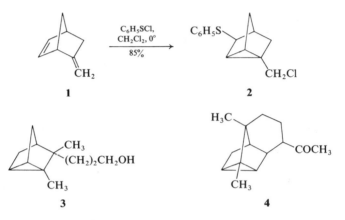

[1] D. Heissler and J.-J. Riehl, *Tetrahedron Letters*, 3957 (1979).
[2] D. Heissler and J.-J. Riehl, *Tetrahedron Letters*, **21**, 4707 (1980).

1,3-Benzodithiolylium tetrafluoroborate (1). Mol. wt. 240.05, m.p. 150° (dec.), stable at 25°, soluble in polar solvents.

Preparation:[1]

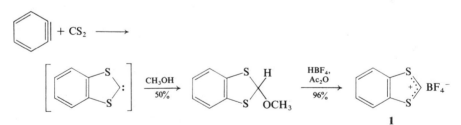

α,β-Unsaturated aldehydes.[2] Lithium trialkylalkynylborates are alkylated by **1** to the vinylboranes **2** and **3** in a ratio of ~5:1. The products are separated as the corresponding hydrolysis products **4** and **5**, protected derivatives of α,β-enals.

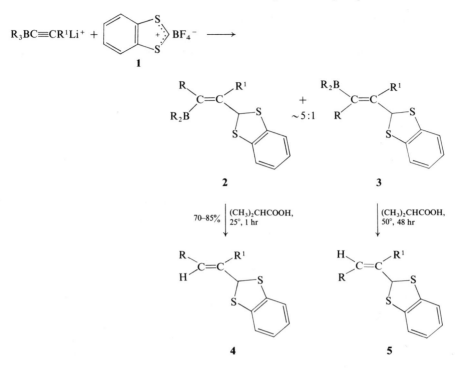

Trimethylamine oxide (**6**, 624; **7**, 507) or alkaline hydrogen peroxide selectively oxidizes **2** and **3** at the C—B bond to give protected 3-keto aldehydes (**6**).

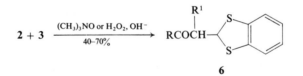

[1] J. Nakayama, K. Fujiwara, and M. Hoshino, *Bull. Chem. Soc. Japan*, **49**, 3567 (1976).
[2] A. Pelter, P. Rupani, and P. Stewart, *J.C.S. Chem. Comm.*, 164 (1981).

1,3-Benzodithiolylium perchlorate, 8, 34.

Caution:[1] A serious explosion of the perchlorate has been reported. The perchlorate is almost as sensitive to friction as mercury fulminate. However, the corresponding tetrafluoroborate is safe to handle and is chemically similar.

2,4,6-Triphenylpyrylium perchlorate also has explosive properties.

[1] A. Pelter, *Tetrahedron Letters*, **22**, No. 22, i (1981).

Benzylamine, $C_6H_5CH_2NH_2$. Mol. wt. 107.16, m.p. 10°, b.p. 184–185°.

Claisen rearrangement of allyloxyanthraquinones.[1] Benzylamine is an effective catalyst for rearrangement of **1** to **2**. The yield is increased, the reaction proceeds at

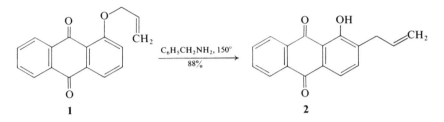

lower temperatures, and the allylic double bond is not isomerized. However, the amine does not effect rearrangement of 2-allyloxyanthraquinone. N,N-Dimethyl-benzylamine effects rearrangement of **3** to **4** as the major product.

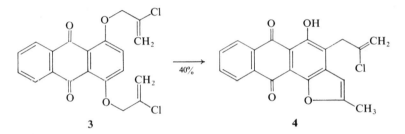

These studies were undertaken in connection with a possible route to unsymmetrically functionalized anthraquinones for intermediates to rhodomycinones.

[1] C. M. Wong, R. Singh, K. Singh, and H. Y. P. Lam, *Can. J. Chem.*, **57**, 3304 (1979).

Benzylchlorobis(triphenylphosphine)palladium(II), 8, 35–36; **9,** 41–42.

Coupling of allyl bromides and allylstannanes. This reaction is possible by catalysis with benzylchlorobis(triphenylphosphine)palladium, which is probably converted into bis(triphenylphosphine)palladium(0), the actual catalyst. This Pd(0) complex is more effective than tetrakis(triphenylphosphine)palladium(0). Zinc chloride in certain cases can function as a catalyst alone or in combination with the Pd(II) catalyst.[1]

Example:

$$(CH_3)_2C{=}CHCH_2Br + (CH_2{=}CHCH_2)_4Sn \xrightarrow[81\%]{\substack{Pd(II), ZnCl_2 \\ THF}} (CH_3)_2C{=}CHCH_2CH_2CH{=}CH_2$$

[1] J. Godschalx and J. K. Stille, *Tetrahedron Letters*, **21**, 2599 (1980).

3-Benzyl-5-(2-hydroxyethyl)-4-methyl-1,3-thiazolium chloride, 6, 38–39, 289; **7,** 16–17. Detailed directions are available for preparation of this thiazolium salt,[1] which is also commercially available (Fluka). Examples of use of this catalyst in combination with triethylamine for acyloin condensation are cited. The heterocyclic furoin can also be prepared with this catalyst. Synthesis of benzoins, however, requires use of a related catalyst, one substituted with an N-methyl or N-ethyl group in place of N-benzyl.

   2-Hydroxycyclopentenones.[2] A new synthesis involves addition of aldehydes to 1-acetoxy-3-butene-2-one catalyzed with this salt (**1**), followed by base-catalyzed condensation of the resulting 1,4-diketone.

   Example:

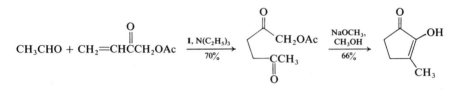

[1] H. Stetter and H. Kuhlmann, *Org. Syn.*, submitted (1980).
[2] H. Stetter and W. Schlenker, *Tetrahedron Letters*, **21**, 3479 (1980).

(−)-Benzylquininium chloride, 8, 430–431.

   Chiral epoxynaphthoquinones. Pluim and Wynberg[1] have prepared a number of optically active epoxides of 2-alkyl- and 2,3-dialkyl-1,4-naphthoquinones by oxidation with 30% H_2O_2, aqueous NaOH, and benzylquininium chloride. Enantiomeric excesses of 45% can be realized, and these can be improved by crystallization. The authors also report that the most satisfactory method for preparation of 2-alkyl-1,4-naphthoquinones is that of Jacobsen (**5,** 17; **8,** 18).

[1] H. Pluim and H. Wynberg, *J. Org.*, **45**, 2498 (1980).

3-Benzylthiazolium bromide, (**1**).

   RCHO → RCOOCH₃.[1] This salt, in CH_3OH containing $N(C_2H_5)_3$, catalyzes a redox reaction in which aldehydes are oxidized to methyl esters, and certain organic compounds such as acridine (**2**) and phenazine are reduced.

   Examples:

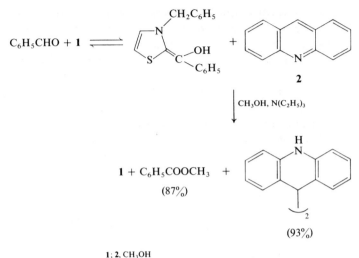

$$C_6H_5CHO + 1 \rightleftharpoons \quad + \quad \textbf{2}$$

1 + C_6H_5COOCH_3 +
(87%)

(93%)

$$CH_3(CH_2)_5CHO \xrightarrow[67\%]{1;2,\ CH_3OH} CH_3(CH_2)_5COOCH_3$$

[1] H. Inoue and K. Higashiura, *J.C.S. Chem. Comm.*, 549 (1980).

Benzyl(triethyl)ammonium permanganate, 9, 43.

Oxidation of amines.[1] Tertiary amines are oxidized by this reagent (2 equivalents) to amides in 70–95% yield. The oxidation of secondary amines is less selective; amides are obtained in 35–50% yield accompanied by carboxylic acids and esters.

Examples:

$$CH_3CH_2CH_2CH_2N(n\text{-}Bu)_2 \xrightarrow[93\%]{C_6H_5CH_2(C_2H_5)_3\overset{+}{N}MnO_4{}^-} CH_3CH_2CH_2\overset{\overset{\displaystyle O}{\|}}{C}N(n\text{-}Bu)_2$$

$$C_6H_5CH_2N(CH_3)_2 \xrightarrow[70\%]{} C_6H_5\overset{\overset{\displaystyle O}{\|}}{C}N(CH_3)_2$$

$$C_6H_5CH_2NHCH_3 \longrightarrow C_6H_5\overset{\overset{\displaystyle O}{\|}}{C}NHCH_3 + C_6H_5CHO + C_6H_5COOH$$

(36%)   (33%)   (27%)

Oxidation of arylmethyl groups.[2] The reagent oxidizes an acetoxyarylmethyl group to the corresponding carboxylic acid in moderate yield.

Examples:

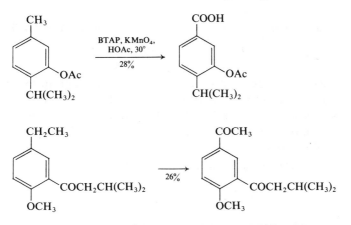

[1] H.-J. Schmidt and H. J. Schäfer, *Angew. Chem. Int. Ed.*, **20**, 109 (1981).
[2] R. Sangaiah and G. S. Krishna Rao, *Synthesis*, 1018 (1980).

Benzyltrimethylammonium fluoride, 6, 44.

Enol carbonates and carbamates. Olofson and Cuomo[1] have extended the regiospecific alkylation of ketones (**6**, 44) to a regiospecific preparation of enol carbonates and carbamates. The enol carbonates are prepared by reaction of silyl enol ethers with alkyl fluoroformates (ROCCl cannot be substituted). Enol carbamates are obtained from the same substrates by use of carbamoyl fluorides.

Examples:

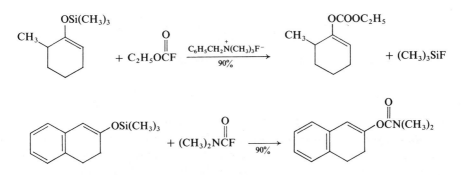

[1] R. A. Olofson and J. Cuomo, *Tetrahedron Letters*, **21**, 819 (1980).

2,2′-Bipyridinium chlorochromate, 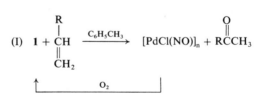 . Mol. wt. 292.6. The reagent

$$ClCrO_3^-$$

is prepared and used in the same way as pyridinium chlorochromate (**6**, 498–499). It is a somewhat milder oxidant than PCC and offers the advantage that excess reagent and by-products can be removed by a dilute acid wash. The complex of CrO_3 and 2,2′-bipyridine has also been prepared. It also is a milder oxidant than the corresponding Collins reagent.[1]

[1] F. S. Guziec, Jr., and F. A. Luzzio, *Synthesis*, 691 (1980).

Bis(acetonitrile)chloronitropalladium(II) (1). Mol. wt. 269.97.
   Preparation:

$$(CH_3CN)_2PdCl_2 \xrightarrow[\text{quant.}]{\substack{AgNO_2, \\ CH_3CN}} (CH_3CN)_2PdClNO_2$$

$$\textbf{1}$$

Oxidation of 1-alkenes to methyl ketones.[1] This Pd catalyst allows air oxidation of 1-alkenes to alkyl methyl ketones in yields of about 350% (based on Pd). The oxidation is also possible under nitrogen (about 90% isolated yield), but then **1** is not functioning as a catalyst (equation I). 2-Alkenes can be oxidized slowly in this way but a number of products are formed.

$$\text{(I)} \quad \textbf{1} + \underset{\overset{\|}{CH_2}}{\overset{R}{\underset{|}{CH}}} \xrightarrow{C_6H_5CH_3} [PdCl(NO)]_n + R\overset{O}{\overset{\|}{C}}CH_3$$

$$\uparrow \underline{\qquad\qquad O_2 \qquad\qquad} |$$

[1] M. A. Andrews and K. P. Kelly, *Am. Soc.*, **103**, 2894 (1981).

Bis(acetonitrile)dichloropalladium(II), 7, 21–22; **8,** 39; **9,** 44.
   Rearrangement of allylic acetates (**9,** 44). Grieco *et al.*[1] have found that this rearrangement involves complete transfer of chirality. Thus the allylic acetate **1** in the presence of this Pd(II) salt rearranges to the allylic acetate **2** in 91% yield. One point of interest is that **2** is convertible by known reactions into 12-methyl-$PGF_{2\alpha}$.

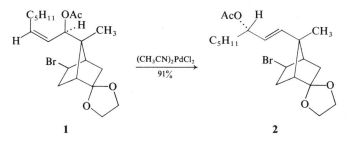

1   **2**

[1] P. A. Grieco, T. Takigawa, S. L. Bongers, and H. Tanaka, *Am. Soc.*, **102**, 7587 (1980).

Bis(benzonitrile)dichloropalladium(II), 5, 31–32; **6**, 45–47; **9**, 44–45.

Cope rearrangement.[1] Stoichiometric amounts of $PdCl_2$ are known to promote the Cope rearrangement of strained cyclic 1,5-dienes. Now catalytic amounts of $PdCl_2(C_6H_5CN)_2$ have been shown to promote the rearrangement of certain alkyl-substituted acyclic 1,5-dienes at 25°.[2]

Examples:

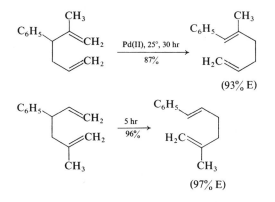

S → N Claisen rearrangement.[3] This Pd(II) salt permits the rearrangement of S-allylthioimidates to N-allylthioamides. Tetrakis(triphenylphosphine)palladium and various other metal salts are inactive. The rearrangement is inhibited by a substituent at the 2-position of the allyl group.

Example:

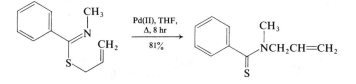

[1] S. J. Rhoads and N. R. Raulens, *Org. React.*, **22**, 1 (1975).
[2] L. E. Overman and F. M. Knoll, *Am. Soc.*, **102**, 865 (1980).
[3] Y. Tamaru, M. Kagotani, and Z. Yoshida, *J. Org.*, **45**, 5221 (1980).

Bis(benzoyloxy)borane, $(C_6H_5CO_2)_2BH$. Mol. wt. 254.04. The borane is prepared from benzoic acid and $BH_3 \cdot THF$ (*caution:* H_2 is evolved).

Deoxygenation of carbonyl compounds (**6**, 98; **7**, 54; **8**, 79–80).[1] This easily prepared borane is as effective as catecholborane for reduction of tosylhydrazones of carbonyl compounds to the corresponding methylene compounds.

[1] G. W. Kabalka and S. T. Summers, *J. Org.*, **46**, 1217 (1981).

Bis[1,2-bis(diphenylphosphino)ethane]palladium, $((C_6H_5)_2PCH_2CH_2P(C_6H_5)_2)_2Pd$
(**1**). Mol. wt. 1959.36. The reagent is prepared by reduction of $PdCl_2$ with hydrazine hydrate in the presence of the phosphine (DIPHOS).

Cyclopentanone synthesis. The palladium reagent (**1**), as well as tetrakis(triphenylphosphine)palladium, promotes a 1,3-alkyl shift from oxygen to carbon with no allylic inversion. Two typical examples are formulated.[1]

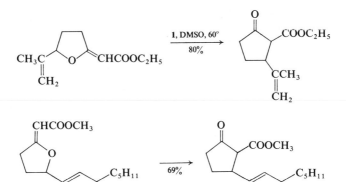

This rearrangement may be related to the rearrangement of allyloxypyridines to N-allylpyridones by Pd(0) (equation I).[2]

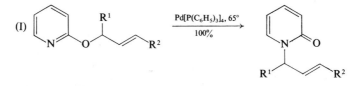

[1] B. M. Trost, T. A. Runge, and L. N. Jungheim, *Am. Soc.*, **102**, 2840 (1980).
[2] G. Balavoine and F. Guibe, *Tetrahedron Letters*, 3949 (1979).

μ-Bis(cyanotrihydroborato)tetrakis(triphenylphosphine)dicopper(I), $\{[(C_6H_5)_3P]_2Cu-BH_3CN\}_2$ **(1).** Mol. wt. 1255.94. The complex is obtained by treatment of $CuSO_4$ in sequence with excess $P(C_6H_5)_3$ and $NaCNBH_3$ in C_2H_5OH. Structure.[1]

Reductions.[2] In contrast to $NaCNBH_3$, the complex **1** reduces acid chlorides only to aldehydes under neutral conditions in 50–90% yield. However, under acidic conditions, **1** reduces carbonyl groups to alcohols (70–80% yield), although diaryl ketones are reduced in low yields. A useful feature of this reagent is that reduction of ketones is more stereoselective than that with $NaCNBH_3$. Thus reduction of 4-*t*-butylcyclohexanone gives a 94:6 mixture of *trans*:*cis* isomeric alcohols in 80% yield.

[1] K. N. Melmed, T. Li, J. J. Mayerle, and S. J. Lippard, *Am. Soc.*, **96**, 69 (1974).
[2] R. O. Hutchins and M. Markowitz, *Tetrahedron Letters*, **21**, 813 (1980).

Bis(1,5-cyclooctadiene)nickel(0), Ni(COD)$_2$, **4**, 33–35; **5**, 34–35; **7**, 428–429; **9**, 45–46.

β,γ-Unsaturated ketones.[1] π-Allylnickel halides (**4**, 33) react with 2-pyridyl-carboxylic acid esters to form a mixture of β,γ- and α,β-unsaturated ketones in which the former isomer predominates (equation I). 2-Pyridylbenzoate can be used

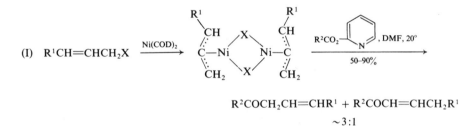

(I) $R^1CH{=}CHCH_2X$

$$R^2COCH_2CH{=}CHR^1 + R^2COCH{=}CHCH_2R^1$$
$$\sim 3:1$$

as a component in this reaction, but other derivatives of benzoic acid fail to give useful yields of ketones.

Diaryl ketones.[2] In the presence of this complex (1 equivalent) S-(2-pyridyl) aryl thioates (1 equivalent), prepared from aryl carboxylic acids, undergo reductive homocoupling to diaryl ketones (equation I).

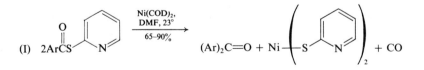

[1] M. Onaka, T. Goto, and T. Mukaiyama, *Chem. Letters*, 1483 (1979).
[2] T. Goto, M. Onaka, and T. Mukaiyama, *Chem. Letters*, 51 (1980).

Bis(dibenzylideneacetone)palladium [Pd(dba)$_2$], **9**, 46. This material can be obtained by treatment of dibenzylideneacetone with Na$_2$Pd$_2$Cl$_4$ and sodium acetate.[1]

Allylic alkylation.[2] This Pd(0) complex, in combination with a small quantity of bis(diphenylphosphino)ethane, is more effective than Pd[P(C$_6$H$_5$)$_3$]$_4$ for alkylation of allylic acetates with the anion of dimethyl malonate. It also permits use of sodium cyclopentadienide as a nucleophile.

[1] M. F. Rettig and P. M. Maithis, *Inorg. Syn.*, **17**, 135 (1977).
[2] J. C. Fiaud and J. L. Malleron, *Tetrahedron Letters*, **21**, 4437 (1980).

Bis(dimethylamino)methoxymethane, 5, 71–73; **7**, 41; **9**, 76.

α-Methylene-γ-butyrolactones.[1] The reagent has been used in a recent synthesis of aromatin (**1**) for methylenation of a γ-butyrolactone.[2]

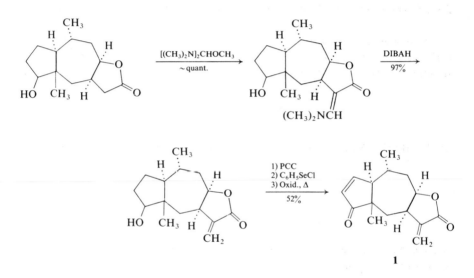

[1] F. E. Ziegler and J.-M. Fang, *J. Org.*, **46**, 825 (1981).
[2] Method of R. H. Mueller and M. Thompson, unpublished work.

Bis-(3-dimethylaminopropyl)phenylphosphine, C$_6$H$_5$P[(CH$_2$)$_3$N(CH$_3$)$_2$]$_2$ (**1**), **5**, 36–39.

Macrocyclic lactones. Ireland and Brown[1] have adapted the Eschenmoser contraction of sulfides to a synthesis of five- and six-membered lactones. An example is formulated in equation (I). A hydroxy thioamide is esterified to give a chloro ester, which is then treated in sequence with NaI and phosphine **1**. The method can also be used for preparation of macrocyclic lactones under high-dilution

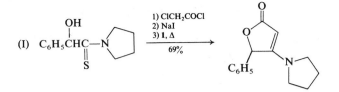

techniques. In fact diplodialide A (**2**) was synthesized by this sulfide contraction, although in only moderate yield.

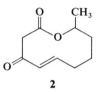

2

[1] R. E. Ireland and F. R. Brown, Jr., *J. Org.*, **45**, 1868 (1980).

1,3-Bis(dimethylphosphono)-2-propanone (1). Mol. wt. 274.15, b.p. 156–158/0.08 mm.

Preparation:

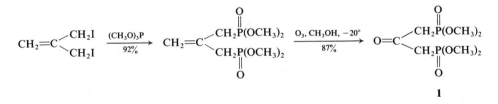

1

Macrocyclic ketones.[1] This reagent has been used for cycloolefination of acetylenic dialdehydes (Wittig-Horner reaction) as a route to muscone, exaltone, and civetone. A synthesis of the last ketone (**3**) is formulated (equation I).

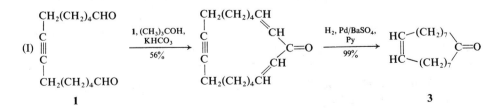

This route to cyclic ketones was suggested by the "rigid group principal".[2] As expected from this rule, dialdehydes such as **1**, because of constraint imposed by the triple bond, undergo cycloolefination more efficiently than saturated dialdehydes.

[1] G. Büchi and H. Wüest, *Helv.*, **62**, 2661 (1979).
[2] W. Baker, *Ind. Chim. Belg.*, **17**, 633 (1952).

trans-2,3-Bis(diphenylphosphine)-[2.2.1]-5-bicycloheptene (Norphos), 9, 48–49.

Asymmetric coupling of a vinyl halide with a Grignard reagent. 1-Phenylethylmagnesium chloride couples with vinyl bromide in the presence of a nickel catalyst (**1**) composed of $NiCl_2$ and (−)-norphos to give (S)-3-phenyl-1-butene (**2**) in 67% enantiomeric excess.[1]

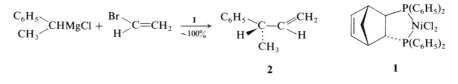

2   **1**

[1] H. Brunner and M. Pröbster, *J. Organometal. Chem.*, **209**, Cl (1981).

2,2′-Bis(diphenylphosphine)-1,1′-binaphthyl (BINAP) (1).

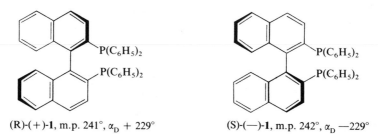

(R)-(+)-**1**, m.p. 241°, α_D + 229°   (S)-(−)-**1**, m.p. 242°, α_D −229°

This dissymmetric phosphine is prepared by standard reactions from (+)-1,1′-binaphthyl and is then resolved as a chiral Pd(II) complex.

Asymmetric hydrogenation.[1] Rh(I) complexed with (R)- or (S)-**1** catalyzes the asymmetric hydrogenation of prochiral α-(acylamino)acrylic acids, $R^1CHC=C$-(COOH)NHCOR², to optically active derivatives of (R)- or (S)-alanine (85–100% ee).

[1] A. Miyashita, A. Yasuda, H. Takaya, K. Toriumi, T. Ito, T. Souchi, and R. Noyori, *Am. Soc.*, **102**, 7932 (1980).

[1,2-Bis(diphenylphosphine)ethane]nickel(II) chloride (Nickel chloride–Diphos), $NiCl_2$–$(C_6H_5)_2PCH_2CH_2P(C_6H_5)_2$ (1).

Alkylation of allylic alcohols. The alkylation of allylic alcohols by Grignard reagents in the presence of nickel–phosphine[1,2] catalysts has been shown to be stereospecific in the case of the *cis*- and *trans*-4-methyl-2-cyclohexene-1-ols (**2**)[3]. The hydroxyl group is replaced with inversion, but the regioselectivity is different. In

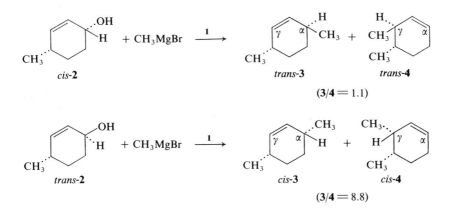

$(3/4 = 1.1)$

$(3/4 = 8.8)$

the case of *cis*-**2**, the replacement occurs about equally at the α- and γ-positions. In the case of *trans*-**2**, α-substitution is markedly favored over γ-substitution, possibly for steric reasons.

[1] H. Felkin, E. Jampel-Costa, and S. Swierczewski, *J. Organometal. Chem.*, **134**, 265 (1977).
[2] G. Consiglio, F. Morandini, and O. Piccolo, *Helv.*, **63**, 987 (1980).
[3] G. Consiglio, F. Morandini, and O. Piccolo, *Am. Soc.*, **103**, 1846 (1981).

1,1′-Bis(diphenylphosphine)ferrocene (dppf) (**1**). Mol. wt. 498.51, m.p. 183–184°.
Preparation[1]:

Coupling of allyl ethers and phenylmagnesium bromide.[2] $NiCl_2$(dppf) uniquely catalyzes the coupling of phenylmagnesium bromide with either isomeric ether **1** or **2** to give the terminal alkene **3** as the major product. On the other hand, $PdCl_2$(dppf) is by far the best catalyst for coupling of either **1** or **2** to give the nonterminal alkene **4**. The regioselectivity of both reactions is independent of the nature of the leaving group, which can be OC_6H_5, OTHP, OH, or Cl.

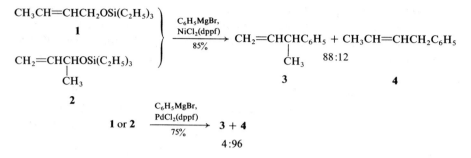

Acetylenic ketones.[3] $PdCl_2(dppf)$ is generally the most satisfactory catalyst for synthesis of acetylenic ketones by carbonylation of aryl and vinyl halides in the presence of a terminal acetylene.

Examples:

$$C_6H_5I + CO + HC\equiv CC_6H_5 \xrightarrow[85\%]{\substack{PdCl_2(dppf), \\ N(C_2H_5)_3}} C_6H_5COC\equiv CC_6H_5$$

$$C_6H_5CH=CHBr + CO + HC\equiv CC_6H_5 \xrightarrow[80\%]{} C_6H_5CH=CHCOC\equiv CC_6H_5$$

[1] J. J. Bishop, A. Davison, M. L. Katcher, D. W. Lichtenberg, R. E. Merrill, and J. C. Smart, *J. Organometal. Chem.*, **27**, 241 (1971).
[2] T. Hayashi, M. Konishi, K. Yokota, and M. Kumada, *J.C.S. Chem. Comm.*, 313 (1981).
[3] T. Kobayashi and M. Tanaka, *J.C.S. Chem. Comm.*, 333 (1981).

Bis(ethylthio)acetic acid (1). This α-keto acid equivalent is converted into the dianion by 2 equivalents of potassium bis(trimethylsilyl)amide in THF at 0°.[1] The dianion is readily alkylated in high yield by a wide variety of electrophiles. The products are converted into α-keto acids by NBS hydrolysis (**4**, 216) or into methyl 2,2-bismethoxycarboxylates by treatment with I_2–CH_3OH (**6**, 238).

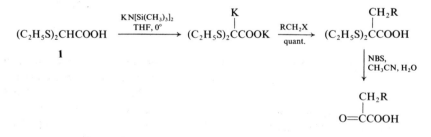

[1] G. S. Bates and S. Ramaswamy, *Can. J. Chem.*, **58**, 716 (1980).

2,4-Bis(4-methoxyphenyl)-1,3,2,4-dithiadiphosphetane-2,4-disulfide (1), **8**, 327; **9**, 49–50. Supplier: Fluka, Aldrich.

$RCOOR' \rightarrow RCH_2OR'$. A general method for this conversion[1] involves thionation of an ester to form a thionoester, which is then reduced with Raney nickel (equation I). Esters of primary and secondary alcohols are reduced in this way in

$$\text{(I)} \quad \underset{\displaystyle RCOR'}{\overset{\displaystyle O}{\|}} \quad \xrightarrow{\ \mathbf{1}\ } \quad \underset{\displaystyle RCOR'}{\overset{\displaystyle S}{\|}} \quad \xrightarrow{\ \text{Raney Ni}\ } \quad RCH_2OR'$$

about 60–70% yield. The limiting step is thionation, which can be retarded by steric factors, by electron-withdrawing groups conjugated to the carbonyl group, or by an ether group.

[1] S. L. Baxter and J. S. Bradshaw, *J. Org.*, **46**, 831 (1981).

Bis[methoxy(thiocarbonyl)] disulfide, $\underset{\displaystyle (CH_3OCS)_2}{\overset{\displaystyle S}{\|}}$ **(1)**. Mol. wt. 214.34, m.p. 23.5°. The disulfide is prepared by reaction of potassium methyl xanthate, CH_3OCSS, with chloramine-T at 25°.[1]

α-Alkylidene-γ-butyrolactones.[2] The disulfide **1** can be used for synthesis of either (E)- or (Z)-α-alkylidene-γ-butyrolactones. The former products are obtained by reaction of γ-butyrolactone with 2.2 equivalents of LDA followed by addition of **1** and then an aldehyde. The latter products are obtained when the same reaction is conducted in the presence of an added metal salt such as CuI or $ZnCl_2$.

Examples:

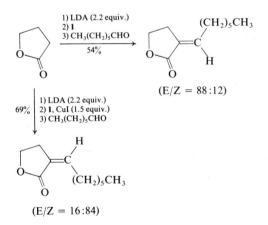

[1] E. J. Stout, B. S. Shasha, and W. M. Doane, *J. Org.*, **39**, 562 (1974); G. Bulmer and F. G. Mann, *J. Chem. Soc.*, 674 (1945).

[2] K. Tanaka, N. Tamura, and A. Kaji, *Chem. Letters*, 595 (1980).

Bis(3-methyl-2-butyl)borane (Disiamylborane), 1, 57–59; **3,** 22; **4,** 37; **5,** 39–41; **6,** 62; **8,** 41.

$RCH{=}CH_2 \rightarrow RCH_2CHO.$ This reaction can be carried out by hydroboration followed by oxidation of the resulting organoborane with pyridinium chlorochromate (PCC). In first reports of the transformation $BH_3:S(CH_3)_2$ was used, but this hydroboration is not regiospecific in the case of terminal alkenes. Use of disiamylborane is superior since this borane is highly selective in hydroboration of both alkenes and dienes.[1]

Examples:

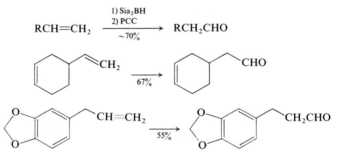

[1] H. C. Brown, S. U. Kulkarni, and C. G. Rao, *Synthesis*, 151 (1980); *J. Organometal. Chem.*, **172,** C20 (1979).

1,2-Bis(methylthio)-1,3-butadiene (1). Mol. wt. 146.27, b.p. 50°/0.25 mm.
    Preparation:

$$CH_3SCH_2C{\equiv}CCH_2SCH_3 \xrightarrow{KOC(CH_3)_3,\ DMSO} $$

1

Diels–Alder reactions[1]:

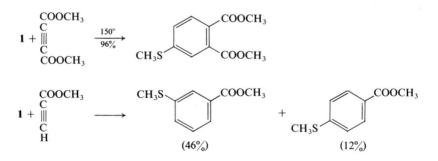

[1] M. E. Garst and P. Arrhenius, *Syn. Comm.*, **11,** 481 (1981).

Bis(*o*-nitrophenyl) phenylphosphonate, $C_6H_5\overset{O}{\underset{\|}{P}}(OC_6H_4NO_2\text{-}o)_2$ (1), 9, 20–21.

Amides and peptides. The preparation of these substances with **1** as the coupling reagent (**9**, 20–21) has been simplified by use of phase-transfer conditions (Bu_4NHSO_4, KOH, H_2O–CH_2Cl_2). The potassium salt of the acid can be prepared *in situ* either from the acid or the ethyl ester. In either case yields of amides are 70–95%.[1]

[1] Y. Watanabe and T. Mukaiyama, *Chem. Letters*, 285 (1981).

N,N-Bis(2-oxo-3-oxazolidinyl)phosphordiamidic chloride (1). Mol. wt. 254.6, m.p. 191–193°.

Preparation:

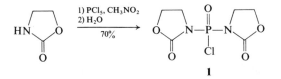

Amides; esters.[1] The reagent reacts with carboxylic acids in the presence of triethylamine to form an intermediate **a** that reacts with amines or alcohols to form amides or esters.

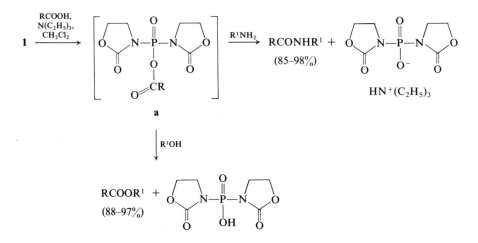

[1] J. Diago-Meseguer, A. L. Palomo-Coll, J. R. Fernandez-Lizarbe, and A. Zugaza-Bibao, *Synthesis*, 547 (1980).

Bis(2,4-pentanedionato)nickel(II) [Ni(acac)$_2$], **5**, 471; **6**, 417; **7**, 250; **9**, 51–52.

Michael additions.[1] Conjugate addition of β-dicarbonyl compounds to various Michael acceptors is catalyzed efficiently by Ni(acac)$_2$. Yields are usually higher than those obtained with traditional bases.

Examples:

$$CH_3COCH_2COCH_3 + CH_2{=}CHCCH_3 \xrightarrow[90\%]{\substack{1\% \text{ Ni(acac)}_2, \\ \text{dioxane}}} \underset{\underset{COCH_3}{|}}{\overset{\overset{COCH_3}{|}}{CHCH_2CH_2CCH_3}}$$

$$CH_3OOCCH_2COOCH_3 + C_6H_5CH{=}CHNO_2 \xrightarrow[62\%]{} \underset{\underset{COOCH_3}{|}}{\overset{\overset{COOCH_3}{|}}{CHCH(C_6H_5)CH_2NO_2}}$$

[1] J. H. Nelson, P. N. Howells, G. C. DeLullo, and G. L. Landen, *J. Org.*, **45**, 1246 (1980).

Bis(phenylthio)methane, 7, 25.

Ketone synthesis. Dialkylation of bis(phenylthio)methane is possible if the anion is prepared with *n*-BuLi–TMEDA in hexane at 0°. Highest yields are obtained with primary alkyl iodides.[1]

$$(C_6H_5S)_2CHR \xrightarrow[60-95\%]{\substack{1)\, n\text{-BuLi, TMEDA} \\ 2)\, R'I}} (C_6H_5S)_2C{\Big\langle}{\substack{R \\ R'}}$$

[1] D. J. Ager, *Tetrahedron Letters*, **21**, 4763 (1980).

Bis(tetra-*n*-butylammonium)dichromate, 9, 53.

Oxidation.[1] Benzylic and allylic alcohols are oxidized to the corresponding ketones or aldehydes in 80–90% yield. Unactivated alcohols resist attack by this oxidant.

[1] E. Santaniello and P. Ferraboschi, *Syn. Comm.*, **10**, 75 (1980).

Bis(tricaprylylmethyl)ammonium   nitrosodisulfonate, ·ON[SO$_3^-$ N$^+$CH$_3$(C$_8$H$_{17}$)$_3$]$_2$ **(1)**. The salt is obtained *in situ* from Fremy's salt, ·ON(SO$_3^-$ K$^+$)$_2$ (**1**, 940; **2**, 347–348; **4**, 411; **5**, 562) and 2 equivalents of Aliquat 336 (**4**, 28; **5**, 460; **6**, 404–406; **7**, 380) as a slurry in benzene.

Oxidation of a water-insoluble phenol.[1] The phenol **2** is not oxidized by Fremy's salt in aqueous ether, methanol, or DMF, but it is oxidized by **1** in benzene or under phase-transfer conditions. The product **3** is a well-known precursor to α-tocopherol (**4**).

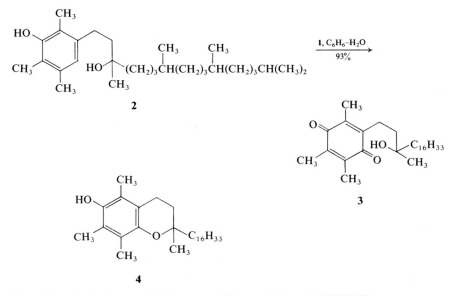

2

3

4

[1] G. L. Olson, H.-C. Cheung, K. Morgan, and G. Saucy, *J. Org.*, **45**, 803 (1980).

Bis(2,2,2-trifluoroethoxy)triphenylphosphorane, $(C_6H_5)_3P(OCH_2CF_3)_2$, (**1**). Mol. wt. 460.36, hygroscopic, stable in solution at 25° for several weeks. The phosphorane is prepared by reaction of triphenylphosphine dibromide with sodium 2,2,2-trifluoroethoxide.

Alkylation or acylation of ketones, sulfides, and amines.[1] This reagent generally reacts with alcohols or carboxylic acids to form 2,2,2-trifluoroethyl ethers or esters in satisfactory yields, except in the case of alcohols prone to dehydration. The reaction of these ethers provides a simple synthesis of unsymmetrical sulfides (equation I). A similar reaction can be used for preparation of secondary amines or amides (equation II). Enolate anions (generated from silyl enol ethers with KF) can be alkylated or acylated with **a** or **b** (equation III). Use of Grignard reagents in this type of coupling results in mediocre yields.

(I) $ROH \xrightarrow{\textbf{1}} \left[(C_6H_5)_3P\begin{array}{c} OCH_2CF_3 \\ OR \end{array} \right] \xrightarrow{R^1SH}$

a

$$RSR^1 + (C_6H_5)_3P{=}O + CF_3CH_2OH$$

(49–95%)

(II) $RCOOH \xrightarrow{\textbf{1}}$ $\left[(C_6H_5)_3P \underset{OCOR}{\overset{OCH_2CF_3}{<}} \right]$ $\xrightarrow{R^1NH_2}$

$$\mathbf{b}$$

$$RCONHR^1 + (C_6H_5)_3P{=}O + CF_3CH_2OH$$
$$(65-80\%)$$

(III) $C_6H_5\overset{\overset{\displaystyle OSi(CH_3)_3}{|}}{C}{=}CH_2$ $+ (C_6H_5)_3P \underset{OCOC_6H_5}{\overset{OCH_2CF_3}{<}}$ $\xrightarrow[82\%]{KF}$ $C_6H_5\overset{\overset{\displaystyle O}{||}}{C}CH_2COC_6H_5$

[1] T. Kubota, S. Miyashita, T. Kitazume, and N. Ishikawa, *J. Org.*, **45**, 5052 (1980).

Bistrimethylsilyl ether (Hexamethyldisiloxane), 9, 55.

Silylation of carboxylic acids.[1] This reagent in the presence of sulfuric acid silylates carboxylic acids in 75–95% isolated yield. The actual reagent may be bis(trimethylsilyl) sulfate, $[(CH_3)_3Si]_2SO_4$. Several other acids are less effective as catalysts.

[1] H. Matsumoto, Y. Hoshino, J. Nakabayashi, T. Nakano, and Y. Nagai, *Chem. Letters*, 1475 (1980).

1,3-Bis(trimethylsilyloxy)-1,3-butadiene (1), 7, 27–28.

Anthraquinones.[1] A regioselective synthesis of polyhydroxyanthraquinones is based on Diels-Alder reactions of **1** with chloro-substituted naphthoquinones. An example is the synthesis of 1,6-dihydroxyanthraquinone (**3**) from 3-chlorojuglone (**2**). Analogous syntheses are possible by use of vinylogous ketene acetals related to **1**.

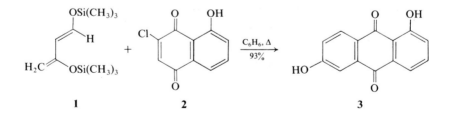

| | | |
| 1 | 2 | 3 |

[1] C. Brisson and P. Brassard, *J. Org.*, **46**, 1810 (1981).

1,2-Bis(trimethylsilyloxy)cyclobutene,   (1). Mol. wt. 230.45, b.p.

75–76°/10–11mm. Preparation.[1]

[2 + 2] Cycloaddition to enones.[2] The reagent undergoes photochemical cyclo-addition to cyclopentenones and cyclohexenones to form *cis-anti-cis*-cycloadducts selectively. The products can be converted into *cis*-hydrindanes and *cis*-decalones.

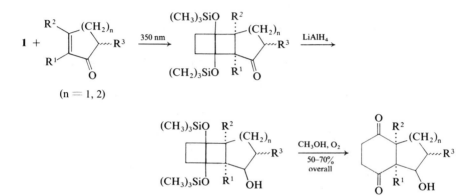

(n = 1, 2)

[1] K. Rühlmann, *Synthesis*, 242 (1971).
[2] M. Van Audenhove, D. de Keukeleire, and M. Vandewalle, *Tetrahedron Letters*, **21**, 1979 (1980).

2,5-Bis(trimethylsilyloxy)furanes (1).

Preparation:

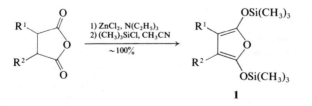

1

Reactions.[1] These silyloxyfuranes undergo Diels-Alder reactions much more readily than furane itself **(6,** 161). The reaction provides a route to *p*-quinones and hydroquinones (equation I).

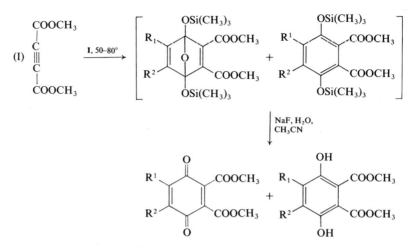

These furanes can be used for synthesis of bislactones or γ-hydroxybutenolides by reaction with aldehydes or ketones in the presence of TiCl$_4$ (equations II and III).

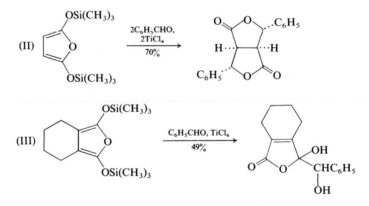

[1] P. Brownbridge and T. H. Chan, *Tetrahedron Letters*, **21**, 3423, 3427, 3431 (1980).

1,3-Bis(trimethylsilyloxy)-1-methoxy-1,3-butadiene (1), 9, 54–55.

3-Hydroxyhomophthalates. Complete details are available for construction of benzene rings by condensation of **1**, as the dianion equivalent of methyl acetoacetate, with various 1,3-dicarbonyl equivalents.[1] The reaction has been extended to condensation of **1** with a 1,3,5-tricarbonyl unit to form 3-hydroxyhomophthalates. Thus when **1** (2 equivalents) is treated with trimethyl orthoformate (1 equivalent) in the presence of TiCl$_4$, dimethyl 3-hydroxy-homophthalate (**2**) is formed in 68% yield. The reaction is believed to involve formation of the 1,3,5-tricarbonyl equivalent **a**, which then condenses with **1** to form **2**.

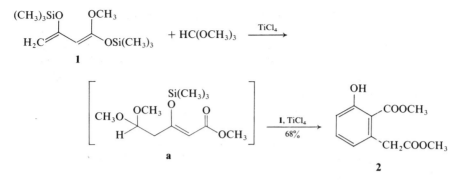

A variation of this reaction was used to synthesize sclerin (**3**), a natural product believed to be biosynthesized from a penta-β-carbonyl precursor (equation I).[2]

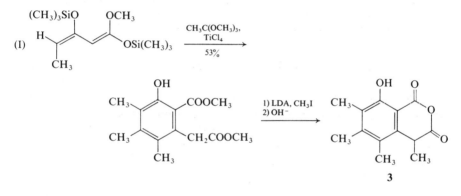

[1] T.-H. Chan and P. Brownbridge, *Am. Soc.*, **102**, 3534 (1980).
[2] T.-H. Chan and P. Brownbridge, *J.C.S. Chem. Comm.*, 20 (1981).

Bis(triphenylphosphine)copper(I) tetrahydroborate, 8, 47; **9,** 57.

Reduction of acid chlorides to aldehydes. Two laboratories[1,2] have published details for this reduction. One difficulty is the large quantities of reagent needed for preparative-scale reactions. For large-scale reductions the Rosenmund reaction is preferable.

Reduction of tosyl- and trisylhydrazones.[3] The reagent (1 equivalent) reduces tosylhydrazones of ketones to alkanes in yields of 60–85% (GLC). The reduction is subject to steric hindrance and so is not effective with the tosylhydrazone of camphor. Tosylhydrazones of aldehydes are reduced in moderate yield (about 50%). The reagent does not reduce tosylhydrazones of aromatic or α,β-unsaturated aldehydes or ketones.

Trisylhydrazones of aliphatic aldehydes and ketones are reduced by this reagent in 40–50% yield.

Reduction of aldehydes and ketones.[4] In the presence of hydrogen chloride (or a Lewis acid), the reagent reduces carbonyl compounds to alcohols in high yield and with high stereoselectivity. However, the reduction of hindered ketones requires a strong Lewis acid ($AlCl_3$). Aldehydes are reduced so much more readily than ketones that selective reductions are possible. The reagent is also useful for reduction of α,β-enals to allylic alcohols.

Reduction of aromatic azides.[5] Aromatic azides can be reduced to anilines by this reagent in $CHCl_3$ at 25°. The rate is accelerated by electron-withdrawing substituents but is reduced by electron-attracting groups. Even so, the presence of an azide group does not interfere with use of the reagent for reduction of acid chlorides to aldehydes.

[1] G. W. J. Fleet and P. J. C. Harding, *Org. Syn.*, submitted (1980).
[2] T. N. Sorrell and P. S. Pearlman, *J. Org.*, **45**, 3449 (1980).
[3] G. W. J. Fleet, P. J. C. Harding, and M. J. Whitcombe, *Tetrahedron Letters*, **21**, 4031 (1980).
[4] G. W. J. Fleet and P. J. C. Harding, *Tetrahedron Letters*, **22**, 675 (1981).
[5] S. J. Clarke, G. W. J. Fleet, and E. M. Irving, *J. Chem. Res. (S)*, 17 (1981).

9-Borabicyclo[3.3.1]nonane (9-BBN), 2, 31; **3**, 24–29; **4**, 41; **5**, 46–47; **6**, 62–64; **7**, 29–31; **8**, 47–49; **9**, 57–58.

Methylenecycloalkanes.[1] A cycloalkene can be converted into a methylene-cycloalkane by the following steps: hydroboration with 9-BBN and carbonylation–reduction[2] to give a β-(cycloalkylmethyl)-9-BBN derivative, which is then allowed to react with an aldehyde[3] to form the final product.

Example:

(75%)

Steroid side chain.[4] The key step in a method for stereocontrolled addition of the side chain to 17-keto steroids is hydroboration of a 17(20)-(Z)-ethylidene steroid (**1**), which proceeds selectively to give **2**, with the desired natural configuration at C_{17} and C_{20}. The product reacts with most alkylating reagents in rather low yield, possibly because of steric factors; however alkylation with the anion of chloroacetonitrile (potassium 2,6-di-*t*-butyl-4-methylphenoxide) in THF gives the nitrile **3** in 60–70% yield. One added attraction of this route is that 9-BBN reacts preferentially with a 17(20)-double bond in the presence of a 5(6)-double bond.

For another approach to natural steroid side chains using the ene reaction, *see* Ethylaluminum dichloride, this volume.

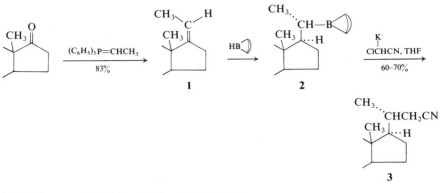

1   **2**

3

[1] H. C. Brown and T. M. Ford, *J. Org.*, **46**, 647 (1981).

[2] H. C. Brown, T. M. Ford, and J. L. Hubbard, *J. Org.*, **45**, 4067 (1980).

[3] M. M. Midland, A. Tramontano, and S. A. Zderic, *J. Organometal. Chem.*, **156**, 203 (1978).

[4] M. M. Midland and G. C. Kevon, *J. Org.*, **46**, 229 (1981).

Borane–Dimethyl sulfide, 4, 124, 191; **5**, 47; **8**, 49–50.

1-Arylnaphthalide lignans.[1] A novel synthesis of these lignans involves *in situ* generation of an isobenzofurane (**a**) from a substrate such as **1**, which is trapped by a Diels-Alder reaction with methyl acetylenedicarboxylate. The adduct aromatizes to a naphthol **2** in the presence of an acid. Selective reduction of the 3-

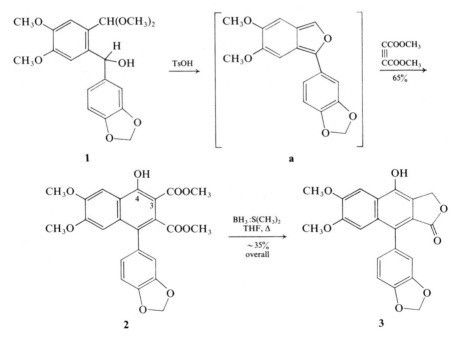

1   **a**

2   **3**

methoxycarbonyl group is achieved with borane–dimethyl sulfide in refluxing THF. The reduction is noteworthy because aromatic esters are reduced only very slowly by $BH_3:S(CH_3)_2$. The positive result is associated with the adjacent hydroxyl group. The product (3) is the natural lignan diphyllin. The same scheme was used to synthesize six other natural lignans.

[1] H. P. Plaumann, J. C. Smith, and R. Rodrigo, *J.C.S. Chem. Comm.*, 354 (1980).

Borane–1,4-Oxathiane, $BH_3:S$ ⟨O⟩ (1). Mol. wt. 118.00, m.p. 11–15°. The complex is obtained when BH_3 is passed into 1,4-oxathiane at 25°.

Hydroboration.[1] This complex of borane is more reactive than borane–dimethyl sulfide (4, 124, 191; 5, 47) and has the added advantage that after hydroboration, 1,4-oxathiane can be removed easily as the water-soluble sulfoxide.

[1] H. C. Brown and A. K. Mandal, *Synthesis*, 153 (1980).

Boron tribromide, 1, 66–67; 2, 33–34; 3, 30–31; 4, 42; 5, 49; 6, 64–65; 9, 61.

Cleavage of aryl methyl ethers (3, 30–31).[1] This reaction is conveniently effected with the stable complex $BBr_3:S(CH_3)_2$, m.p. 108°. The complex $BCl_3:S(CH_3)_2$, m.p. 90°,[2] is recommended for cleavage of a methylenedioxy group (4, 43).

[1] P. G. Williard and C. B. Fryble, *Tetrahedron Letters*, **21**, 3731 (1980).
[2] M. Schmidt and H. D. Block, *Ber.*, **103**, 3705 (1970).

Boron trifluoride, 1, 68–69; 3, 32–33; 5, 51–52; 7, 31.

Diels-Alder reactions (6, 65–66). 2-Methoxy-5-methylbenzoquinone shows no regioselectivity in reactions with piperylene or isoprene. These cycloadditions are catalyzed by both BF_3 and $SnCl_4$, but they favor different adducts in each case. The difference is believed to arise from different types of complexes with the quinone.[1]

Example:

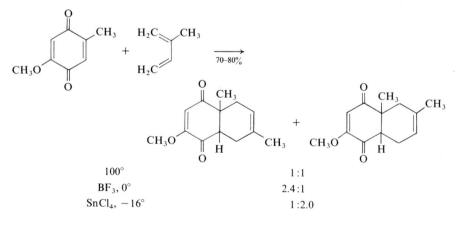

	100°	1:1
	BF_3, 0°	2.4:1
	$SnCl_4$, −16°	1:2.0

Rearrangement of γ-trimethylsilyl alcohols. In the presence of $BF_3\cdot 2HOAc$ *t*-alcohols substituted by a γ-silyl group undergo a pinacol-like rearrangement to give an alkene with loss of the silyl group. The rearrangement involves a hydride or an alkyl shift.[2]

Examples:

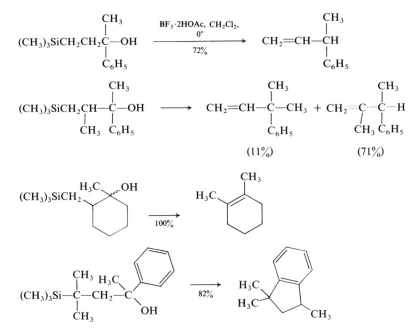

[1] J. S. Tou and W. Reusch, *J. Org.*, **45**, 5012 (1980).
[2] I. Fleming and S. K. Patel, *Tetrahedron Letters*, **22**, 2321 (1981).

Boron trifluoride–Dimethyl sulfide.

Debenzylation.[1] Benzyl ethers are cleaved in high yield by this combination, which is superior to BF_3 etherate and ethanethiol (**9**, 63) for substrates containing a carbonyl group or a Michael acceptor, both of which can react with ethanethiol.

[1] K. Fuji, T. Kawabata, and E. Fujita, *Chem. Pharm. Bull. Japan*, **28**, 3662 (1980).

Boron trifluoride–Ethanedithiol, 7, 33; 9, 63.

Deacetylation.[1] Treatment of the acetylpyrrole **1** with BF_3 etherate and ethane-dithiol in acetic acid gives the pyrrole **2** in quantitative yield. This reaction was first observed with the copper(II) complex of the acetylporphyrin **3**, which was converted to **4** in this way. The proposed mechanism involves protonation of the dithio-ketal *in situ*.

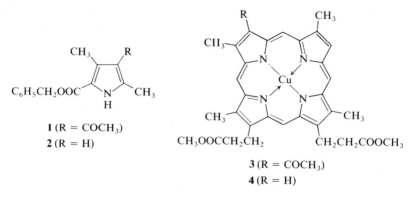

1 (R = COCH₃)
2 (R = H)

3 (R = COCH₃)
4 (R = H)

Cleavage of styrenes. ² The double bond of styrenes when substituted in the β-position by at least one electron-withdrawing group (NO₂, CN, COOC₂H₅) is cleaved by a hard Lewis acid and ethanethiol to give dithioacetals of benzaldehydes. The corresponding aliphatic compounds are cleaved by this system in lower yield.

Examples:

$$C_6H_5CH=C\begin{smallmatrix}CN\\COOC_2H_5\end{smallmatrix} \xrightarrow[88\%]{\substack{BF_3\cdot(C_2H_5)_2O,\\C_2H_5SH}} C_6H_5CH(SC_2H_5)_2$$

$$C_6H_5CH=C\begin{smallmatrix}NO_2\\CH_3\end{smallmatrix} \xrightarrow[48\%]{} C_6H_5CH(SC_2H_5)_2$$

¹ K. M. Smith and K. C. Langry, *J.C.S. Chem. Comm.*, 283 (1981).
² K. Fuji, T. Kawabata, M. Node, and E. Fujita, *Tetrahedron Letters*, **22**, 875 (1981).

Boron trifluoride etherate, 1, 70–72; **2**, 35–36; **3**, 33; **4**, 44–45; **5**, 54–55; **6**, 65–67; **7**, 31–32; **8**, 51–52; **9**, 64–65.

Diels-Alder reactions with **p-quinones** (**6**, 65–66).¹ The orientation of Diels-Alder reactions of 6-methoxy-1-vinyl-3,4-dihydronaphthalene (**1**) with *p*-quinones is subject to reversal by addition of BF₃ etherate (1.3 equivalent). Thus the thermal reaction with 2,6-dimethyl-*p*-benzoquinone (**2**) results in exclusive formation of **3**, whereas the catalyzed reaction leads predominately to the isomer **4**. The adduct **3** is stable to base, but the *syn, cis*-isomer **4** on treatment with Na₂CO₃ is converted to the more stable *anti, trans*-isomer **5**.

This orientation reversal is noted also with 2,5-dimethyl-*p*-benzoquinone and 2-acetyl-*p*-benzoquinone, but not with 6-methoxy-*p*-toluquinone. The orientation observed in the thermal reactions is unexpected, since Diels-Alder reactions with simple dienophiles give the two possible adducts in approximately equal amounts.

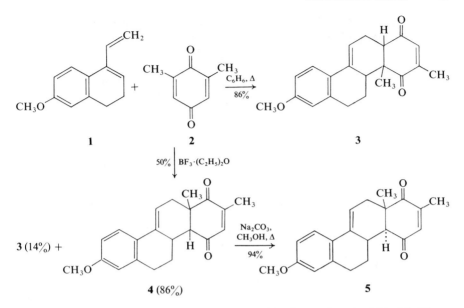

This orientation also obtains in cycloaddition with citraconic anhydride (**6**). These results are difficult to rationalize; in any case the reaction of **1** with quinones such as **2** provides a route to D-homosteroids with substituents at either C_{13} or C_{14}. In fact, the adduct **5** was converted to *dl*-estrone methyl ether.

6

Cuprate conjugate additions. One step in a recent synthesis of (+)-modhephene (**3**), a natural sesquiterpene with a (3.3.3)propellane skeleton, involved conjugate addition of lithium dimethylcuprate to **1**. The desired reaction proved difficult

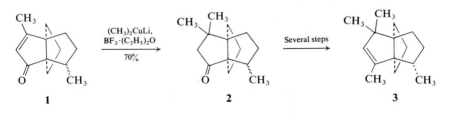

because of steric factors, but was finally achieved with catalysis by boron trifluoride etherate.[2]

A different route to **3** also involved a conjugate addition of a cuprate to a hindered enone, **4 → 5**, which was possible with BF_3 etherate as catalyst.[3]

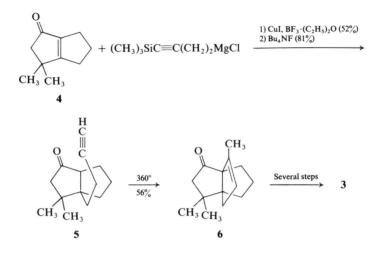

Use of BF_3 etherate as catalyst was suggested by the report of Yamamoto and Maruyama[4] that $n\text{-}C_4H_9Cu$ complexed with BF_3 undergoes conjugate addition to α,β-unsaturated acids and esters, a reaction that has not been observed with R_2CuLi, even in the presence of BF_3 etherate.

Cyclopentenones (6, 67–68). Details of the Lewis acid catalyzed decomposition of β,γ-unsaturated diazo ketones to form cyclopentenones have been published. Similar decomposition of γ,δ-unsaturated diazo ketones to β,γ-unsaturated cyclohexenones is possible, but yields are significantly lower.[5]

This decomposition can be used to initiate cyclization of di- and triunsaturated diazo ketones to bi- and tricyclic ketones. However, the cyclization is only possible if the β,γ-double bond is di- or trisubstituted (equation I). When the β,γ-double bond

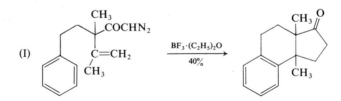

is monosubstituted, only the cyclopentenone is formed (in low yield). Moreover, the stereochemistry of the β,γ-double bond does not control the stereochemistry of the ring fusion obtained in cyclization. Formation of *cis*-fused products is highly favored as in the cyclization shown in equation (II).[6]

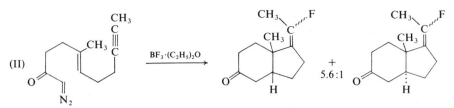

(II)

α-Methylene-γ-butyrolactones.[7] In the presence of this Lewis acid α-alkoxycarbonylallylsilanes react with acetals; the products can be converted into α-methylene-γ-butyrolactones by dealkylation with iodotrimethylsilane. The free aldehydes react with the silanes in the presence of $TiCl_4$, but yields are low.

Example:

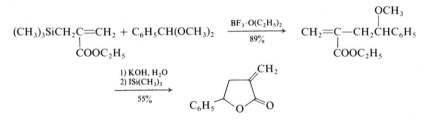

trans-2-Alkyltetrahydrofurane-3-carbaldehydes.[8] A new stereoselective synthesis of these compounds is outlined in equation (I). The synthesis involves the isomerization of a 4,7-dihydro-1,3-dioxepine to a 4,5-dihydrodioxepine with a ruthenium hydride complex followed by a Lewis acid catalyzed 1,3-alkyl migration. Both steps proceed in high yield; the second rearrangement is also highly stereoselective.

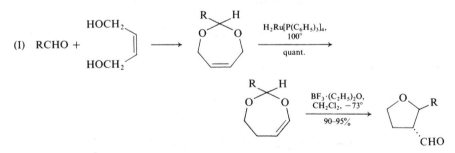

[1] J. Das, R. Kubela, G. A. MacAlpine, Ž. Stojanac, and Z. Valenta, *Can. J. Chem.*, **57**, 3308 (1979).

[2] A. B. Smith III, and P. J. Jerris, *Am. Soc.*, **103**, 194 (1981).

[3] H. Schostarez and L. A. Paquette, *Am. Soc*, **103**, 722 (1981).

[4] Y. Yamamoto and K. Maruyama, *Am. Soc.*, **100**, 3240 (1978).

[5] A. B. Smith III, B. H. Toder, S. J. Branca, and R. K. Dieter, *Am. Soc.*, **103**, 1996 (1981).

[6] A. B. Smith III, and R. K. Dieter, *Am. Soc.*, **103**, 2009, 2017 (1981).

[7] A. Hosomi, H. Hashimoto, and H. Sakurai, *Tetrahedron Letters*, **21**, 951 (1980).

[8] H. Suzuki, H. Yashima, T. Hirose, M. Takahashi, Y. Moro-oka, and T. Ikawa, *Tetrahedron Letters*, **21** 4927 (1980); *see also* H. D. Scharf and H. Frauenrath, *Ber.*, **113**, 1472 (1980).

Bromine, 3, 34; **4,** 46–47; **5,** 55–57; **6,** 70–73; **7,** 33–35; **8,** 52–53; **9,** 65–66.

OH → Br.[1] A primary or secondary alcohol can be converted to the alkyl phenyl selenide by reaction with phenyl selenocyanate and tri-*n*-butylphosphine (7, 252–253). Reaction with Br_2 and triethylamine (7, 34–35) replaces the SeC_6H_5 group by Br.

Overall yields are generally about 60%. In the case of secondary alcohols both steps proceed with about 95% inversion of configuration; the net effect is >90% retention of configuration.

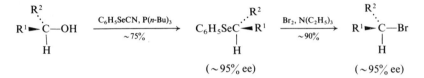

[1] M. Sevrin and A. Krief, *J.C.S. Chem. Comm.*, 656 (1980).

Bromomethylenetriphenylphosphorane, $(C_6H_5)_3P=CHBr$ (**1**). Mol. wt. 355.23. The phosphorane can be generated *in situ* by treatment of bromomethyltriphenyl-phosphonium bromide with potassium *t*-butoxide in THF at −78°.

Terminal alkynes.[1] The phosphorane reacts with aldehydes to form (Z)- and (E)-1-bromoalkenes. In the presence of excess base 1-alkynes are formed in

$$RCHO \xrightarrow{\quad 1 \quad} RCH=CHBr \xrightarrow[35-80\%]{KOC(CH_3)_3} RC\equiv CH$$

satisfactory yield. This reaction is also useful for preparation of conjugated terminal enynes.

[1] M. Metsumoto and K. Kuroda, *Tetrahedron Letters*, **21**, 4021 (1980).

Bromomethyl methyl ether (Methoxymethyl bromide), $BrCH_2OCH_3$. Mol. wt. 124.97, b.p. 87°. Supplier: Aldrich.

α-Methylenation of a γ-lactone.[1] A two-step preparation of an α-methylene-γ-lactone (**3**) involves alkylation of the enolate of the γ-lactone (**1**) with bromomethyl methyl ether under the conditions of Herrmann and Schlessinger.[2] The product (**2**) is treated with anhydrous $KOH-KOC(CH_3)_3$ to effect elimination of methanol and hydrolysis to the acrylate anion; acid quench then affords the desired **3**.

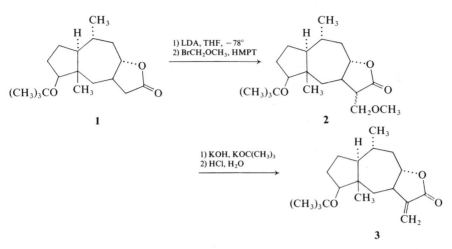

1

1) LDA, THF, −78°
2) BrCH₂OCH₃, HMPT

2

1) KOH, KOC(CH₃)₃
2) HCl, H₂O

3

¹ P. T. Lansbury, D. G. Hangauer, Jr., and J. P. Vacca, *Am. Soc.*, **102**, 3964 (1980).
² J. L. Herrmann and R. H. Schlessinger, *J.C.S. Chem. Comm.*, 711 (1973).

N-Bromosuccinimide, 1, 78–80; **2**, 40–42; **3**, 34–36; **4**, 49–53; **5**, 65–66; **6**, 74–76; **7**, 37–40; **8**, 54–56; **9**, 70–72.

α,β-*Enone fragmentation.* The fragmentation of α,β-enones to ring-enlarged alkynones has usually been conducted via the tosylhydrazone of the α,β-epoxy ketone (Eschenmoser fragmentation, **2**, 419–422¹). The most serious drawback to this method is the difficulty in carrying out epoxidation of sterically hindered α,β-enones. This reaction has been circumvented in a new fragmentation reaction based on a method for regeneration of ketones from the tosylhydrazone by NBS and

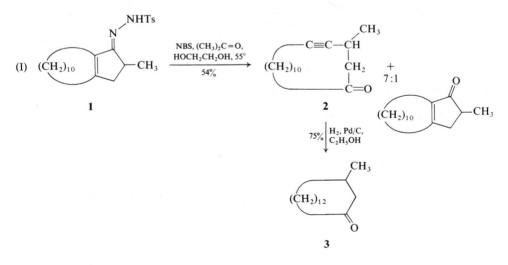

methanol (**6**, 74–75). The vinylogous fragmentation is facilitated by use of a more bulky alcohol such as *sec*-butyl alcohol or ethylene glycol in place of methanol. A synthesis of muscone (**3**) by this method is shown in equation (I).[2]

Brominative cyclization. Treatment of either costunolide (**1**) or deydrosaussurea lactone (**2**) with NBS in aqueous acetone at 20° results in the cyclized bromo lactones **3**, **4**, and **5**, in approximately the same ratio from both substrates.[3]

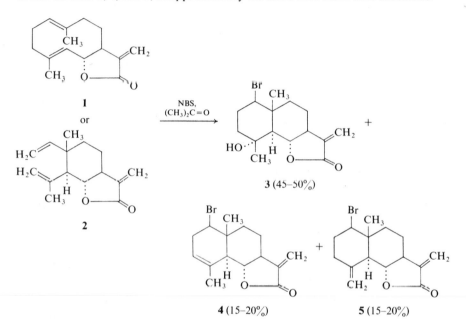

[1] For improvements *see* G. A. MacAlpine and J. Warkentin, *Can. J. Chem.*, **56**, 308 (1978).
[2] C. Fehr, G. Ohloff, and G. Büchi, *Helv.*, **62**, 2655 (1979).
[3] T. C. Jain and C. M. Banks, *Can. J. Chem.*, **58**, 447 (1980).

***trans*-Bromo-*o*-tolylbis(triethylphosphine)nickel(II),** $[(C_2H_5)_3P]_2Ni\genfrac{}{}{0pt}{}{C_6H_4CH_3\text{-}o}{Br}$ (**1**).

Mol. wt. 466.06, m.p. 102–103° (dec.). The complex is prepared by reaction of *o*-tolylmagnesium bromide with dibromobis(triethylphosphine)nickel(II); yield 67%.[1]

Ullmann reaction. During a study of the synthesis of biaryls with nickel catalysts (**4**, 33; **6**, 654), Tsou and Kochi[2] found that nickel(II) complexes such as **1** readily yield biaryls on treatment with an aryl halide and that they are important intermediates in the radical-chain process of nickel-catalyzed Ullmann reactions.

Halogen exchange with aryl and vinyl halides.[3] This nickel complex is a particularly active catalyst for exchange between haloarenes or vinylic halides and inorganic halide salts. In the case of the vinylic halides the exchange is particularly rapid and no *cis–trans* isomerization is involved. The inorganic halide salt can also be replaced by another aryl halide.

[1] D. G. Morrell and J. K. Kochi, *Am. Soc.*, **97**, 7262 (1965).
[2] T. T. Tsou and J. K. Kochi, *Am. Soc.*, **101**, 7547 (1979).
[3] T. T. Tsou and J. K. Kochi, *J. Org.*, **45**, 1930 (1980).

Bromotrimethylsilane, 9, 73–74.

Acid bromides.[1] Acid chlorides can be converted to acid bromides by reaction with bromotrimethylsilane (75–95% yield). Acid iodides can be obtained by use of iodotrimethylsilane.

Alkyl bromides.[2] The reagent cleaves oxiranes of type of example 1 exclusively to primary alkyl bromides. Four-membered cyclic ethers also react readily, but five-membered cyclic ethers require long periods of reflux.

Examples:

$$CH_3CH-CH_2 \xrightarrow[61\%]{BrSi(CH_3), -60°} CH_3CH-CH_2Br \quad (OSi(CH_3)_3)$$

$$\square O \xrightarrow[62\%]{-60°} BrCH_2CH_2CH_2OSi(CH_3)_3$$

$$\text{(cyclic)} O \xrightarrow[82\%]{\Delta} Br(CH_2)_4OSi(CH_3)_3$$

$$C_6H_5CHOC_2H_5 \xrightarrow[93\%]{\Delta} C_6H_5CHBr \quad (CH_3)$$

Glycosyl bromides.[3] Glycosyl bromides can be prepared conveniently and in high yield by reaction of bromotrimethylsilane with anomeric glycosyl acetates. Ester groups at other positions are stable under the mild conditions ($25°$, $HCCl_3$, ~ 40 min) employed. Methyl glycosides can also be used, but 1,2-O-isopropylidenes react very slowly.

[1] A. H. Schmidt, M. Russ, and D. Grosse, *Synthesis*, 216 (1981).
[2] H. R. Kricheldorf, G. Morber, and W. Regel, *Synthesis*, 383 (1981).
[3] J. W. Gillard and M. Israel, *Tetrahedron Letters* **22**, 513 (1981).

Bromotrimethylsilane–Cobalt(II) bromide.

α-Glucosylation.[1] 2,3,4,6-Tetra-O-benzyl-α-D-glucopyranose (**1**) is converted in one step to the α-D-glucoside by reaction with an alcohol in CH_2Cl_2 in the presence of $BrSi(CH_3)_3$ (1 equivalent), $CoBr_2$ (1 equivalent), $(n\text{-}C_4H_9)_4NBr$ (1 equivalent), and 4 Å molecular sieves (equation I).

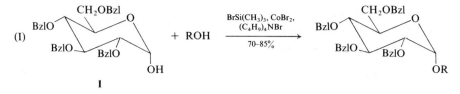

(I)

¹ N. Morishima, S. Koto, C. Kusuhara, and S. Zen, *Chem. Letters*, 427 (1981).

Bromotriphenylphosphonium bromide, 1, 1247–1249; **2,** 446; **3,** 320–322; **4,** 555; **5,** 729–731; **6,** 647–648; **7,** 407.

Carbodiimides. Even unstable carbodiimides can be obtained by dehydration of ureas with this reagent and triethylamine in cold methylene chloride (equation I).[1]

$$(I)\quad R^1NHCNHR^2 + (C_6H_5)_3\overset{+}{P}BrBr^- \xrightarrow[\text{}]{N(C_2H_5)_3,\ CH_2Cl_2,\ 0°} R^1NH\overset{\overset{+}{O}P(C_6H_5)_3}{\overset{|}{C}}=NR^2Br^- \xrightarrow[30-90\%]{}$$

$$R^1N=C=NR^2$$

N-Tosylketenimines.[2] Sulfimides (**1**) are converted to N-tosylketenimines (**2**) on reaction with $(C_6H_5)_3PBr_2$ and $N(C_2H_5)_3$. These ketenimines form [2 + 2] cycloadducts (**3**) with Schiff bases. The method is applicable to preparation of N-tosylazetidin-2-imines related to cephalosporins.

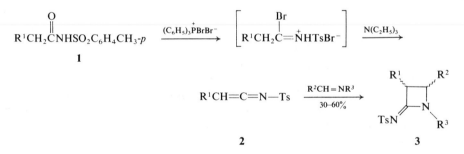

¹ C. Palomo and R. Mestres, *Synthesis*, 373 (1981).
² A. Van Camp, D. Goossens, M. Moya-Portuguez, J. Marehand-Brynaert, and L. Ghosez, *Tetrahedron Letters*, **21**, 3081 (1980).

Butane-1,4-bis[triphenylphosphonium] dibromide (1). Mol. wt. 740.51, m.p. 293°.
Preparation[1]:

$$(C_6H_5)_3P + Br(CH_2)_4Br \xrightarrow[80\%]{250°} (C_6H_5)_3\overset{+}{P}(CH_2)_4\overset{+}{P}(C_6H_5)_32Br^-$$

1

2,3-Dialkyl-1,3-cyclohexadienes.[2] The diylide of **1** [KOC(CH$_3$)$_3$, tetraglyme] reacts with α-dicarbonyl compounds to form 2,3-dialkyl-1,3-cyclohexadienes, albeit in modest yields. Tetraglyme is used as solvent, since the products are rather volatile and are isolated by distillation.

Examples:

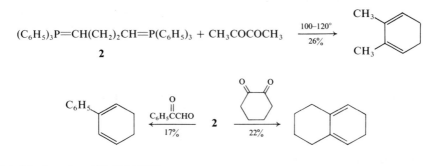

$$(C_6H_5)_3P{=}CH(CH_2)_2CH{=}P(C_6H_5)_3 + CH_3COCOCH_3 \xrightarrow[26\%]{100-120°}$$

2

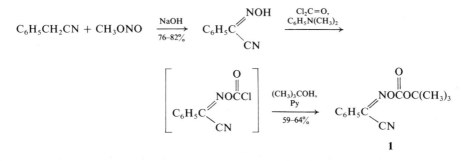

[1] A. Mondon, *Ann.*, **603**, 115 (1957).
[2] K. B. Becker, *Synthesis*, 238 (1980).

2-*t*-Butoxycarbonyloxyimino-2-phenylacetonitrile (1). Mol. wt. 246.26, m.p. 84–86°, gradually evolves CO$_2$ on storage.

Preparation:

$$C_6H_5CH_2CN + CH_3ONO \xrightarrow[76-82\%]{NaOH} \overset{NOH}{\underset{CN}{C_6H_5C}} \xrightarrow{\underset{C_6H_5N(CH_3)_2}{Cl_2C=O,}}$$

$$\left[\overset{\overset{O}{\parallel}}{\underset{CN}{C_6H_5C}}\overset{NOCCl}{} \right] \xrightarrow[59-64\%]{\underset{Py}{(CH_3)_3COH,}} \underset{CN}{C_6H_5C}\overset{\overset{O}{\parallel}}{NOCOC(CH_3)_3}$$

1

BOC-Amino acids. The reagent, in combination with 1.5 equivalent of triethylamine, converts amino acids into the BOC derivative at 25° in 65–100% yield. The by-product, 2-hydroxyimino-2-phenylacetonitrile, is easily removed by extraction into an organic solvent (ethyl acetate–hexane).[1]

Chemists at Merck Sharp & Dohme have reported the preparation of the BOC derivatives of a number of amino acids by use of this reagent in 60–95% yield.[2]

[1] M. Itoh, D. Hagiwara, and T. Kamiya, *Bull. Chem. Soc. Japan*, **50**, 718 (1977); *Org. Syn.*, **59**, 95 (1979).
[2] W. J. Paleveda, F. W. Holly, and D. F. Veber, *Org. Syn.* submitted (1980).

t-**Butylamine, 1**, 84; **2**, 43.

Demethylation of $\overset{\displaystyle O}{\overset{\|}{\diagup}}POCH_3$**.** Phosphate and phosphonate esters are partially dealkylated by tertiary amines. *t*-Butylamine is superior for this cleavage, and it is very selective. Only O—CH$_3$ bonds are affected: ethyl esters are stable to the amine for several weeks. A benzyl group can be cleaved under forcing conditions.[1]

Examples:

$$CH_3\overset{O}{\overset{\|}{P}}(OCH_3)_2 + (CH_3)_3CNH_2 \xrightarrow{25°,\ 10\ min} CH_3\underset{\overset{|}{OCH_3}}{\overset{\overset{\displaystyle O}{\|}}{P}}{-}O^- \ CH_3\overset{+}{N}H_2C(CH_3)_3$$

$$C_6H_5\underset{\overset{\|}{O}}{\overset{\overset{\displaystyle OCH_3}{|}}{P}}(CH_2)_2COOCH_3 + (CH_3)_3CNH_2 \longrightarrow$$

$$C_6H_5\underset{\overset{\|}{O}}{\overset{\overset{\displaystyle O^-}{|}}{P}}(CH_2)_2COOCH_3CH_3\overset{+}{N}H_2C(CH_3)_3$$

This selectivity permits use of a methyl protecting group in phosphate triester synthesis of nucleotides. Removal is effected by heating solutions of the protected nucleotide in *t*-butylamine at reflux (46°) for 15 hours.[2]

[1] M. D. M. Gray and D. J. H. Smith, *Tetrahedron Letters*, **21**, 859 (1980)
[2] D. J. H. Smith, K. K. Ogilvie, and M. F. Gillen, *Tetrahedron Letters*, **21**, 861 (1980).

t-**Butyldimethylchlorosilane, 4**, 57–58, 176–177; **5**, 74–75; **6**, 78–79; **7**, 59; **8**, 77; **9**, 77.

Transsilylation. Several reagents have been recommended for preparation of *t*-butyldimethylsilyl ethers by transsilylation. These include allyl-*t*-butyldimethyl-silane[1] and *t*-butyldimethylsilyl enol ethers of pentane-2,4-dione and methyl aceto-acetate,[2] both prepared with *t*-butyldimethylchlorosilane and imidazole. Unlike the reaction of *t*-butyldimethylchlorosilane with alcohols, which requires a base catalyst, these new reagents convert alcohols to silyl ethers under slightly acidic conditions (TsOH) in good yield. The trimethylsilyl ethers of pentane-2,4-dione and methyl acetoacetate convert alcohols to trimethylsilyl ethers at room temperature even with no catalyst. The former reagent is also useful for silylation of nucleotides.[3]

[1] T. Morita, Y. Okamoto, and H. Sakurai, *Tetrahedron Letters*, **21**, 835 (1980).
[2] T. Veysoglu and L. A. Mitscher, *Tetrahedron Letters*, **22**, 1299, 1303 (1981).
[3] C. F. Bigge and M. P. Mertes, *J. Org.*, **46**, 1994 (1981).

$$\underset{\overset{|}{CH_3}}{\overset{\overset{CH_3}{|}}{}}$$

t-**Butyldimethyliodosilane,** $I\overset{\overset{CH_3}{|}}{\underset{\underset{CH_3}{|}}{Si}}C(CH_3)_3$ (**1**). Mol. wt. 242.18, unstable. The reagent is

generated *in situ* by treatment of *t*-butyldimethyl(phenylseleno)silane with 0.5 equivalent of iodine in CH_3CN.

Allylic alcohols from epoxides.[1] Reaction of **1** with epoxides and then DBN gives reasonable yields of allylic alcohols as the *t*-butyldimethylsilyl ether.

Examples:

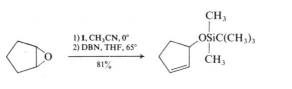

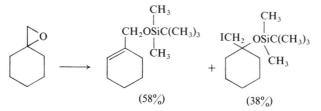

(58%)   (38%)

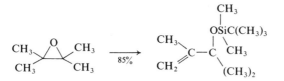

[1] M. R. Detty, *J. Org.*, **45**, 924 (1980).

t-**Butyldimethylsilyl   trifluoromethanesulfonate,** $(CH_3)_3C\overset{\overset{CH_3}{|}}{\underset{\underset{CH_3}{|}}{Si}}OSO_2CF_3$. Mol. wt.

207.23, b.p. 180°. The reagent is obtained by reaction of *t*-butyldimethylchlorosilane with trifluoromethanesulfonic acid (89% yield).

Silylation.[1] This triflate is a very reactive silylating reagent. It was used successfully to prepare 1,4-bis(*t*-butyldimethylsilyloxy)-2,5-di-*t*-butylbenzene (88% yield).

[1] R. F. Stewart and L. L. Miller, *Am. Soc.*, **102**, 4999 (1980).

***t*-Butyl hydroperoxide, 1,** 88–89; **2,** 49–50; **3,** 37–38; **5,** 75–77; **6,** 81–82; **7,** 43–45; **8,** 62–64; **9,** 78–79. Absolute reagent can be obtained from 80% commercial material by five consecutive treatments with thoroughly dried 3-Å molecular sieves (Fluka).[1]

Asymmetric epoxidation of allylic alcohols.[2] Epoxidation of allylic alcohols with *t*-butyl hydroperoxide in the presence of titanium(IV) isopropoxide as the metal catalyst and either diethyl D- or diethyl L-tartrate as the chiral ligand proceeds in ⩾90% stereoselectivity, which is independent of the substitution pattern of the allylic alcohol but dependent on the chirality of the tartrate. Suggested standard conditions are 2 equivalents of anhydrous *t*-butyl hydroperoxide with 1 equivalent each of the alcohol, the tartrate, and the titanium catalyst. Lesser amounts of the last two components can be used for epoxidation of reactive allylic alcohols, but it is important to use equivalent amounts of these two components. Chemical yields are in the range of 70–85%.

Examples:

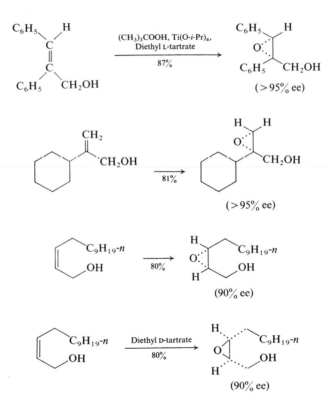

Asymmetric epoxidation is a key step in a synthesis of (+)-disparlure (**1**), the sex attractant of the gypsy moth, as outlined in equation (I).[3]

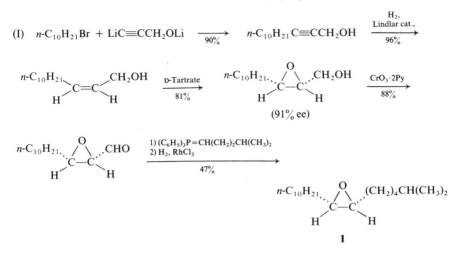

(I) n-$C_{10}H_{21}Br$ + $LiC≡CCH_2OLi$ $\xrightarrow{90\%}$ n-$C_{10}H_{21}C≡CCH_2OH$ $\xrightarrow[96\%]{\text{H}_2,\ \text{Lindlar cat.,}}$

1

[1] H. Langhals, E. Fritz, and I. Mergelsberg, *Ber.*, **113**, 3662 (1980).

[2] T. Katsuki and K. B. Sharpless, *Am. Soc.*, **102**, 5974 (1980).

[3] B. E. Rossiter, T. Katsuki, and K. B. Sharpless, *Am. Soc.*, **103**, 464 (1981).

t-Butyl hydroperoxide–Diaryl selenides, 9, 79.

Selective oxidations. *t*-Butyl hydroperoxide (70%) in combination with dimesityl diselenide is useful for oxidation of hydroxyl groups in the presence of —SC_6H_5 or —SeC_6H_5 groups.[1]

Examples:

$$n\text{-}C_4H_9\text{—}\underset{\underset{OH}{|}}{CH}\text{—}\underset{\underset{SeC_6H_5}{|}}{CH}C_4H_9\text{-}n \xrightarrow[90\%]{t\text{-BuOOH, (ArSe)}_2} n\text{-}C_4H_9\underset{\underset{O}{\|}}{C}\text{—}\underset{\underset{SeC_6H_5}{|}}{CH}C_4H_9\text{-}n$$

$$CH_3\underset{\underset{SeC_6H_5}{|}}{C}=CHCH_2OH \xrightarrow{64\%} CH_3\underset{\underset{SeC_6H_5}{|}}{C}=CHCHO$$

[1] M. Shimizu, H. Urabe, and I. Kuwajima, *Tetrahedron Letters*, **22**, 2183 (1981).

t-Butyl hydroperoxide–Selenium dioxide, 8, 64–65; 9, 79–80.

Allylic oxidation. Germacrane-type sesquiterpene lactones such as epitulipin-olide (**1**) are oxidized by *t*-butyl hydroperoxide and a catalytic amount of SeO_2 regio- and stereoselectively at C_{14} to allylic alcohols (**2**) containing a (Z)-double bond.[1]

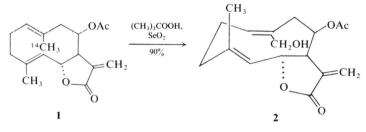

1   2

[1] M. Haruna and K. Ito, *J.C.S. Chem. Comm.*, 483 (1981).

t-Butyl hydroperoxide–Vanadyl acetylacetonate, 2, 287; **4**, 346; **5**, 75–76; **7**, 43–44; **9**, 81–82.

   Stereoselective epoxidation of an allylic alcohol (**5**, 75–76; **9**, 81–82). The antibiotic methyl pseudomonate A (**2**) is the β-epoxide of methyl pseudomonate C. Epoxidation of methyl pseudomonate C with *m*-chloroperbenzoic acid in CH_2Cl_2

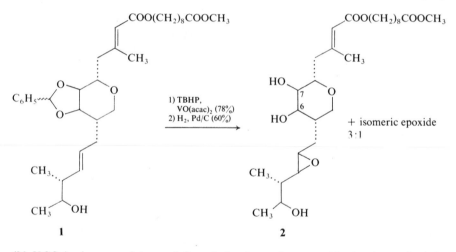

1   2

($NaHCO_3$) gives a mixture of **2** and the isomeric α-epoxide in the ratio 1:1. Epoxidation with TBHP/VO(acac)$_2$ gives the two epoxides in the ratio 1.5:1. This reaction becomes more stereoselective if the 6,7-*cis*-diol unit is protected as the benzylidene derivative (**1**). In this reaction the two epoxides are formed in the ratio 3:1, presumably because the 6,7-diol can no longer form a complex with the metal catalyst.[1]

[1] A. P. Kozikowski, R. J. Schmiesing, and K. L. Sorgi, *Tetrahedron Letters*, **22**, 2059 (1981).

t-Butyl hypochlorite, 1, 90–94; **2**, 50; **3**, 38; **4**, 58–60; **5**, 77–78; **6**, 82.

   Selenoxides: telluroxides. Diaryl or alkyl aryl selenides and diaryl or dialkyl tellurides are oxidized to oxides in 40–95% yield by treatment with *t*-butyl hypochlorite followed by alkaline hydrolysis ($NaHCO_3$).[1] The by-product is *t*-butyl alcohol. NCS can also be used, but separation of the by-product succinimide is sometimes tedious.

[1] M. R. Detty, *J. Org.*, **45**, 274 (1980).

t-Butylisonitrile, 2, 50–51; 9, 82.[1]

Amino acid synthesis. Ugi's amino acid synthesis consists of condensation of a carbonyl compound, a carboxylic acid, an amine, and an isonitrile.[2]

Joullié and co-workers[3] have found that even labile thiophene-, furane-, and tetrahydrofuranecarboxaldehydes can be used in this condensation. More recently they[4] have employed this reaction in a total synthesis of (+)-furanomycin (**2**), an antibiotic from *S. threomyceticus*, which also clarified the configuration (equation I). The starting chiral material (**1**) is derived from α-D-glucose, from which the overall yield of **2** is 6%.

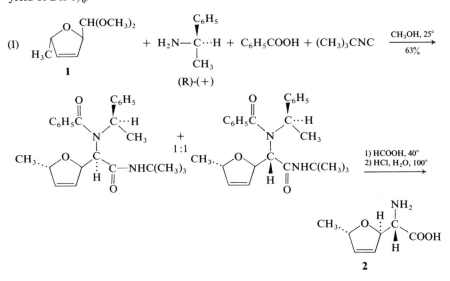

[1] Preparation: I. Ugi, R. Meyr, M. Lipinski, F. Bodesheim, and F. Rosendahl, *Org. Syn. Coll. Vol.*, **5**, 300 (1973).
[2] I. Ugi, *Angew. Chem. Int. Ed.*, **14**, 61 (1975).
[3] H. R. Divanford, Z. Lysenko, P.-C. Wang, and M. M. Joullié, *Syn. Comm.*, **8**, 269 (1978).
[4] M. M. Joullie, P.-C. Wang, and J. E. Semple, *Am. Soc.*, **102**, 887 (1980).

t-Butyl lithioacetate, 4, 306; 5, 371; 6, 84.

Chain extension.[1] An alternative to the malonic ester synthesis involves alkylation of *t*-butyl lithioacetate. Yields are improved by preparation of the enolate with LiICA (**4**, 306) at −78°. Alkylation proceeds in highest yields at −78 to −35°

and in THF–HMPT as solvent. Alkyl iodides are generally superior to bromides. The final step involves thermolysis at 100° in the presence of *p*-TsOH.

Example:

$$\text{LiCH}_2\text{COOC(CH}_3)_3 \xrightarrow[\substack{80-85\%}]{\substack{1) \text{ C}_8\text{H}_{17}\text{I, THF–HMPT} \\ 2) 100°, \text{ H}^+}} \text{C}_8\text{H}_{17}\text{CH}_2\text{COOH}$$

[1] W. Bos and H. J. J. Pabon, *Rec. trav.*, **99**, 141 (1980).

n-Butyllithium, 1, 95–96; **2**, 51–53; **4**, 60–63; **5**, 78; **6**, 85–91; **7**, 45–47; **8**, 65–66; **9**, 83–87.

Metalated enamines. Metalation–alkylation of the β,γ-unsaturated amine (**1**) occurs regioselectively to give enamines such as **2**. This reaction was used in a short synthesis of N-methyl-14α-morphinan (**5**), which contains the ring system of the morphine alkaloids, as outlined in equation (I). The product was cyclized, reduced, and converted into the *cis*-fused immonium salt **3**, which differs from the morphinan **5** only by a methylene bridge. Reaction of diazomethane with **3** gave an aziridinium salt, which was cleaved with lithium chloride to produce **4**. Ring closure to **5** was effected with aluminum chloride. Surprisingly, the reaction of diazomethane with **3** gave **5** directly in 15% yield.[1]

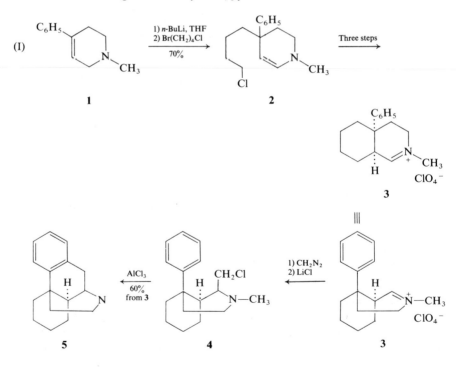

o-*Lithiation of arenesulfonic acids.* Lithium arenesulfonates are lithiated exclusively at the *ortho*-position by *n*-butyllithium in THF at 0°.[2] The products can react with various electrophiles. Since the sulfonic acid group can be replaced by hydrogen by acid hydrolysis, this *ortho*-lithiation can be used to obtain substituted arenes that are not readily available.

Example:

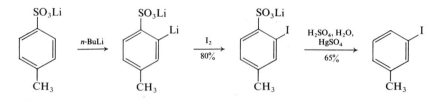

Indole synthesis. Fuhrer and Gschwend[3] have converted N-pivaloylaniline into the dilithio derivative **a** by reaction with 2 equivalents of *n*-butyllithium and have reported that *ortho*-substituted derivatives are obtained in good yield by reaction of **a** with an electrophile. The same species can be prepared somewhat more efficiently from an *o*-bromo-N-pivaloylaniline (**1**) by bromine–lithium exchange and N-deprotonation with methyllithium and *t*-butyllithium (equation I). This dilithium reagent can be used for synthesis of indoles.[4] Thus it reacts with a biselectrophile

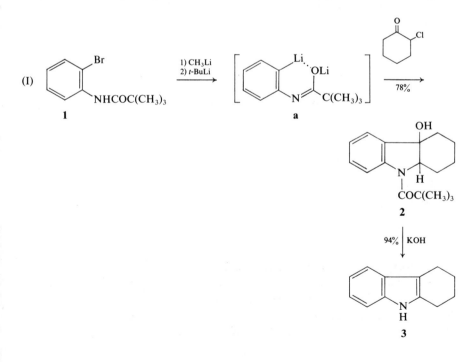

such as α-chlorocyclohexanone to give **2**, which is converted on hydrolysis with anhydrous KOH into tetrahydrocarbazole **3**.

An enedione can also be used as the biselectrophile as formulated for a synthesis of the methyl ester of indole-2-acetic acid (equation II).

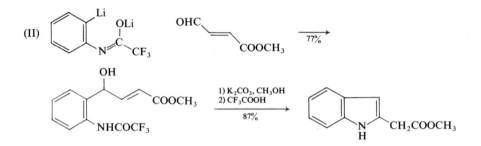

2,4-Dienones.[5] The cleavage by *n*-BuLi of an Se—C bond with generation of a furfuryllithium has been used for homologation of aldehydes to 2,4-dienones (equation I).

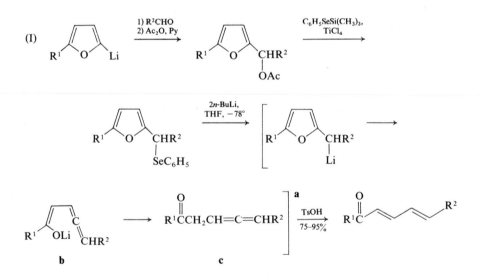

Trimethylsilyloxyallenes; α,β-unsaturated ketones.[6] 1-Trimethylsilylpropargylic alcohols (**1**) on treatment with a catalytic amount of *n*-butyllithium in THF undergo Brook rearrangement[7] to give trimethylsilyloxyallenes (**2**). If 1 equivalent of *n*-butyllithium is used and hexane is the solvent the intermediate (**b**) can be alkylated to afford α,β-unsaturated ketones (**3**) after acidic work-up.

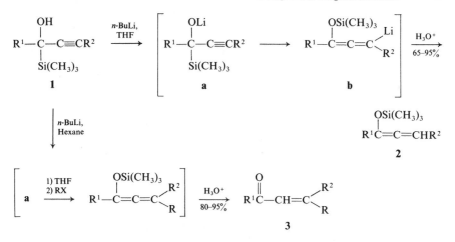

Secondary thiols.[8] A new preparation involves dithiolane cleavage with *n*-butyllithium to thiones, which are known to be reduced by the same reagent to *sec*-thiols by β-hydride transfer. Yields are 78–90% from saturated thioketals. The method is less useful for preparation of aryl substrates owing to competitive metalation of the aromatic ring.

[1] D. A. Evans, C. H. Mitch, R. C. Thomas, D. M. Zimmerman, and R. L. Robey, *Am. Soc.*, **102**, 5955 (1980).

[2] G. D. Figuly and J. C. Martin, *J. Org.,* **45**, 3728 (1980).

[3] W. Fuhrer and H. W. Gschwend, *J. Org.*, **44**, 1133 (1979).

[4] P. A. Wender and A. W. White, *Tetrahedron Letters,* **22**, 1475 (1981).

[5] I. Kuwajima, S. Hashino, T. Tanaka, and M. Shimizu, *Tetrahedron Letters*, **21**, 3209 (1980); *see also* K. Atsumi and I. Kuwajima, *Am. Soc.*, **101**, 2208 (1979).

[6] I. Kuwajima and M. Kato, *Tetrahedron Letters*, **21**, 623 (1980).

[7] A. G. Brook, *Accts. Chem. Res.*, **7**, 77 (1974).

[8] S. R. Wilson and G. M. Georgiadis, *Org. Syn.*, submitted (1980).

n-Butyllithium–Magnesium bromide.

Epoxide cleavage.[1] Reaction of *o*-bromo- and *o*-iodostyrene oxides in THF containing suspended MgBr₂ (2 equivalents) with *n*-butyllithium at −78° results in benzocyclobutenols.

Examples:

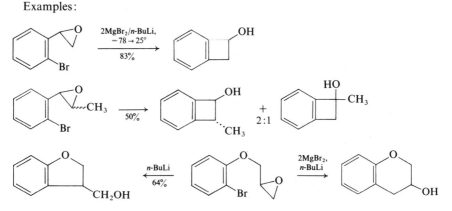

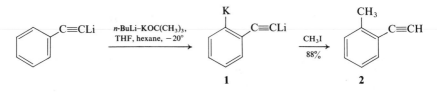

The last example illustrates a difference between the reaction of such substrates with *n*-butyllithium alone[2] and with *n*-butyllithium and magnesium bromide.

[1] K. L. Dhawan, B. D. Gowland, and T. Durst, *J. Org.*, **45**, 922 (1980).
[2] C. K. Bradsher and D. C. Reames, *J. Org.*, **43**, 3800 (1978).

n-Butyllithium–Potassium t-butoxide, 5, 552; **8**, 67; **9**, 87.

Dimetallation of phenylacetylene. This dimetallation is not possible with *n*-butyllithium, but is possible with the complex base *n*-BuLi–KOC(CH$_3$)$_3$ (2:1 equivalents) in THF and hexane at $-20°$. The dimetallated product **1** reacts as

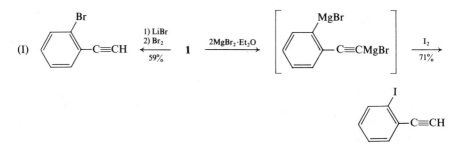

expected with methyl iodide and dimethyl disulfide. However, the reaction of **1** with chlorotrimethylsilane (1 equivalent) results in comparable amounts of phenylacetylene and the two possible monotrimethylsilyl derivatives of **1**.[1]

The dianion can be converted into the *o*-haloacetylenes (equation I).[2]

[1] H. Hommes, H. D. Verkruijsse, and L. Brandsma, *J.C.S. Chem. Comm.*, in press.
[2] H. Hommes, H. D. Verkruijsse, and L. Brandsma, *Tetrahedron Letters*, **22**, 2495 (1981).

n-**Butyllithium–Tetramethylethylenediamine**, **2**, 403; **3**, 284; **4**, 485–489; **5**, 678–680; **7**, 47–48; **8**, 67–69.

α,α′-Dianions of ketones. The α,α′-dianions **1** of acetone can be prepared by treatment of potassioacetone (prepared with KH in ether, 25°) with equivalent amounts of *n*-butyllithium and TMEDA at 0° for 5 minutes. This dianion can be monoalkylated or monoacylated much more readily than the monoanion. The ability of aliphatic ketones to form dianions is fairly general. This useful method, however, is probably limited to symmetrical ketones. Some useful reactions of these dianions are formulated in equations (I) and (II).[1]

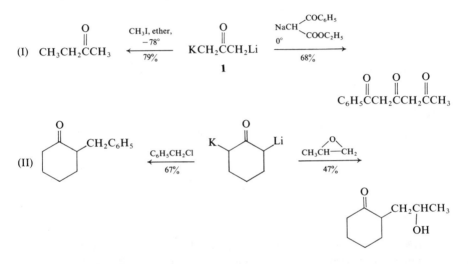

The same technique can be used to form the 1,3,5-trianion of 2,4-pentanedione. This anion is also monoalkylated at C_1 (equation (III).

$$\text{(III)} \quad \underset{\underset{Na}{|}}{\text{LiCH}_2\overset{O}{\overset{\|}{\text{C}}}\text{CHCH}\overset{O}{\overset{\|}{\text{C}}}\text{CH}_2\text{Li}} \quad \xrightarrow[62\%]{\text{C}_6\text{H}_5\text{CH}_2\text{Cl, 0°}} \quad \text{C}_6\text{H}_5\text{CH}_2\text{CH}_2\overset{O}{\overset{\|}{\text{C}}}\text{CH}_2\overset{O}{\overset{\|}{\text{C}}}\text{CH}_3$$

Dilithiation of α,β-unsaturated amides. Two laboratories[2,3] have reported that either *n*-butyllithium or *sec*-butyllithium in combination with TMEDA can dimetalate α,β-unsaturated secondary amides and that the second metalation occurs selectively at the β′-position even when the γ-position bears a proton. A similar preference has been noted in aromatic systems.[4] This dimetalation has been developed particularly in the case of N-*t*-butylmethacrylamide (**1**) and leads to a reagent (**2**) equivalent to the dianion of methacrylic acid in reactions with various electrophiles.

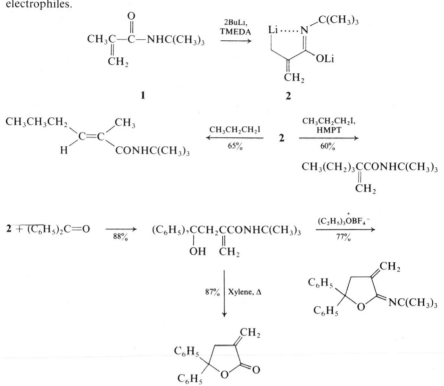

ortho-**Lithiation of benzamides** (**9**, 5–6). Use of dilithio derivatives of benzamides for the synthesis of anthraquinones has been extended to a synthesis of aklavinone (**1**), the aglycone of an 11-deoxyanthracycline antibiotic.[5]

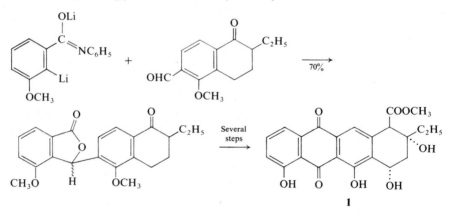

[1] J. S. Hubbard and T. M. Harris, *Am. Soc.*, **102**, 2110 (1980).

[2] P. Beak and D. J. Kempf, *Am. Soc.*, **102**, 4550 (1980).

[3] J. J. Fitt and H. W. Gschwend, *J. Org.*, **45**, 4259 (1980).

[4] H. W. Gschwend and H. R. Rodriguez, *Org. React.*, **26**, 1 (1979).

[5] A. S. Kende and J. P. Rizzi, *Tetrahedron Letters*, **22**, 1779 (1981).

sec-**Butyllithium, 5**, 78–79; **9**, 87–88.

Directed metalation. ortho-Metalation of benzamides has been used for a one-pot route to polycyclic anthraquinones and heterocyclic benzoquinones.[1]

Example:

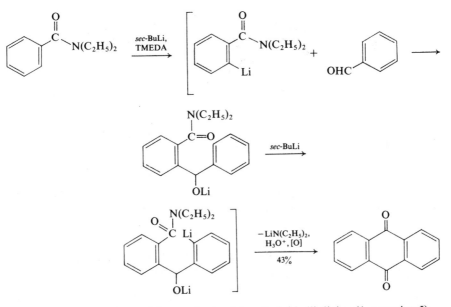

The method was also used for synthesis of the alkaloid elliplicine (**1**, equation I).

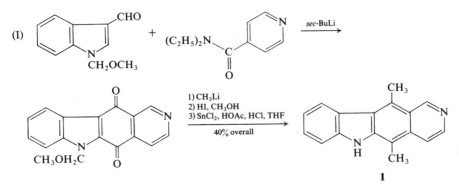

[1] M. Watanabe and V. Snieckus, *Am. Soc.*, **102**, 1457 (1980).

t-**Butyllithium, 1**, 96–97; **5**, 79–80; **7**, 47; **8**, 70–72; **9**, 89.

Aryllithiums. Halogen–metal exchange to form an aryllithium from aryl halides was first demonstrated by Gilman and co-workers[1] and has since been shown to be relatively independent of the type of substituents.[2] This exchange has been used in a general synthesis of benzocyclobutenes and of isoquinolines.[3]

Example:

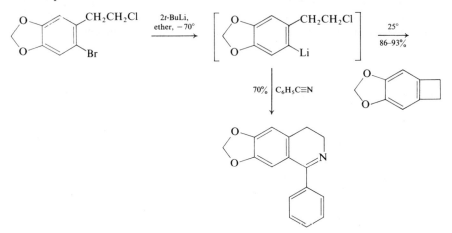

α-Keto dianions. α-Keto dianions (**1**) can be obtained by metal–halogen exchange of lithium enolates of α-bromo ketones with *t*-butyllithium in ether.[4] One

interesting use of these dianions is in the preparation of α-trimethylsilyl ketones (equation I).

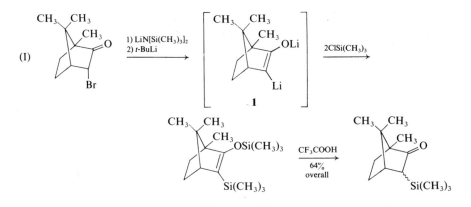

Lithium trialkylborohydrides. The reaction of *t*-butyllithium with trialkylboranes provides a convenient route to lithium trialkylborohydrides.[5]

[1] R. G. Jones and H. Gilman, *Org. React.*, **6**, 339 (1951).

[2] W. E. Parham, Y. Sayed, and L. D. Jones, *J. Org.*, **39**, 2053 (1974).

[3] D. J. Jakiela, P. Helquist, and L. D. Jones, *Org. Syn.*, submitted (1981).

[4] C. J. Kowalski, M. L. O'Dowd, M. C. Burke, and K. W. Fields, *Am. Soc.*, **102**, 5411 (1980).

[5] H. C. Brown, G. W. Kramer, J. L. Hubbard, and S. Krishnamurthy, *J. Organometal. Chem.*, **188**, 1 (1980).

C

Carboethoxycyclopropyltriphenylphosphonium tetrafluoroborate, 5, 90–91.

Pyrrolizidines. Reaction of sodium succinimide with Fuch's reagent (1) leads to the pyrrolizidine 2 in high yield. The product is converted into the alkaloid isoretro-

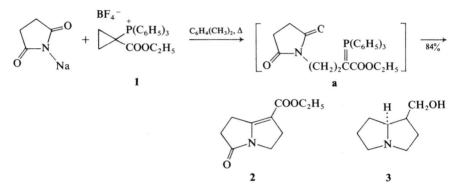

1

a

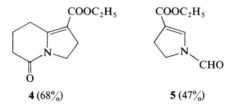

2

3

necanol (3) by catalytic hydrogenation followed by reduction with lithium aluminum hydride.[1,2]

The salt reacts with the anions of glutarimide and diformylimide to form the indolizidine 4 and the pyrrolidine 5, respectively.

4 (68%)

5 (47%)

[1] J. M. Muchowski and P. H. Nelson, *Tetrahedron Letters,* **21,** 4585 (1980).
[2] W. Flitsch and P. Wernsmann, *Tetrahedron Letters,* **22,** 719 (1981).

(Carboethoxymethylene)triphenylphosphorane, 1, 112–114; **2,** 60. Suppliers: Aldrich, Fluka.

α-Allenic esters. These esters can be prepared by reaction of an acid chloride and the phosphorane in the presence of triethylamine (1 equivalent).[1] The synthesis involves *in situ* generation of a ketene followed by a Wittig reaction. If a phosphonium salt is used, 2 equivalents of base is required.

Example:

$$(C_6H_5)_3P{=}CHCOOC_2H_5 + CH_3CH_2COCl \xrightarrow[\substack{67-75\%}]{\substack{N(C_2H_5)_3, \\ CH_2Cl_2, 20°}}$$

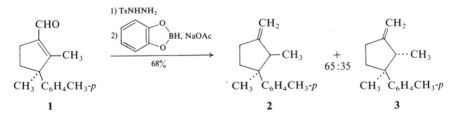

[1] R. W. Lang and H.-J. Hansen, *Helv.*, **63**, 438 (1980); *Org. Syn.*, submitted (1980).

Carbonylchlorobis(triphenylphosphine)iridium, $[(C_6H_5)_3P]_2Ir(CO)Cl$ **(1).** Mol. wt. 781.25, m.p. 215°. Supplier: Alfa.

Transfer hydrogenation of α,β-enones. The polystyrene-supported reagent[1] is a more efficient catalyst than the homogeneous reagent[2] for transfer hydrogenation of α,β-enones by formic acid. In addition, the immobilized complex can be used repeatedly. Other supported catalysts prepared from $[(C_6H_5)_3P]_2Rh(CO)Cl$, $[(C_6H_5)_3P]_3RhCl$, and $[(C_6H_5)_3P]_3RuCl_2$ are less efficient than the homogeneous counterparts.

[1] J. Azran, O. Buchman, and J. Blum, *Tetrahedron Letters*, **22**, 1925 (1981).
[2] J. Blum, Y. Sasson, and S. Iflah, *Tetrahedron Letters*, 1015 (1972).

Catecholborane, 4, 25, 69–70; **5,** 100–101; **6,** 33–34; **7,** 54–55; **8,** 79–80; **9,** 97–99.

Deoxygenation (**6,** 98, **7,** 54; **8,** 79–80; **9,** 97–99). Taber and Anthony[1] have used Kabalka's deoxygenation reaction for a stereoselective synthesis of (±)-laurene (**2**) from the unsaturated aldehyde (**1**). Some other methods are somewhat less selective.

Actually models indicate that steric effects are relatively slight in **1**, with the aryl group exerting a greater effect than the methyl group on hydride delivery. Greater stereoselectivity should be possible in more biased systems.

[1] D. F. Taber and J. M. Anthony, *Tetrahedron Letters*, **21**, 2779 (1980).

Cerium(IV) ammonium nitrate (CAN), 1, 120–121; **2,** 63–65; **3,** 44–45; **4,** 71–74; **5,** 101–102; **6,** 99; **7,** 55–56; **9,** 99.

Cyclopropanes.[1] Cyclopropanes substituted by two electron-withdrawing substituents $(COOCH_3)$ can be prepared by reaction of an alkene such as **1** with diazomethane to form a pyrazoline (**2**). Treatment of **2** with 3–10 mole % of CAN in acetone at 0° results in loss of N_2 and formation of a cyclopropane (3) alone or mixed with an alkene 4. Only 3 is formed when R = $COOCH_3$, $COCH_3$, C_2H_5, or p-$CH_3OC_6H_4CH_2$. When R is C_6H_5 or a substituted phenyl group 3 and 4 are both formed. When the pyrazoline is substituted with only one $COOCH_3$ group, no reaction takes place even at 60°.

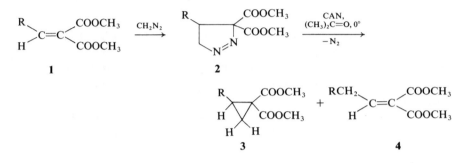

Naphtho[2,3-c]pyrane-5,10-quinones.[2] The dimethoxynaphthalenes 1 and 3 are oxidized to quinones, as expected, with Ag(II) oxide, but on oxidation with CAN undergo oxidative cyclization.

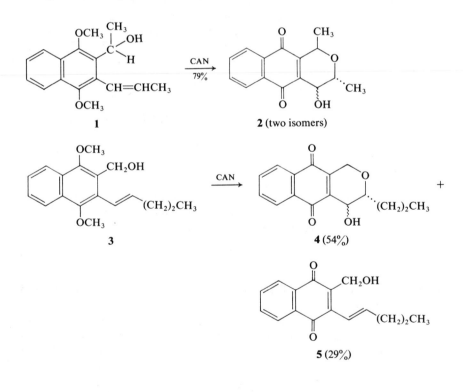

Hydrolysis of 1,3-dithianes (**4**, 74). The cleavage of 1,3-dithianes by CAN requires 4 equivalents of the salt for high yields because of the mechanism involved (equation I).[3]

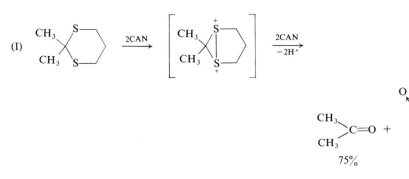

(I)

75%   60%

Oxidative cleavage of alkyl and silyl ethers.[4] CAN catalyzes the oxidative cleavage of alkyl and silyl ethers to carbonyl compounds with sodium bromate in yields usually of 75–95% (equation I).

(I) R^1, R^2CHOR^3 + NaBrO$_3$ $\xrightarrow[\text{75–95\%}]{\text{CAN, H}_2\text{O,}\atop\text{CH}_3\text{CN, 80°}}$ R^1, R^2C=O

R^1, R^2 = H, R, Ar

R^3 = CH$_3$, C$_2$H$_5$,

Si(CH$_3$)$_3$, Si(CH$_3$)$_2$C(CH$_3$)$_3$

$>$*CHNO$_2$ → $>$C=O.* This transformation can be conducted by treatment of nitro compounds with CAN and a base (triethylamine, equation I).[5]

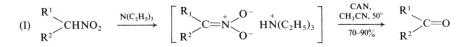

(I)

[1] J. Martelli and R. Grie, *J.C.S. Chem. Comm.*, 355 (1980).

[2] T. A. Chorn, R. G. F. Giles, P. R. K. Mitchell, and I. R. Green, *J.C.S. Chem. Comm.*, 534 (1981).

[3] H.-J. Cristau, B. Chabaud, R. Labandinière, and H. Christol, *Syn. Comm.*, **11**, 423 (1981).

[4] G. A. Olah, B. G. B. Gupta, and A. P. Fung, *Synthesis*, 897 (1980).

[5] G. A. Olah and B. G. B. Gupta, *Synthesis*, 44 (1980).

Cesium fluoride, 7, 57–58; **8,** 81–82; **9,** 100.

Cross-aldol and Michael reactions.[1] CsF is an effective catalyst in reactions of silyl enol ethers with aldehydes and ketones to form α,β-unsaturated ketones.

Michael adducts can also be obtained with this catalyst from the reaction of α,β-unsaturated ketones with silyl enol ethers.

Examples:

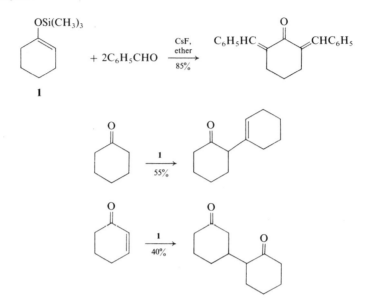

CsF in the presence of tetraalkoxysilanes also effects Michael addition of ketones to α,β-unsaturated ketones, esters, and nitriles. Presumably the enolate is generated and is converted by $Si(OR)_4$ into the silyl enol ether, which reacts *in situ*.[2]

Examples:

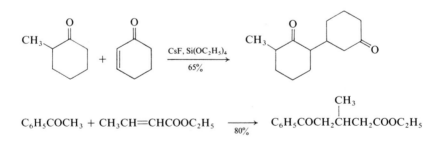

$$C_6H_5COCH_3 + CH_3CH{=}CHCOOC_2H_5 \xrightarrow[80\%]{} C_6H_5COCH_2\overset{\overset{\displaystyle CH_3}{|}}{C}HCH_2COOC_2H_5$$

Acid scavenger.[3] Cesium fluoride is recommended as a scavenger for HF and HBF_4 in place of triethylamine in reactions involving base-sensitive substrates. An example is the esterification formulated in equation (I). No racemization occurs in the transesterification of equation (II), although it is extensive when triethylamine is used.

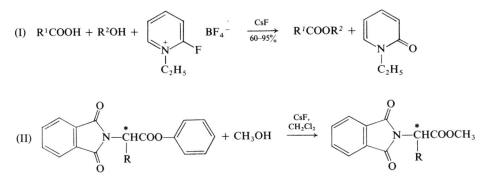

(I) $R^1COOH + R^2OH +$ ... BF_4^- $\xrightarrow[60-95\%]{CsF}$ $R^1COOR^2 +$ ...

(II) ... $+ CH_3OH$ $\xrightarrow{CsF, CH_2Cl_2}$ ...

o-Quinodimethanes.[4] Reaction of the quaternary ammonium salts of **1** with F^- at 25° induces a 1,4-elimination to give o-xylylene (**a**), which undergoes rapid cyclo-addition to alkenes or alkynes.[5]

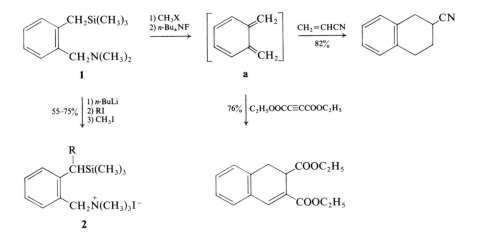

One attractive feature of this method is that α-substituted o-xylylenes are readily obtained by 1,4-elimination reactions with products (**2**) of alkylation of **1**. An interesting example is the synthesis of **4** from **3**.

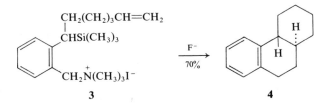

The reaction provides the key step in a stereoselective synthesis of estrone methyl ether (**6**) from **5**.[6]

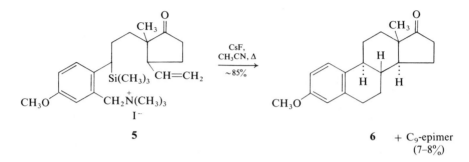

5

6   + C$_9$-epimer
(7–8%)

Magnus and co-workers[7] have used a related desilylation for a novel synthesis of 11α-hydroxyestrone methyl ether (**8**) from the epoxide (**7**) via an *o*-xylylene intermediate.

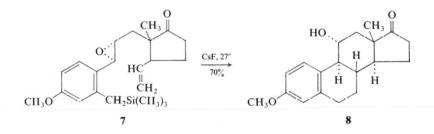

7

8

[1] J. Boyer, R. J. P. Corriu, R. Perz, and C. Réyé, *J. Organometal. Chem.*, **184**, 157 (1980).
[2] J. Boyer, R. J. P. Corriu, R. Perz, and C. Réyé, *J.C.S. Chem. Comm.*, 122 (1981).
[3] S. Shoda and T. Mukaiyama, *Chem. Letters*, 391 (1980).
[4] Y. Ito, M. Nakatsuka, and T. Saegusa, *Am. Soc.*, **102**, 863 (1980).
[5] W. Oppolzer, *Synthesis*, 793 (1978), has reviewed intramolecular cycloaddition reactions of *o*-quinodimethanes.
[6] Y. Ito, M. Nakatsuka and T. Saegusa, *Am. Soc.*, **103**, 476 (1981).
[7] S. Djuric, T. Sarkar, and P. Magnus, *Am. Soc.*, **102**, 6885 (1980).

Cesium fluoroxysulfate, CsSO$_4$F (1). Mol. wt. 247.97, stable solid at $-10°$, stable in CH$_3$CN for about a day. The reagent is prepared by reaction of cesium sulfate in water with fluorine.[1]

Caution: The salt can detonate on contact with metal.

Fluorination of aromatics. The reagent reacts with toluene to form benzyl fluoride as the major product ($\sim 65\%$ yield). It is also useful for fluorination of phenols[2] and of alkyl ethers of phenols[3]; the *ortho*-isomer is formed as the major product. Reactions with this reagent thus differ from those with xenon difluoride, which generally favors formation of *para*-isomers.

[1] E. H. Appelman, L. J. Basile, and R. C. Thompson, *Am. Soc.*, **101**, 3384 (1978).
[2] D. P. Ipe, C. D. Arthur, R. E. Winans, and E. H. Appelman, *Am. Soc.*, **103**, 1964 (1981).
[3] S. Stavber and M. Zupan, *J.C.S. Chem. Comm.*, 148 (1981).

Chloramine-T, 4, 75, 445–446; **5**, 104; **7**, 58; **8**, 83; **9**, 101–102.

Chlorohydrins. Chloramine-T reacts with olefins in an acidic medium to form chlorohydrins in 40–75% isolated yield. The reaction involves a *trans*-addition. Terminal olefins give a mixture of Markovnikov and anti-Markovnikov adducts in a ratio of 4:1.[1]

Telluroxide elimination.[2] Tellurides are converted to alkenes by reaction with chloramine-T, presumably via the adduct **a** (equation I). This elimination proceeds in low yield with *t*-butyl hydroperoxide.

(I) $RCH_2CH_2TeC_6H_5 + p\text{-}CH_3C_6H_4SO_2NClNa \xrightarrow{\text{THF}}$ $\left[\begin{array}{c} RCH_2CH_2TeC_6H_5 \\ \| \\ NSO_2C_6H_4CH_3 \end{array} \right]$

$$\mathbf{a}$$

$$\xrightarrow[66-93\%]{\Delta} RCH{=}CH_2 + C_6H_5TeNHSO_2C_6H_4CH_3$$

[1] B. Damin, J. Garapon, and B. Sillion, *Synthesis*, 362 (1981).
[2] T. Otsubo, F. Ogura, H. Yamaguchi, H. Higuchi, Y. Sakata, and S. Misumi, *Chem. Letters*, 447 (1981).

Chloranil, 1, 125–127; **2**, 66–67; **3**, 46; **4**, 75–76.

Oxidative coupling of furanes with quinones. Furanes do not react with naphthoquinone under usual conditions. However, Bridson and co-workers[1] have effected this addition in the presence of a high-potential quinone to oxidize the initial adduct to a 2-furyl-1,4-naphthoquinone and displace the equilibrium. The reaction with 2-methoxyfurane is particularly interesting because the adduct can be converted to a naphthacenequinone (equation I).

[1] J. N. Bridson, S. M. Bennett, and G. Butler, *J.C.S. Chem. Comm.*, 413 (1980).

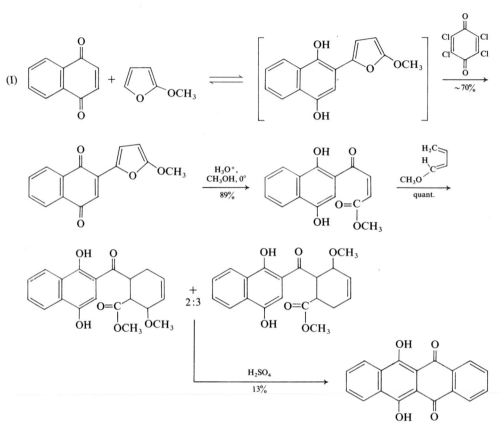

Chlorine–Chlorosulfuric acid, 8, 83.

α-Chlorination of carboxylic acids.[1] This reaction is possible with a combination of chlorine and oxygen (2:1) with chlorosulfonic acid as solvent and acidic catalyst. The function of O_2 is to scavenge free radicals and thereby suppress chlorination at other positions, particularly the β-position. Formulas and yields of some α-chloro acids prepared in this way are listed.

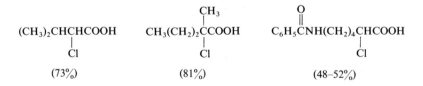

(73%) (81%) $(48–52\%)$

Analogous α-bromination[2] and α-iodination[3] of carboxylic acids have been reported. Iodination, unlike chlorination or bromination, is markedly affected by alkyl substitution at the β-position. All three reactions are believed to involve halogenation of a ketene intermediate, formed by dehydration with chlorosulfonic acid.

[1] Y. Ogata, T. Sugimoto, and M. Inaishi, *Org. Syn.*, **59**, 20 (1979); Y. Ogata, T. Harada, and T. Sugimoto, *Can. J. Chem.*, **55**, 1268 (1977).
[2] Y. Ogata and T. Sugimoto, *J. Org.*, **43**, 3684 (1978).
[3] Y. Ogata and S. Watanabe, *J. Org.*, **45**, 2831 (1980).

α-**Chloroallyllithium,** $[CH_2 \cdots CH \cdots CHCl]^- Li^+$ **(1).** The anion is generated *in situ* in THF by deprotonation of allyl chloride with LDA at $-78°$.

α-*Alkylation of allyl chloride.*[1] This anion is alkylated regioselectively ($>97\%$) at the α-position by primary alkyl bromides in yields of 70–90%. Yields from secondary bromides are only $<10\%$. The products undergo S_N2 displacement at the allylic position to afford stereoselectively (E)-alkenes. An example is the synthesis of the *A. leucotreta* sex pheromone **(2)** (equation I).

(I) $1 + Br(CH_2)_6Br \xrightarrow[88\%]{} CH_2{=}CHCH(CH_2)_6Br \xrightarrow[80\%]{(n\text{-}Pr)_2CuLi}$
                                                            |
                                                            Cl

$$\underset{CH_3(CH_2)_3}{\overset{H}{\diagdown}} C{=}C \underset{H}{\overset{(CH_2)_6Br}{\diagup}} \xrightarrow[76\%]{NaOAc,\ HOAc} \underset{CH_3(CH_2)_3}{\overset{H}{\diagdown}} C{=}C \underset{H}{\overset{(CH_2)_6OAc}{\diagup}}$$

(E/Z = 85:15)   **2** (E/Z = 83:17)

This α-alkylation–cuprate displacement sequence is also applicable to anions of methallyl and crotyl chloride and is a promising route to more complex systems.

[1] T. L. Macdonald, B. A. Narayan, and D. E. O'Dell, *J. Org.*, **45**, 1504 (1981).

μ-**Chlorobis(η-2,4-cyclopentadien-1-yl)(dimethylaluminum)-μ-methylenetitanium   (1),** **8**, 83–84, extremely sensitive to air and moisture.

Esters → vinyl ethers. The original publication (**8**, 83–84) mentioned that ethyl acetate reacts with **1** to give ethyl isopropenyl ether. More recent work[1] has shown that this conversion of esters to vinyl ethers is general and proceeds in high yields. The reaction rate is dramatically increased by the presence of donor ligands (THF, C_5H_5N). The solvent is benzene or toluene. Ketone methylenation is possible, and actually the rate is four times that of ester methylenation. Isolated yields are cited in the examples.

Examples:

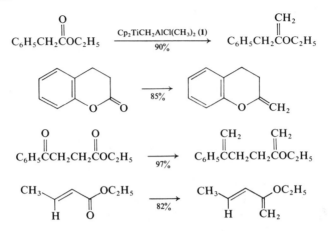

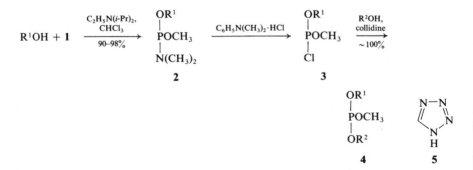

¹ S. H. Pine, R. Zahler, D. A. Evans, and R. H. Grubbs, *Am. Soc.*, **102**, 3270 (1980).

Chloro-N,N-dimethylaminomethoxyphosphine, $CH_3OP(Cl)N(CH_3)_2$ **(1).** Mol. wt. 137.06, b.p. 40–42/13 mm. The phosphine is prepared by reaction of dimethylamine with CH_3OPCl_2 (71% yield).

*Polynucleotides.*¹ These compounds are now generally prepared by use of a bifunctional phosphitylating agent such as *o*-chlorophenylphosphorodichloridite (**6,** 114–115) or methoxydichlorophosphine, CH_3OPCl_2. The intermediate nucleoside phosphites from these reagents tend to be unstable. This difficulty can be alleviated by use of **1,** which reacts with suitably protected nucleosides to form stable phosphoramidates **2** in good yield. These products can be activated for formation of a dinucleotide phosphite (**4**) by treatment with a weak acid such as N,N-dimethylaniline hydrochloride or 1*H*-tetrazole (**5**).

$$R^1OH + 1 \xrightarrow[90-98\%]{\substack{C_2H_5N(i\text{-}Pr)_2, \\ CHCl_3}} \underset{\textbf{2}}{\overset{OR^1}{\underset{N(CH_3)_2}{\mid}} POCH_3} \xrightarrow{C_6H_5N(CH_3)_2 \cdot HCl} \underset{\textbf{3}}{\overset{OR^1}{\underset{Cl}{\mid}} POCH_3} \xrightarrow[\sim 100\%]{\substack{R^2OH, \\ collidine}}$$

$$\underset{\textbf{4}}{\overset{OR^1}{\underset{OR^2}{\mid}} POCH_3} \qquad \underset{\textbf{5}}{\underset{H}{\overset{N-N}{\diagdown\diagup}}}$$

¹ S. L. Beaucage and M. H. Caruthers, *Tetrahedron Letters*, **22**, 1859 (1981).

2-(Chloroethoxy)carbene, $ClCH_2CH_2O\ddot{C}H$ **(1),** 7, 214; 9, 83.

Preparation[1]:

$$ClCH_2CH_2OH + HCl + (CH_2O)_n \xrightarrow[68\%]{} ClCH_2CH_2OCH_2Cl \xrightarrow{LiTMP} 1$$

3-Cyclopentenols.[2] 1,3-Dienes can be converted into 3-cyclopentenols in two steps: *syn*-addition of **1** to give a mixture of four vinylcyclopropyl ethers, which undergo ether cleavage and a 1,3-sigmatropic rearrangement to afford usually a single 3-cyclopentenol on exposure to *n*-butyllithium in THF–hexane–HMPT (25–50°).

Examples:

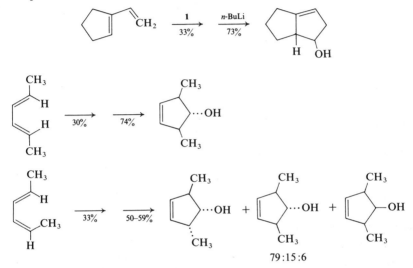

79:15:6

[1] G. N. Barber and R. A. Olofson, *Tetrahedron Letters*, 3783 (1976).
[2] R. L. Danheiser, C. Martinez-Davila, R. J. Auchus, and J. T. Kadonaga, *Am. Soc.*, **103**, 2443 (1981).

Chloroiodomethane, $ClCH_2I$. Mol. wt. 176.40, b.p. 108–109°. The reagent can be prepared by reaction of CH_2Cl_2 with NaI in refluxing DMF (5–10 hours); the yield is about 65% (based on NaI).[1]

Addition to terminal alkenes.[2] In the presence of a radical initiator, ICH_2Cl adds to terminal alkenes (80°, 6 hours) to give 1-chloro-3-iodoalkanes in 80–90% yield (equation I). Only polymers form from a similar reaction of styrene with ICH_2Cl; addition to internal alkenes is not useful either.

$$(I) \quad RCH{=}CH_2 + ICH_2Cl \xrightarrow[80-90\%]{AIBN, 60°} \underset{\underset{I}{|}}{R}CHCH_2CH_2Cl$$

Some useful transformations of the products are outlined in equation (II).

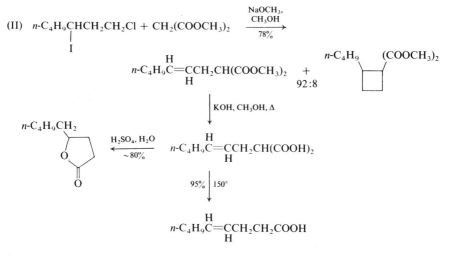

(II) n-C$_4$H$_9$CHCH$_2$CH$_2$Cl + CH$_2$(COOCH$_3$)$_2$ $\xrightarrow[78\%]{\text{NaOCH}_3, \text{CH}_3\text{OH}}$

[1] S. Miyano and H. Hashimoto, *Bull. Chem. Soc., Japan*, **44**, 2864 (1971).
[2] S. Miyano, H. Hokari, Y. Umeda, and H. Hashimoto, *Bull. Chem. Soc., Japan*, **53**, 770 (1980).

Chloromethylcarbene, $\begin{matrix} \text{Cl} \\ \text{CH}_3 \end{matrix}$C:. The carbene is generated by treatment of 1,1-dichloroethane with base (**6**, 530).

α-Methyl-α,β-enones. The carbene adds to the trimethylsilyl enol ethers of cycloalkanones to give a mixture of (Z)- and (E)-chloromethylcyclopropanes. The mixture when heated in toluene or xylene eliminates ClSi(CH$_3$)$_3$ to form the higher ring α-methylcycloalkenone. This step can also be effected in refluxing methanol containing triethylamine.

Examples:

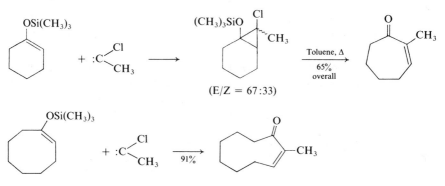

The same reaction is applicable to open-chain carbonyl compounds.[2]
Example:

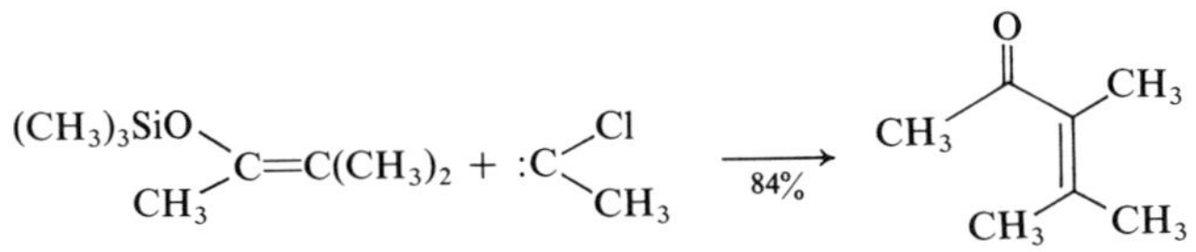

[1] L. Blanco, P. Amice, and J.-M. Conia, *Synthesis*, 289 (1981).
[2] L. Blanco, P. Amice, and J.-M. Conia, *Synthesis*, 291 (1981).

Chloromethyldiphenylsilane, $CH_3(C_6H_5)_2SiCl$ **(1).** Mol. wt. 232.77, b.p. 295°. Supplier: Petrarch.

Silylation.[1] Lithium enolates of esters and lactones undergo α-silylation with this reagent rather than the more usual O-silylation encountered with other reagents. Addition of HMPT results in considerable O-silylation. The reagent effects only O-silylation of ketones.

Examples:

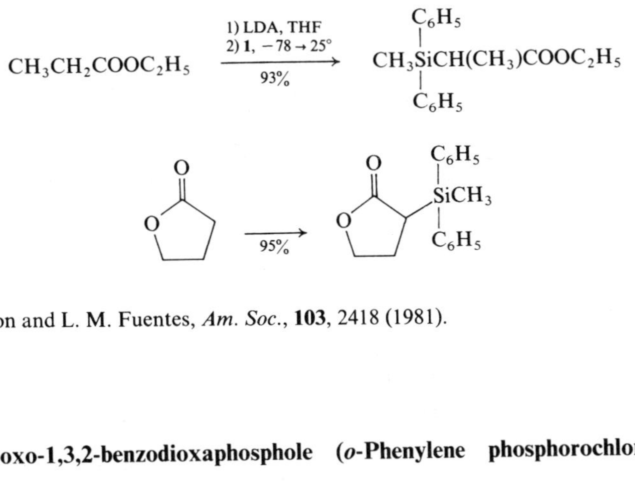

[1] G. L. Larson and L. M. Fuentes, *Am. Soc.*, **103**, 2418 (1981).

2-Chloro-2-oxo-1,3,2-benzodioxaphosphole (*o*-Phenylene **phosphorochloridate),** **1,** 837; **3,** 223.

Amides. The reagent **(1)** can be used to prepare amides.[1] The first step in a one-pot reaction involves formation of **2,** which reacts with tetrabutylammonium carboxylates through several intermediates, such as **a** and **b,** to form an amide.

[1] M. Wakselman and F. Acher, *Tetrahedron Letters*, **21**, 2705 (1980).

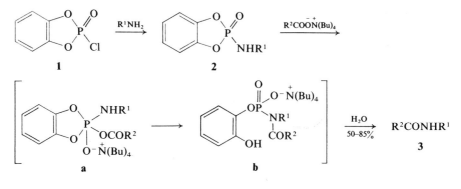

1   **2**

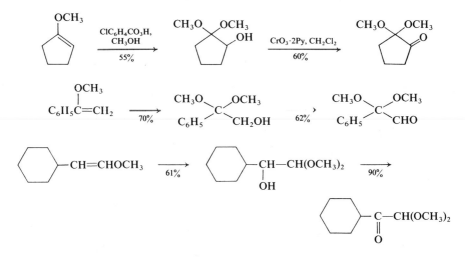

a   **b**

m-Chloroperbenzoic acid, **1**, 135–139; **2**, 68–69; **3**, 49–50; **6**, 110–114; **7**, 62–64; **8**, 97–102; **9**, 108–110.

Monoacetals of α-dicarbonyl compounds. Frimer's epoxidation of enol ethers (**8**, 97) has been extended to a synthesis of monoacetals of α-dicarbonyl compounds.[1]
Examples:

α-Hydroxy esters (*cf.*, **5**, 113; **8**, 100–101). Alkyl trimethylsilyl ketene ketals are oxidized by this peracid to α-hydroxy esters (equation I).[2]

(I) $RCH=C\begin{smallmatrix}OC_2H_5\\OSi(CH_3)_3\end{smallmatrix}$ $\xrightarrow[70-85\%]{ClC_6H_4CO_3H}$ $RCH(OH)COOC_2H_5$

3-Alkyl ethers of digitoxigenin. The Williamson ether synthesis fails when applied to digitoxigenin (**1**). In fact, only three successful methods are known for preparation of the 3-alkyl ethers. The 3-methyl ether can be obtained in 66% yield with diazomethane catalyzed by boron trifluoride etherate, but this reaction is not general for more complicated diazo compounds. The other two methods proceed from α-iododigitoxigenin (**2**), prepared from the alcohol with triphenylphosphine diiodide. This iodide is converted into the methyl ether (55% yield) when solvolyzed in methanol containing silver tetrafluoroborate. Olefinic material is formed as a byproduct. The most general method is reaction of **2** with *m*-chloroperbenzoic acid (**8**, 98–99) in the presence of an alcohol. The yield of various ethers range from 35 to 55%. The reaction probably involves solvolysis of an iodoso intermediate.[3]

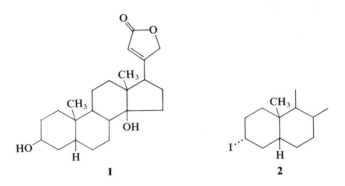

1   **2**

Oxidative desulfenylation.[4] β-Alkylthio carbonyl compounds are converted to α,β-unsaturated carbonyl compounds by oxidation with *m*-chloroperbenzoic acid or sodium periodate at room temperature.

$$R^1CCH_2CSC_2H_5 \xrightarrow[60-90\%]{m\text{-}ClC_6H_4CO_3H} R^1CCH=C\diagdown_{R^3}^{R^2}$$

[1] F. Huet, A. Lechevallier, and J. M. Conia, *Syn. Comm.*, **10**, 83 (1980).
[2] G. M. Rubottom and R. Marrero, *Syn. Comm.*, **11**, 505 (1981).
[3] R. Greenhouse and J. M. Muchowski, *Can. J. Chem.*, **59**, 1025 (1981).
[4] T. Nishio and Y. Omote, *Synthesis*, 392 (1980).

5-Chloro-1-phenyltetrazole, 2, 319–320.

Conversion of phenolic OH to H.[1] This reagent is superior to several related heterocycles for this reaction. Hydrogenolysis of the resulting 1-phenyltetrazolyl ether is best conducted using Pd/C and hydrazine as the H-donor in C_6H_6–C_2H_5OH–H_2O (7:3:2).

[1] I. D. Entwistle, B. J. Hussey, and R. A. W. Johnstone, *Tetrahedron Letters*, **21**, 4747 (1980).

N-Chlorosuccinimide, 1, 139; **2**, 69–70; **5**, 127–129; **6**, 115–118; **7**, 65; **8**, 103–105; **9**, 111.

Oxidative cleavage of **t-***butyldimethylsilyl ethers.*[1] These ethers are usually cleaved with tetra-*n*-butylammonium fluoride in aprotic solvents. The cleavage is also possible with NCS (1.1 equivalent) in DMSO (75–90% yield). A tetrahydropyranyl group is unaffected by NCS under these conditions.

Alkylsulfenyl chlorides.[2] Sulfenyl chlorides are usually prepared by reaction of thiols with chlorine or sulfuryl chloride. NCS can also be used. NBS can be used to obtain sulfenyl bromides, but these substances are extremely sensitive to heat.

[1] R. J. Batten, A. J. Dixon, R. J. K. Taylor, and R. F. Newton, *Synthesis*, 234 (1980).
[2] H. Seliger and H.-H. Gortz, *Syn. Comm.*, **10**, 175 (1980).

Chlorosulfonyl isocyanate (CSI), 1, 117–118; **2**, 70; **3**, 51–53; **4**, 90–94; **5**, 132–136; **6**, 122; **7**, 65–66; **8**, 105–106.

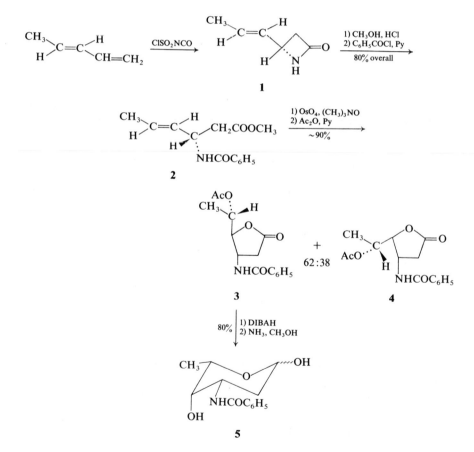

3-Amino hexoses.[1] A short synthesis of DL-daunosamine, the sugar component in some anthracyclines, starts with the addition of CSI to (E)-1,3-pentadiene to give the azetidinone **1**, which is readily converted to the β-amino ester **2**. Hydroxylation of the double bond of **2** with osmium tetroxide and trimethylamine oxide (this volume) is accompanied by lactonization to give two products, separated by chromatography of the corresponding acetates, **3** and **4**. Reduction of the carbonyl group of **3** followed by ammonolysis gives the N-benzoate of DL-daunosamine (**5**).

[1] F. H. Hauser and R. P. Rhee, *J. Org.*, **46**, 227 (1981).

Chlorothexylborane (1). Mol. wt. 132.44. One synthesis involves the reaction of thexylborane with 1 equivalent of hydrogen chloride in ether.[1] This reagent can also be prepared by reaction of 2,3-dimethyl-2-butene with monochloroborane–dimethyl sulfide (equation I).[2]

(I) $(CH_3)_2C=C(CH_3)_2 + H_2BCl \cdot S(CH_3)_2 \xrightarrow[98\%]{CH_2Cl_2}$

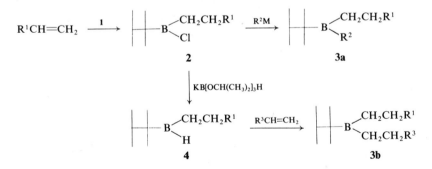

Unsymmetrical ketones.[1,3] This reagent hydroborates 1-alkenes with high regioselectivity to give alkylchlorothexylboranes (**2**). The reaction can be used for a general synthesis of ketones by two routes: The chlorine in **2** is replaced by a second alkyl group by reaction with an alkyllithium or a Grignard reagent to give a dialkylthexylborane (**3a**),[1] or the chlorine is reduced with potassium triisopropoxy-

borohydride (**5**, 565). The resulting borane **4** is used to hydroborate a second alkene to give a dialkylthexylborane (**3b**).[3] The thexylborane group of **3** is then replaced by a carbonyl group by the cyanidation reaction (**6**, 535). The method has great flexibility and is applicable to any alkene as well as to 1-alkynes. Overall yields range from 60 to 80%.

[1] G. Zweifel and N. R. Pearson, *Am. Soc.*, **102**, 5919 (1980).
[2] H. C. Brown, J. A. Sikowski, S. U. Kulkarni, and H. D. Lee, *J. Org.*, **45**, 4540 (1980).
[3] S. U. Kulkarni, H. D. Lee, and H. C. Brown, *J. Org.*, **45**, 4542 (1980).

Chlorotrimethylsilane, 1, 1232; **2**, 435–438; **3**, 310–312; **4**, 537–539; **5**, 709–713; **6**, 626–628; **7**, 66–67; **8**, 107–109; **9**, 112–113.

Vinylsilanes (**8**, 491–492); *allylic alcohols* (**9**, 340). Details are available for conversion of a ketone to a vinylsilane in which the C—Si bond has replaced the C=O group (enesilylation). The reaction affords the less substituted vinylsilane in the case of unsymmetrical ketones. The paper includes details for use of vinylsilanes for cyclopentenone annelation by Friedel–Crafts acylation with acryloyl chlorides and subsequent cyclization of pentadienyl cations (**9**, 498–499).[1]

Details for the use of enesilylation for 1,2-transposition of ketones (**8**, 289–290) have been published.[2] The paper includes the conversion of vinylsilanes to rearranged allylic alcohols. The trimethylsilyl group directs the ene reaction with singlet oxygen exclusively to the α-position. A typical reaction is shown in equation (I).

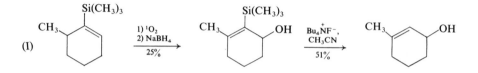

Trimethylsilyl esters. These esters can be prepared in 60–90% yield by reaction of carboxylic acids with chlorotrimethylsilane in 1,2-dichloroethane at 65°.[3]

Silylation of tertiary alcohols. This reaction is possible with $ClSi(CH_3)_3$ and $N(C_2H_5)_3$ if a catalytic amount of DMSO, HMPT, DBU, or imidazole is present. The first two substances are more effective than the amine catalysts.[4]

[1] L. A. Paquette, W. E. Fristad, D. S. Dime, and T. R. Bailey, *J. Org.*, **45**, 3017 (1980).
[2] W. E. Fristad, T. R. Bailey, and L. A. Paquette, *J. Org.*, **45**, 3028 (1980).
[3] H. H. Hergott and G. Simchen, *Synthesis*, 626 (1980).
[4] R. G. Visser, H. J. T. Bos, and L. Brandsma, *Rec. Trav.*, **99**, 70 (1980).

Chlorotrimethylsilane–Lithium bromide.

$ROH \rightarrow RBr$.[1] This transformation can be carried out with $ClSi(CH_3)_3$ and LiBr (or NaBr) in refluxing acetonitrile (93–100% yield). The reaction is more rapid than that with $BrSi(CH_3)_3$ itself, possibly owing to catalysis by Br^-.

The transformation is also possible with hexamethyldisilane and pyridinium bromide perbromide as the source of bromine. This reaction, however, is slower in the case of primary and secondary alcohols. Benzylic, allylic, and tertiary alcohols react rapidly. Hence selective reactions are possible. In the case of secondary alcohols, the conversion occurs with 88% inversion.

[1] G. A. Olah, B. G. B. Gupta, R. Malhotra, and S. C. Narang, *J. Org.*, **45**, 1638 (1980).

Chlorotrimethylsilane–Sodium iodide, 9, 251–252.

Deoxygenation of alcohols and ethers.[1] Treatment of alcohols and methyl or trimethylsilyl ethers in acetonitrile with this iodotrimethylsilane equivalent and then zinc (previously activated with aqueous hydrochloric acid) and a little acetic acid results in deoxygenation to alkanes, usually in 60–90% yield. Presumably an alkyl iodide is an intermediate.

Examples:

$$n\text{-}C_{10}H_{21}OH \xrightarrow[\text{80\%}]{\begin{array}{l}\text{1) ClSi(CH}_3)_3\text{, NaI, CH}_3\text{CN, 1 hr}\\\text{2) Zn, HOAc, 6 hr}\end{array}} n\text{-}C_{10}H_{22}$$

$$n\text{-}C_{10}H_{21}OCH_3 \xrightarrow[\text{82\%}]{} n\text{-}C_{10}H_{22} \xleftarrow[\text{82\%}]{} n\text{-}C_{10}H_{21}OSi(CH_3)_3$$

Cleavage of enol and dienol methyl ethers.[2] Treatment of these substrates in aqueous acetonitrile at 20° for 5 minutes gives the aldehydes or ketones in quantitative yield after aqueous work-up or treatment with alumina.

Example:

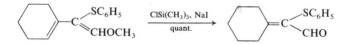

Enol silyl ethers. Enol silyl ethers of aldehydes and ketones are formed rapidly and in high yield by reaction with $(CH_3)_3SiCl$, NaI, and $N(C_2H_5)_3$. The (Z)-isomer is formed preferentially or even exclusively.[3]

Alkenes from β-oxygenated selenides.[4] β-Phenylseleno lactones, ethers, and alcohols are converted into alkenes on treatment with $ClSi(CH_3)_3$ and NaI in CH_3CN. Hydriodic acid (formed by inadvertant hydrolysis) may play a role; this acid can effect this reaction, but in lower yield. This elimination thus reverses cyclofunctionalizations induced by C_6H_5SeX (*cf.* **Na–NH₃, 9**, 26), and in addition provides a stereospecific route to alkenes by way of β-hydroxy selenides.

Example:

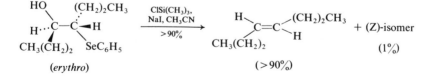

Reduction of α-halo ketones.[5] α-Bromo and α-chloro ketones are reduced to the parent ketone by this reagent in CH_3CN at room temperature in 3–12 hours. Yields are in the range 75–95%.

[1] T. Morita, Y. Okamoto, and H. Sakurai, *Synthesis*, 32 (1981).
[2] Z. Kosarych and T. Cohen, *Tetrahedron Letters*, **21**, 3959 (1980).
[3] P. Cazeam, F. Moulines, O. Laporte, and F. Duboudin, *J. Organometal. Chem.*, **201**, C9 (1980).
[4] D. L. J. Clive and V. N. Kate, *J. Org.*, **46**, 231 (1981).
[5] G. A. Olah, M. Arvanaghi, and Y. D. Vankar, *J. Org.*, **45**, 3531 (1980).

Chlorotrimethylsilane–Zinc.

Deoxygenation of sulfoxides.[1] Sulfoxides are reduced to sulfides in 75–95% yield by reaction with $ClSi(CH_3)_3$ and zinc in THF. The reaction is exothermic and should be maintained below 30°.

[1] A. H. Schmitt and M. Russ, *Ber.*, **114**, 822 (1981).

Chlorotris(triphenylphosphine)rhodium(I), 1, 1252; 2, 248–253; 3, 325–329; 4, 559–562; 5, 736–740; 6, 652–653; 7, 68; 8, 109; 9, 113–114.

Cyclopentanones. The cyclization of 4-pentenals to cyclopentanones catalyzed by Wilkinson's catalyst (**4**, 560) has been improved considerably by modification of the phosphine ligands.[1] Catalysts containing tri-*p*-tolylphosphine, tri-*p*-anisylphosphine, and tris(*p*-dimethylaminophenyl)phosphine are particularly useful. Simple cyclopentanones can be prepared in high yields. The reaction also provides a route to spirocyclic and bicyclic ketones (equations I and II). Unfortunately this method is not applicable to synthesis of larger rings.

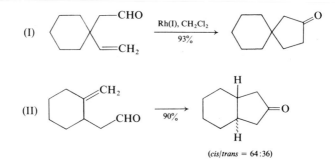

(*cis/trans* = 64:36)

Protection of primary amines (*cf.* **8**, 437). Laguzza and Ganem[2] recommend the diallyl derivatives for protection of primary amines. These compounds are prepared by reaction of the amine with allyl bromide and ethyldiisopropylamine. Deprotection is accomplished by isomerization of the allyl groups to propenyl groups by Wilkinson's catalyst (**5**, 736) in refluxing aqueous acetonitrile. Under these conditions the resulting enamine is hydrolyzed with evolution of propionaldehyde, and the amine is obtained in 65–90% yield.

The method has been used in a synthesis of anticapsin (**3**) from a derivative (**1**) of L-tyrosine.

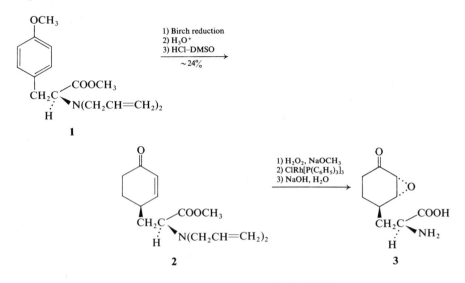

1   **2**   **3**

[1] R. C. Larock, K. Oertle, and G. F. Potter, *Am. Soc.*, **102**, 190 (1980).
[2] B. C. Laguzza and B. Ganem, *Tetrahedron Letters*, **22**, 1483 (1981).

Chromic anhydride, 1, 144–147; **2**, 72–75; **3**, 54–57; **4**, 96–97; **5**, 140–141; **7**, 70; **9**, 115.

γ-Keto-α,β-unsaturated esters.[1] α,β-Unsaturated esters undergo allylic oxidation when treated with excess CrO_3 in Ac_2O–HOAc (1:2) at 15–25° for 30–60 minutes (equation I).

$$\text{(I)} \quad RCH_2CH{=}CHCOOCH_3 \xrightarrow[\substack{50-86\%}]{\substack{CrO_3,\ Ac_2O, \\ HOAc,\ 15-25°}} R\overset{\overset{\displaystyle O}{\|}}{C}CH{=}CHCOOCH_3$$

[1] M. Nakayama, S. Shinke, Y. Matsushita, S. Ohira, and S. Hayashi, *Bull. Chem. Soc. Japan*, **52**, 184 (1979).

Chromic anhydride–Pyridine, $CrO_3 \cdot 2Py$, **1**, 145–146; **2**, 74–75; **3**, 55; **4**, 96; **9**, 121.

$CrO_3 \cdot 2Py$–Celite. Merck chemists[1] have oxidized the alcohol **1** with a modified Ratcliffe reagent. CrO_3 (1 equivalent) and pyridine (2 equivalents) are dissolved with stirring in CH_3CN to give an orange-red solution. Celite is added and after a few minutes a solution of **1** in CH_3CN is added. The resulting aldehyde is isolated as the thioacetal **2**.

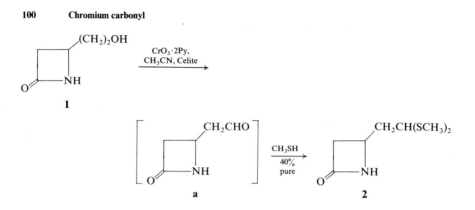

[1] S. M. Schmitt, D. B. R. Johnston, and B. G. Christensen, *J. Org.*, **45**, 1135, 1142 (1980).

Chromium carbonyl, 5, 142–143; **6**, 125–126; **7**, 71–72; **8**, 110; **9**, 117–119.

Arene-chromium tricarbonyl complexes. Semmelhack and Yamashita[1] have used the activating and *meta*-directing effect of the Cr(CO)₃ group to obtain the spiro-[4.5]decenone system of acorenone (**3**) and acorenone B (**4**) from anisole. Treatment

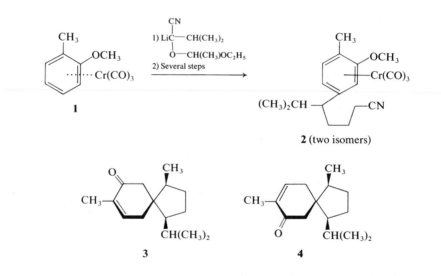

of the complex **1** with a cyanohydrin acetal anion led to substitution almost entirely in the desired *meta*-position. The intermediate **2** was obtained after a few steps as two diastereomers. One of these was eventually converted to **3**, the other into **4**. Overall yields were low because cyclization of either **2a** or **2b** resulted in fused cyclo-hexenones as well as the desired spirohexenone.

[1] M. F. Semmelhack and A. Yamashita, *Am. Soc.*, **102**, 5924 (1980).

Chromium(III) chloride–Lithium aluminum hydride, 8, 110–112; **9,** 119.

Allenic alcohols **(9,** 119).[1] The reaction of propargylic bromides **(1)** with aldehydes or ketones can result in allenic alcohols **(3)** and/or homopropargylic alcohols **(4).** The selectivity depends on the substitution in **1** and **2** and is also

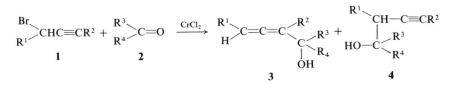

markedly influenced by addition of HMPT, which favors formation of the allenic alcohol.

Reduction of propargylic bromides. The reagent reduces propargylic bromides to allenes as the major products in the presence of a proton donor (water). Addition of HMPT suppresses reduction to acetylenes.[2] Use of a protonic chiral reagent results in optically active allenes. Highest inductions have been observed with (−)-menthol and (−)-borneol (about 20% ee).[3]

[1] P. Place, C. Vernière, and J. Goré, *Tetrahedron*, **37,** 1359 (1981).
[2] F. Delbecq, R. Baudouy, and J. Goré, *Nouv. J. Chim.*, **3,** 321 (1979).
[3] C. Verniere, B. Cazes, and J. Goré, *Tetrahedron Letters*, **22,** 103 (1981).

Chromium(II) perchlorate–Butanethiol.

Dehalogenation. Barton *et al.* **(1,** 148) effected dehalogenation of steroidal β-hydroxy halides with chromium(II) acetate and butanethiol as the proton donor in DMSO. The method is only useful with tertiary halides. A recent improvement that permits reduction of halides of all types uses the ethylenediamine complex of $Cr(ClO_4)_2$ and the tetrahydropyranyl ethers of the β-hydroxy halide. Catalytic amounts of the reducing agent can be used in "indirect electrolysis." The reaction is convenient for preparation of deoxynucleosides.[1]

Example:

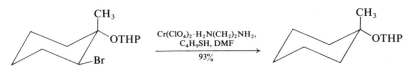

[1] J. Wellmann and E. Steckhan, *Angew. Chem. Int. Ed.*, **19,** 46 (1980).

Cobalt(II) chloride, $CoCl_2$. Mol. wt. 129.84.

Alkylation of 1,3-diketones with allylic and benzylic alcohols. Allylic and benzylic alcohols can alkylate 1,3-diketones and other active hydrogen compounds in the presence of $CoCl_2$. Alcohols are preferred to alkyl halides, since formation of hydrogen halide can lead to further reactions.[1]

Examples:

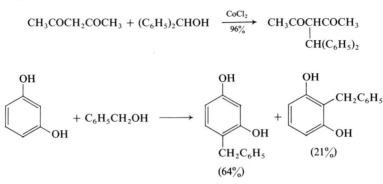

$$CH_3COCH_2COCH_3 + (C_6H_5)_2CHOH \xrightarrow[96\%]{CoCl_2} CH_3COCHCOCH_3$$
$$\qquad\qquad\qquad\qquad\qquad\qquad\qquad\qquad\qquad CH(C_6H_5)_2$$

[1] J. Marquet and M. Moreno-Mañas, *Chem. Letters*, 173 (1981).

Cobalt(II) phthalocyanine (CoPc), **9**, 119–121.

Reduction of nitro groups.[1] The lithium anion of phthalocyaninecobalt(I), Li[Co(I)Pc], selectively reduces aliphatic and aromatic nitro compounds to primary amines at room temperature in 65–95% yield. Double bonds, nitriles, carbonyl groups, and aryl halides are not reduced.

This reaction is particularly useful in the Friedlander quinoline synthesis, the use of which has been limited by the inaccessibility of *o*-aminobenzaldehydes. Thus a one-pot synthesis of quinolines is now possible by reduction of an *o*-nitrobenzaldehyde with the reagent in the presence of an α-methylene ketone.

Examples:

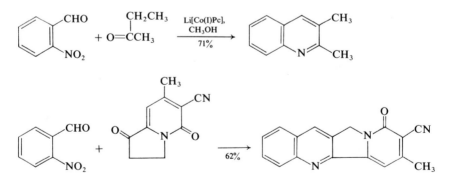

The second example is noteworthy because condensation of *o*-aminobenzaldehyde with the same carbonyl component under Friedlander conditions results in the quinoline in only 20% yield. The product is of interest because it can serve as a precursor to camptothecine (A).

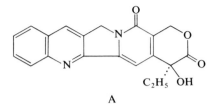

A

Fischli has used a related reagent, Cob(I)alamine,[2] obtained by reduction of vitamin B_{12} with zinc and acetic acid, as a catalyst for reduction with zinc and acetic acid of various unsaturated systems such as α,β-unsaturated nitriles,[3] esters,[4] and carbonyl compounds,[5] allylic alcohols and amines,[6] and also isolated double bonds.[6]

[1] H. Eckert, *Angew. Chem. Int. Ed.*, **20**, 208 (1981).
[2] A. Fischli, *Helv.*, **61**, 2560 (1978).
[3] A. Fischli, *Helv.*, **62**, 882 (1979).
[4] A. Fischli and D. Suss, *Helv.*, **62**, 48 (1979).
[5] A. Fischli, *Helv.*, **62**, 2361 (1979).
[6] A. Fischli and P. M. Müller, *Helv.*, **63**, 1619 (1980).

Copper(II) acetate, 1, 159; 3, 84.

Oxidation of hydrazides.[1] Acid hydrazides are oxidized to carboxylic acids by several Cu(II) salts, particularly $Cu(OH)_2$ and $Cu(OH)Cl$. The salt can be used in catalytic amounts if oxygen is available to convert Cu(I) to Cu(II). For this purpose the most convenient salt is $Cu(OAc)_2$ in THF or CH_3OH (equation I).

$$(I) \quad RCONHNH_2 + O_2 \xrightarrow{Cu(OAc)_2} RCOOH + N_2 + H_2O$$

Methyl esters are obtained only in low yields in a reaction conducted in CH_3OH. They are obtained most satisfactorily by use of excess $CuCl_2$ and $NaOCH_3$ in THF. Presumably the oxidant is $Cu(OCH_3)Cl$ (equation II).

$$(II) \quad RCONHNH_2 + 4Cu(OCH_3)Cl \longrightarrow RCOOCH_3 + 4CuCl + N_2 + 3CH_3OH$$

[1] J. Tsuji, T. Nagashima, N. T. Qui, and H. Takayanagi, *Tetrahedron*, **36**, 1311 (1980).

Copper(II) acetate–Iron(II) sulfate.

Fragmentation of α-alkoxy hydroperoxides. Schreiber[1] has reported a short synthesis of the macrolide ($\pm$)-recifeiolide (**2**) in which the key step involves fragmentation of the β-alkoxy hydroperoxide **1**, available in two steps from

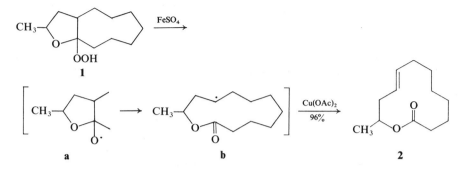

1

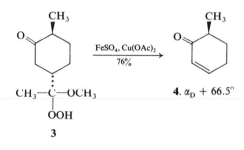

a   **b**   **2**

cyclononanone. Treatment of **1** with $FeSO_4$ and $Cu(OAc)_2$ leads to the macrolide in 96% yield. The paper suggests formation of the intermediate radicals **a** and **b** and an intermediate pseudo six-membered ring formed by chelation in a copper intermediate.

The paper also reports fragmentation by this reaction of the α-alkoxy hydroperoxide **3**, derived from (−)-carvone, as a novel route to the optically active **4**.

CH₃ ... FeSO₄, Cu(OAc)₂ / 76% ... CH₃

4. $α_D + 66.5°$

CH₃—C—OCH₃
|
OOH

3

[1] S. L. Schreiber, *Am. Soc.*, **102**, 6163 (1980).

Copper(I) bromide–Dimethyl sulfide, 6, 225; 8, 117–118.

Purification. Two laboratories have reported procedures for preparation of the pure complex. In one method[1] commercial CuBr is washed with methanol until the effluent is clear. The other group[2] has prepared CuBr by reduction of $CuBr_2$ with Na_2SO_3 as a pure white solid, which was then used to prepare the complex (m.p. 128–130°).

γ-Allylation of α,β-unsaturated acids (7, 82–83). Further study on the γ-allylation of copper dienolates derived from α,β-unsaturated acids reveals that use of $CuBr·S(CH_3)_2$ rather than CuI improves the regioselectivity of α:γ attack to 10:90 and also the yield. The effect of Cu^+ on regioselectivity is essentially limited to allylic halides and vinyl epoxides. No other metal ions have shown this ability to increase the γ-selectivity. Unfortunately some isomerization around the double bond derived from the electrophile can occur. Even with this limitation the reaction is efficient for synthesis of certain isoprenoid 1,5-dienes.[3]

Examples:

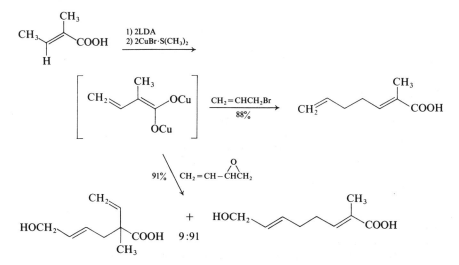

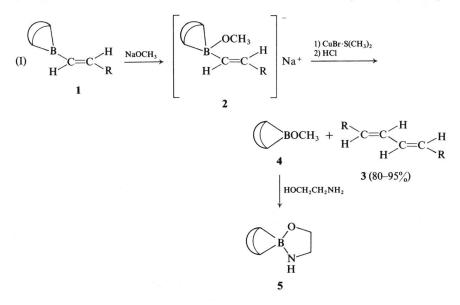

1,3-*[4] *and 1,4-Dienes*[5] *from alkenyldialkylboranes. The borate complexes (**2**), formed by addition of sodium methoxide to alkenyldialkylboranes (**1**), react rapidly with CuBr·S(CH$_3$)$_2$ to give (E,E)-1,3-dienes (**3**) together with **4**, which is precipitated on addition of ethanolamine as the adduct **5**. The synthesis probably involves transmetalation from boron to copper. Indeed, if an allylic halide is added to the above sequence prior to the quenching with HCl, a 1,4-diene (6) is formed (equation II).

$$(II) \quad 2 \xrightarrow[\text{2) CH}_2=\text{CHCH}_2\text{Br}]{\text{1) CuBr·S(CH}_3)_2} \quad \begin{array}{c} R \\ \diagdown \\ H \diagup \end{array} C=C \begin{array}{c} \diagup H \\ \diagdown \\ CH_2CH=CH_2 \end{array} \quad + 3 \quad \text{(trace)}$$

6 (75–95%)

[1] P. G. M. Wuts, *Syn. Comm.*, **11**, 139 (1981).
[2] A. B. Theis and C. A. Townsend, *Syn. Comm.*, **11**, 157 (1981).
[3] P. A. Savu and J. A. Katzenellenbogen, *J. Org.*, **46**, 239 (1981).
[4] J. B. Campbell, Jr., and H. C. Brown, *J. Org.*, **45**, 549 (1980).
[5] H. C. Brown and J. B. Campbell, Jr., *J. Org.*, **45**, 550 (1980).

Copper(II) chloride, 1, 163; **2**, 84–85; **3**, 66; **4**, 105–107; **5**, 158–160; **6**, 139–141; **7**, 79; **8**, 119–120; **9**, 123.

α-Chloro ketones. Ketones are converted to α-chloro ketones by reaction with CuCl$_2$ in DMF (**2**, 85), but the reaction is not regioselective. This drawback is remedied by use of silyl enol ethers as substrates. Use of DMF as solvent is essential. CuCl$_2$ in some cases can be replaced by FeCl$_3$, but yields are usually lower and by-products are formed if an additional double bond is present in the starting material.[1]

Example:

$$(CH_3)_3CC=CH_2 \xrightarrow[65\%]{\text{CuCl}_2, \text{DMF}} (CH_3)_3C\overset{\displaystyle O}{\overset{\|}{C}}CH_2Cl$$
$$\mid$$
$$OSi(CH_3)_3$$

Oxyselenation of alkenes (**8**, 119–120). Details of this reaction have been published. The role of the metal catalyst is considered to be polarization of the Se—CN bond of the selenocyanate, and in fact the pyridine complex of the metal salt is less effective. An episelenonium ion is postulated as the intermediate. CuBr$_2$, CiCl, NiBr$_2$, and NiCl$_2$ are similar to CuCl$_2$ in activity. CrCl$_3$ and CoCl$_2$ show limited activity, whereas halides of many other metals are inactive.

No oxyselenation products are formed from olefins with electron-withdrawing groups (CH$_2$=CHCOCH$_3$, CH$_2$=CHCN, CH$_2$=CHCOOCH$_3$), but acetals of α,β-unsaturated aldehydes do react.[2]

[1] Y. Ito, M. Nakatsuka, and T. Saegusa, *J. Org.*, **45**, 2022 (1980).
[2] A. Toshimitsu, T. Aoai, S. Uemura, and M. Okano, *J. Org.*, **45**, 1953 (1980).

Copper(I) iodide, 1, 169; **2**, 92; **3**, 69–71; **5**, 167–168; **6**, 147; **7**, 81–83; **8**, 121–122; **9**, 124–125.

Copper dienolate of a β,γ-unsaturated acid (*cf.*, **7**, 82–83). The site of alkylation of the dienolate of the acid **1** is markedly dependent on the counterion. The lithium

$$\overset{\displaystyle CH_2}{\underset{\displaystyle }{(CH_3)_3SiCH_2\overset{\|}{C}CH_2COOH}}$$

1

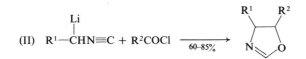

	2	70–80 : 30–20	**3**
LDA; RX			
LDA, CuI; RX	**2**	10–20 : 90–80	

dianion is alkylated preferentially at the α-position and addition of HMPT or TMEDA has no effect on the regioselectivity. In contrast, the copper dianion is alkylated preferentially at the γ-position to give products with the (E)-configuration.

[1] H. Nishiyama, K. Itagaki, K. Takahashi, and K. Itoh, *Tetrahedron Letters*, **22**, 1691 (1981).

Copper(I) oxide, 1, 169–170.

Oxazoles. Activated isonitriles react with selenol esters in the presence of triethylamine and Cu_2O at 25° to form oxazoles, presumably via a β-keto isonitrile (equation I).[1]

(I) $RCOSeCH_3 + C_2H_5OOCCH_2N{\equiv}C \xrightarrow[\substack{60-90\%}]{\substack{Cu_2O,\ N(C_2H_5)_3,\\ THF,\ 25°}}$

This reaction is a modification of Schöllkopf's oxazole synthesis[2] from acid chlorides and α-metalated isonitriles (equation II).

(II) $R^1{-}\overset{\displaystyle Li}{\underset{\displaystyle }{C}}HN{\equiv}C + R^2COCl \xrightarrow[60-85\%]{}$

[1] A. P. Kozikowski and A. Ames, *Am. Soc.*, **102**, 860 (1980).
[2] R. Schröder, U. Schöllkopf, E. Blume, and I. Hoppe, *Ann.*, 533 (1975).

Copper(II) sulfate, 1, 164; **2**, 89; **5**, 162; **6**, 141; **8**, 125.

Dehydration of alcohols. Anhydrous $CuSO_4$ is recommended as a general catalyst for this reaction.[1]

[1] R. V. Hoffman, R. D. Bishop, P. M. Fitch, and R. Hardenstein, *J. Org.*, **45**, 917 (1980).

Copper(I) trifluoromethanesulfonate (CuOTf), **5**, 151–152; **6**, 130–133; **7**, 75–76; **8**, 125–126.

   Bicyclo(3.2.0)heptane-2-ols (2); cyclopentenones.[1] Photobicyclization of 1,6-heptadiene-3-ols (**1**) in the presence of CuOTf results in bicyclo(3.2.0)heptane-2-ols (**2**) in 70–85% yield. The corresponding ketones readily fragment to ethylene and cyclopentenones (**4**) when heated in a quartz tube at ~600°.

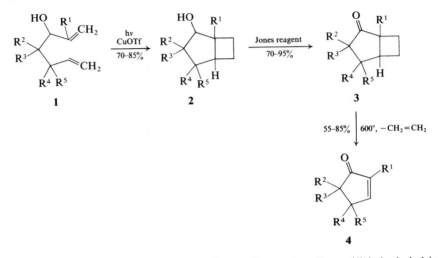

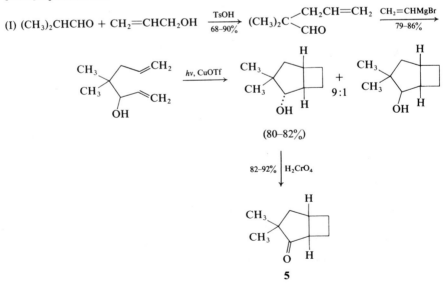

   A general Claisen rearrangement leading to the starting dienes (**1**) is included in an example of this reaction for synthesis of 3,3-dimethyl-*cis*-2-bicyclo-[3.2.0]heptanone (**5**).

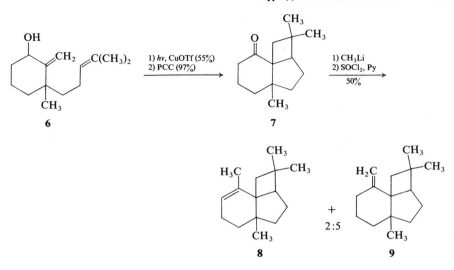

An intramolecular photocyclization catalyzed by copper(I) triflate provided a key step in a total synthesis of the ginseng sesquiterpenes α- and β-panasinsene, (**8**) and (**9**). The unsaturated allylic alcohol (**6**) is cyclized by irradiation at 254 nm in the presence of CuOTf to a mixture of saturated alcohols, which is oxidized to the ketone **7**. The ketone is inert to methylenetriphenylphosphorane, but can be converted into a 2:5 mixture of **8** and **9** by addition of methyllithium followed by dehydration.[2]

Lactonization.[3] Cyclization of the thioester **1** cannot be effected with mercuric trifluoroacetate or copper(I) trifluroacetate. The usual reagents used for this reaction (**6**, 582; **7**, 444) are ineffective, but cyclization is effected with CuOTf complexed with benzene. Two isomeric lactones (**2** and **3**) are obtained in 62% yield. One of these is the acetate of the pyrrolizidine alkaloid crobarbatine. Unfortunately deacetylation of these products is accompanied by further hydrolysis to the pyrrolizidine unit (retronecine).

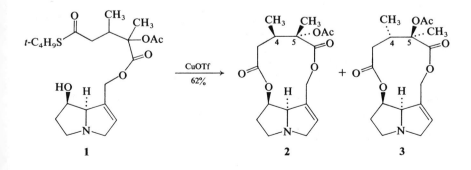

¹ R. G. Salomon, D. J. Coughlin, and E. M. Easler, *Am. Soc.*, **101**, 3961 (1979); R. G. Salomon and S. Ghosh, *Org. Syn.*, submitted (1980).
² J. E. McMurry and W. Choy, *Tetrahedron Letters*, **21**, 2477 (1980).
³ J. Huang and J. Meinwald, *Am. Soc.*, **103**, 861 (1981).

Copper(II) trifluoromethanesulfonate, 5, 152; **7,** 76; **8,** 126–127.

1,4-Diketones (**8,** 126–127). Complete details of the synthesis of 1,4-diketones by oxidative coupling of ketone enolates and trimethylsilyl enol ethers with Cu(OTf)₂ are available.¹ Use of isobutyronitrile is essential for the coupling; it is not only a suitable solvent, but the nitrile group apparently facilitates reduction of the intermediate copper enolate to CuOTf.² When acetonitrile is used by-products containing a nitrile group are formed. 1,4-Diketones are formed only in traces when DMF, DMSO, or HMPT is used.

*Friedel–Crafts acylation.*³ The complex (CuOTf)₂·C₆H₆ (**5,** 151–152) induces acylation of aromatics by selenol esters. Other copper salts (Cu₂O, CuCl₂) are ineffective.

Examples:

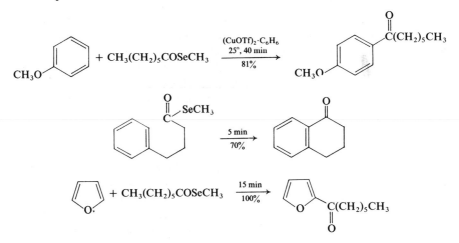

¹ Y. Kobayashi, T. Taguchi, T. Morikawa, E. Tokuno, and S. Sekiguaki, *Chem. Pharm. Bull. Japan*, **28**, 262 (1980).
² Copper(I) triflate complexed with acetonitrile is effective for homogeneous Ullman coupling of aryl bromides: T. Cohen and J. G. Tirpak, *Tetrahedron Letters*, 143 (1975); T. Cohen and I. Cristea, *Am. Soc.*, **98**, 748 (1976).
³ A. P. Kozikowski and A. Ames, *Am. Soc.*, **102**, 860 (1980).

Crown ethers, 4, 142–145; **5,** 152–155; **6,** 133–137; **7,** 76–79; **8,** 128–130; **9,** 126–127.

Wittig–Horner reactions. Addition of catalytic amounts of a crown ether to the Wittig–Horner reactions with diethyl phenylmethanephosphonate permits reactions at a lower temperature and in shorter times than those conventionally used with significantly higher yield (equation I).[1]

(I) $C_6H_5CH_2\overset{\overset{\displaystyle O}{\|}}{P}(OC_2H_5)_2$ + $\underset{R^2}{\overset{R^1}{>}}C=O$ $\xrightarrow[\text{85-99\%}]{\text{NaH, THF, crown ether}}$ $\underset{H}{\overset{C_6H_5}{>}}C=C\underset{R^2}{\overset{R^1}{<}}$

In reactions with aldehydes the (E)-isomer is formed exclusively.

The phosphonate **1** does not react with the aldehyde **2** under usual conditions ($NaOCH_3$, DMF, 20–110°).[2] Addition of a catalytic amount of 15-crown-6 permits the desired condensation in 45% yield.[3] The product (**3**) was used for synthesis of pallescensin-E (**4**).

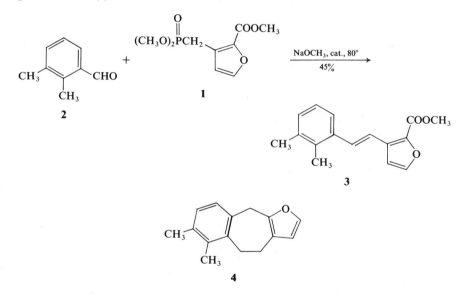

Alkynes.[4] Alkynes are obtained by bisdehydrohalogenation of *vic*-1,2-dibromides or *gem*-dichlorides with solid potassium *t*-butoxide in petroleum ether in the presence of 0.1 mole % of 18-crown-6. Dichlorides obtained from methyl ketones are converted into 1-alkynes only if the 3-position is blocked. Alkynes are also formed from (E)-haloalkenes by *syn*-elimination. Yields of alkynes are >80% for the most part.

N-Alkylation.[5] Exclusive N-alkylation of pyrrole, pyrazole, and related heterocycles is possible with potassium *t*-butoxide as base and 18-crown-6 as catalyst in ether or benzene.

Fulvenes.[6] In the presence of 18-crown-6, potassium cyclopentadienide reacts with ketones to form fulvenes in moderate yields (equation I). Aldehydes only polymerize under these conditions.

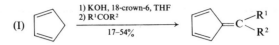

Review.[7] The preparation of various crown ethers as well as derivatives and analogs has been reviewed (218 references).

[1] R. Baker and R. J. Sims, *Synthesis*, 117 (1981).
[2] W. S. Wadsworth, *Org. React.*, **25**, 73 (1977).
[3] R. Baker and R. J. Sims, *Tetrahedron Letters*, **22**, 161 (1981).
[4] E. V. Dehmlow and M. Lissel, *Ann.*, 1 (1980).
[5] W. C. Guida and D. J. Mathre, *J. Org.*, **45**, 3172 (1980).
[6] H. Alper and D. E. Laycock, *Synthesis*, 799 (1980).
[7] J. S. Bradshaw and P. E. Stott, *Tetrahedron*, **36**, 461 (1980).

Cryptates, 5, 156; **6,** 137–138; **8,** 130; **9,** 127.

Alkylation of nonenolizable ketones.[1] Potassium hydride in THF reduces cyclopropyl phenyl ketone; however, in the presence of cryptate [2.2.2] (1 equivalent) the ketone can be alkylated in high yield (equation I).

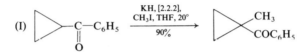

[1] H. Handel, M.-A. Pasquini, and J.-L. Pierre, *Bull. soc.*, II-351 (1980).

Cyanotrimethylsilane, 4, 542–543; **5,** 720–722; **6,** 632–633; **7,** 397–399; **8,** 133; **9,** 127–129. A new synthesis is formulated in equation (I).[1]

$$(I)\ [(CH_3)_3SiO]_2SO_2 + KCN \xrightarrow[96\%]{} (CH_3)_3SiCN$$

α,β-Unsaturated nitriles (**1,** 876–881).[2] A new one-pot conversion of ketones into α,β-unsaturated nitriles is formulated in equation (I). The synthesis is applicable to hindered and to α,β-unsaturated ketones.

α-Hydroxy amides.[3] The reaction of alkyl aryl ketones and HCN usually proceeds in low yield. The α-hydroxy amides can be obtained satisfactorily by use of cyanotrimethylsilane followed by hydrolysis (equation I). The hydrolysis step can proceed in low yield with diaryl ketones.

$$\text{(I)} \quad \text{ArCOR} \xrightarrow[\text{ZnI}_2]{\text{CNSi(CH}_3)_3,} \quad \underset{R}{\overset{OSi(CH_3)_3}{Ar-C-CN}} \xrightarrow[70-90\%]{\text{HCl}} \underset{R}{\overset{OH}{Ar-C-CONH_2}}$$

Review.[4] Synthetic applications of cyanotrimethylsilane, iodotrimethylsilane, azidotrimethylsilane, and methylthiotrimethylsilane have been reviewed (108 references).

[1] W. Kantlehner, E. Haug, and W. W. Mergen, *Synthesis*, 460 (1980).
[2] M. Oda, A. Yamamuro, and T. Watabe, *Chem. Letters*, 1427 (1979).
[3] G. L. Gruenwald, W. J. Brouillette, and J. A. Finney, *Tetrahedron Letters*, **21**, 1219 (1980).
[4] W. C. Groutas and D. Felker, *Synthesis*, 861 (1980).

Cyanotrimethylsilane–Triethylaluminum.

Conjugate addition.[1] In the presence of 1 equivalent of triethylaluminum, cyanotrimethylsilane undergoes conjugate addition to α,β-enones in high yield. The products are converted into β-cyano ketones by acid hydrolysis. The addition is kinetically controlled in toluene at room temperature, but thermodynamically controlled in refluxing THF (equation I).

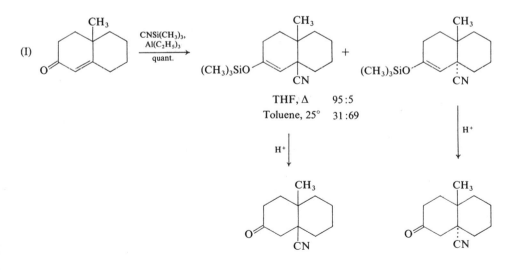

The paper reports one example of 1,6-addition (equation II).

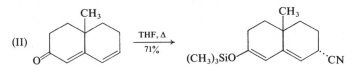

(II)   THF, Δ
                            ———→
                             71%

¹ K. Utimoto, M. Obayashi, Y. Shishiyama, M. Inoue, and H. Nozaki, *Tetrahedron Letters*, **21**, 3389 (1980).

Cyanuric chloride, 3, 72; **5**, 522–523; **6**, 149.

Macrocyclic lactonization. ω-Hydroxy acids are converted into the corresponding lactones by reaction with cyanuric chloride and triethylamine in acetone or acetonitrile at 25°.¹ Isolated yields of 13-, 16-, 17-, and 19-membered lactones are 70, 68, 85, and 33%, respectively. An example is the lactonization of aleuritic acid (**1**) to the lactone acetonide **2**.

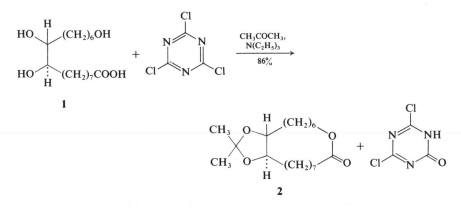

1

2

*β-Lactams.*² β-Lactams can be prepared from an acetic acid derivative and a Schiff base in the presence of cyanuric chloride (1 equivalent) and triethylamine (equation I).

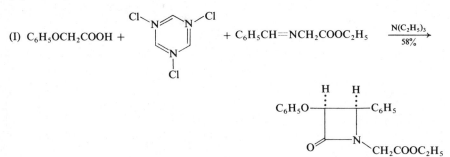

(I) $C_6H_5OCH_2COOH$ +   + $C_6H_5CH=NCH_2COOC_2H_5$ $\xrightarrow[58\%]{N(C_2H_5)_3}$

[1] K. Venkataraman and D. R. Wagle, *Tetrahedron Letters*, **21**, 1893 (1980).
[2] M. S. Manhas, A. K. Bose, and M. S. Khajavi, *Synthesis*, 209 (1981).

Cyanuric fluoride, 5, 171.

Deoxygenation of sulfoxides.[1] This reaction can be carried out with cyanuric fluoride in dioxane (3–9 hours, 65–80% yield). Cyanuric chloride (**3**, 72; **4**, 522; **5**, 687; **6**, 149) can be used in the case of aryl sulfoxides, but alkyl sulfoxides undergo chlorination with this reagent.

[1] G. A. Olah, A. P. Fung, B. G. B. Gupta, and S. C. Narang, *Synthesis*, 221 (1980).

Cyclohexylisopropylaminomagnesium bromide, $C_6H_{11}\underset{\underset{\displaystyle MgBr}{|}}{N}CH(CH_3)_2$ **(1)**. Mol. wt. 244.47. The reagent is prepared from CH_3MgBr and the corresponding amine in THF.

Isomerization of α,β-unsaturated epoxides. This base is superior to LDA or lithium diethylamide for isomerization of α,β-unsaturated epoxides to allylic alcohols with a *trans* double bond.[1,2]

Examples:

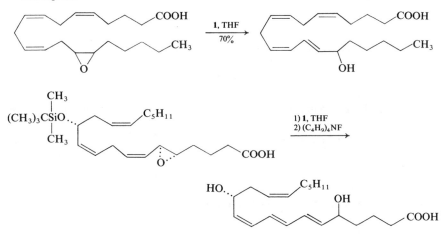

[1] E. J. Corey, A. Marfat, J. R. Falck, and J. O. Albright, *Am. Soc.*, **102**, 1433 (1980).
[2] E. J. Corey, P. B. Hopkins, J. E. Munroe, A. Marfat, and S. Hashimoto, *Am. Soc.*, **102**, 7986 (1980).

(1,5-Cyclooctadiene)bis(methyldiphenylphosphine)iridium(I) hexafluorophosphate, 8, 135–137.

Isomerization of allyl ethers.[1] This complex is particularly useful in carbohydrate chemistry for protection of hydroxy groups by the allyl group. After isomerization to the propenyl ethers, the free hydroxy compound is obtained by treatment with HgCl–HgO.

[1] J. J. Oltvoort, C. A. A. van Boeckel, J. H. de Koning, and J. H. van Boom, *Synthesis*, 305 (1981).

(1,5-Cyclooctadiene)(pyridine)(triphenylphosphine)iridium(I) hexafluorophosphate (1).

Hydrogenation catalyst. In a noncoordinating solvent (CH_2Cl_2) and in the presence of H_2 and an alkene, **1** produces cyclooctane and a very active catalyst in which the metal is complexed with hydrogen and the alkene. It is one of the most effective homogeneous hydrogenation catalysts and hydrogenates even tetra-substituted alkenes at useful rates.[1] It selectively hydrogenates hindered steroid double bonds from the α-face, but does not reduce keto groups, carbon–halogen bonds, or cyclopropanes. A $\Delta^{9(11)}$-steroid is not reduced.[2]

[1] R. J. Crabtree and G. E. Morris, *J. Organometal. Chem.*, **135**, 395 (1977); R. H. Crabtree, H. Felkin, and G. E. Morris, *J. Organometal. Chem.*, **141**, 205 (1977); R. H. Crabtree, *Accts. Chem. Res.*, **12**, 331 (1979).
[2] J. W. Suggs, S. D. Cox, R. H. Crabtree, and J. M. Quirk, *Tetrahedron Letters*, **22**, 303 (1981).

Cyclopentadienylnitrosylcobalt dimer, 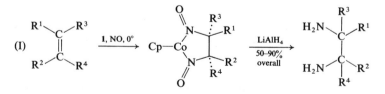 **(1).** Mol wt. 308.1.

Unstable to air. The reagent is prepared by reaction of $CpCo(CO)_2$ with NO in benzene.[1]

vic-*Diamination.* A few years ago Brunner and Loskot[2] reported that **1** in the presence of nitrous oxide forms a *cis*-adduct with norbornene and other strained alkenes. This reaction actually is general, but strained alkenes react more readily. The adducts are reduced to *vic*-diamines by $LiAlH_4$ (equation I). The first step is a stereospecific *cis*-addition, but reduction proceeds only stereoselectively.[3]

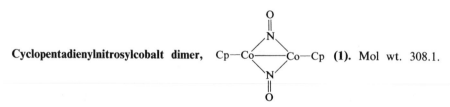

[1] H. Brunner, *J. Organometal. Chem.*, **12**, 517 (1968).
[2] H. Brunner and S. Loskot, *J. Organometal. Chem.*, **61**, 401 (1973).
[3] P. N. Becker, M. A. White, and R. G. Bergman, *Am. Soc.*, **102**, 5676 (1980).

D

Dehydroabietylamine, 1, 183; **2,** 97; **3,** 73. Supplier: Aldrich.

Resolution of alcohols (cf., 1-(naphthyl)ethyl isocyanate, **8,** 356–357). A practical synthesis of the methyl ester of (S)-5-HETE **(1)** from arachidonic acid involves chromatographic separation of the diastereomeric urethanes prepared from the isocyanate derived from dehydroabietylamine (hydrogen chloride and phosgene).[1] Urethanes from other chiral amines are less useful. The urethanes are cleaved with triethylamine and trichlorosilane to give the corresponding pure enantiomeric esters, which can be hydrolyzed by base.

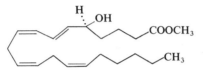

1, (S)-(+)-5-HETE methyl ester

[1] E. J. Corey and S. Hashimoto, *Tetrahedron Letters,* **22,** 299 (1981).

Diacetatobis(triphenylphosphine)palladium(II), 5, 497–498; **6,** 156–157; **7,** 274; **8,** 380–381.

Terminal arylalkynes.[1,2] The coupling between ethynyltrimethylsilane and aryl halides in the presence of this catalyst and triethylamine provides a useful route to terminal alkynes.

Example:

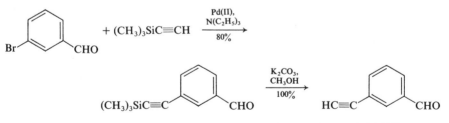

[1] S. Takahashi, Y. Kuroyama, K. Sonogashira, and N. Hagihara, *Synthesis,* 627 (1980).
[2] W. B. Austin, N. Bilow, W. J. Kelleghan, and K. S. Y. Lau, *J. Org.,* **46,** 2280 (1981).

Dialkylaluminum amides, 8, 182; **9,** 177.

β-Amino alcohols. A diethylaluminum amide reacts with epoxides at room temperature to give, after hydrolysis, a β-amino alcohol in 50–85% yield. Dimethyl-aluminum amides are somewhat more reactive.[1]

Examples:

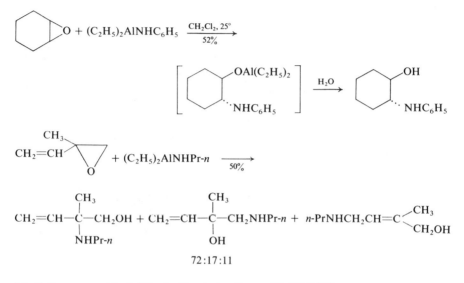

[1] L. E. Overman and L. A. Flippin, *Tetrahedron Letters*, **22**, 195 (1981).

Dialkylboron trifluoromethanesulfonates, 9, 131–132.

Enantioselective aldol condensation. Masamune *et al.*[1] have prepared optically pure β-hydroxy-α-methyl carboxylic acids by aldol condensation with the (S)- and (R)-isomers of the ethyl ketone **1**, prepared in three steps from commercially available (S)- and (R)-mandelic acid. For example, (S)-**1** is converted into the (Z)-boron enolate (**2**), which condenses with propionaldehyde to form a single aldol

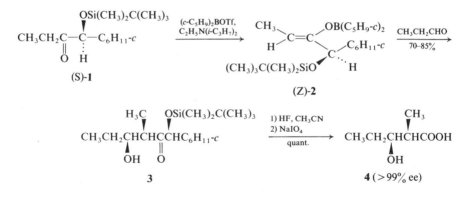

product **3**. Desilylation of **3** followed by oxidation gives acid **4**. Application of the same sequence to (R)-**1** results in the enantiomer of **4**. Thus the aldol products all have the 2,3-*syn* orientation. If the aldehyde has an α-alkyl substituent, 9-BBN triflate is the reagent of choice for formation of the boron enolate.

These highly diastereoselective aldol reactions have been used in a synthesis of 6-deoxyerythronolide B (**5**), which contains 10 asymmetric centers. Four aldol reactions, indicated by dotted lines, were used to construct the carbon framework with overall stereoselection of 85%.[2]

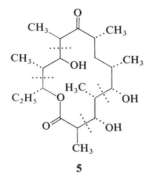

5

Vinyloxyboranes (boron enolates) are obtained in quantitative yield by reaction of silyl enol ethers with dialkylboron triflates in CH_2Cl_2 at $-22°$. The products can be used for stereoselective aldol condensations.[3]

Example:

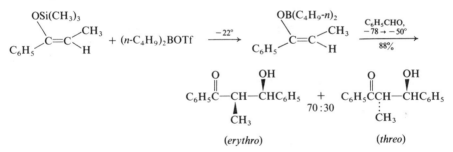

[1] S. Masamune, W. Choy, F. A. J. Kerdesky, and B. Imperiali, *Am. Soc.*, **103**, 1566 (1981).
[2] S. Masamune, M. Hirama, S. Mori, S. A. Ali, and D. S. Garvey, *Am. Soc.*, **103**, 1568 (1981).
[3] M. Wada, *Chem. Letters*, 153 (1981).

2,4-Diamino-1,3-thiazole hydrotribromide, (**1**). Mol. wt. 355.8,

m.p. 198–200° (dec.). The reagent is prepared by reaction of 2,4-diamino-1,3-thiazole hydrobromide with 1 equivalent of Br_2 (60% yield).

Bromination.[1] Ketones of the type $ArCOCH_2R$ are brominated at the methylene position by **1** in 80–95% yield. Phenols are brominated at the *ortho-* and *para-*positions.

[1] L. Fortini, *Synthesis*, 487 (1980).

Diazidotin dichloride (1). Mol. wt. 273.65; explodes on heating.
Preparation[1]:

$$(CH_3)_3SiN_3 + SnCl_4 \longrightarrow Cl_3SnN_3 \cdot (CH_3)_3SiN_3 \xrightarrow{80-90°} Cl_2Sn(N_3)_2 + (CH_3)_3SiCl$$
$$\textbf{1}$$

Imino thioethers.[2] The reagent, prepared *in situ*, converts the dimethyl thioketal of a cyclic ketone (**2**) into the imino thioether **3** in high yield. The reaction can also be carried out with azidotrimethylsilane (1 equivalent), stannic chloride (1 equivalent), and iodine (10 mole %), but yields are lower. This reaction was examined because of the reaction of thioketals with iodine azide (this volume).

$$\textbf{2} \qquad\qquad\qquad \textbf{3}$$

[1] N. Wilberg and K. H. Schmid, *Ber.*, **100**, 748 (1967)
[2] B. M. Trost, M. Vaultier, and M. L. Santiago, *Am. Soc.*, **102**, 7929 (1980).

Diazoacetaldehyde dimethyl acetal (1). Mol. wt. 116.12.
Preparation[1]:

$$NH_2CH_2CH(OCH_3)_2 \xrightarrow[\text{2) H}_2\text{O}]{\text{1) KCN}} H_2NCONHCH_2CH(OCH_3)_2 \xrightarrow{N_2O_4}$$

$$ONCONHCH_2CH(OCH_3)_2 \xrightarrow[\substack{45\% \\ \text{overall}}]{\text{NaOH}} N_2CHCH(OCH_3)_2$$
$$\textbf{1}$$

Cyclopropyl aldehydes.[2] The reagent is a more reactive 1,3-dipole than alkyl diazoacetates. It reacts with various olefins such as styrene, methyl acrylate, and **2** to form pyrazolines **a**, which on photolysis are converted into the cyclopropyl acetals **b**. The free aldehydes (**3**) are obtained by acid hydrolysis in overall yields greater than 80%.

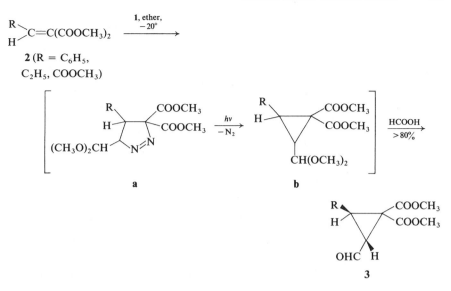

a   **b**

3

The reagent can be used to prepare 3-formylpyrazoles from alkynes (equation I).

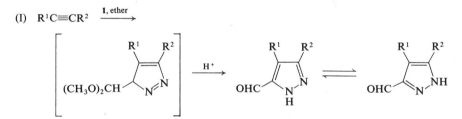

---

[1] Method of W. Kirmse and M. Buschhoff, *Ber.*, **100**, 1491 (1967).
[2] H. Abdallah and R. Grée, *Tetrahedron Letters*, **21**, 2239 (1980).

Dibenzoyl peroxide–Nickel(II) bromide, 9, 136.

Selective oxidation of primary diols. 2,2-Diphenyl-1,4-butanediol (**1**) is oxidized

exclusively to β,β-diphenyl-γ-butyrolactone (**2**) by either $C_6H_5\overset{O}{\overset{\|}{C}}OO\overset{O}{\overset{\|}{C}}C_6H_5$–$NiBr_2$ or trityl tetrafluoroborate (**8**, 525). In general the latter reagent is the more selective, but isolated yields of the lactone are lower (77% for **1** → **2**). Other oxidants show less selectivity. Use of LiBr or $(CH_3)_4NBr$ in place of $NiBr_2$ results in complete loss of selectivity. Presumably the $NiBr_2$ modifies the steric requirement of diols by a reversible association.[1]

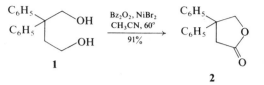

1   **2**

[1] M. P. Doyle, R. L. Dow, V. Bagheri, and W. J. Patrie, *Tetrahedron Letters*, **21**, 2795 (1980).

Dibromoisocyanuric acid,

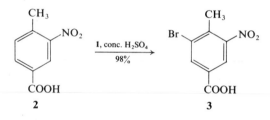

(1). Mol. wt. 285.88, m.p. 307–309° (dec).

The reagent is prepared by reaction of cyanuric acid with 2 equivalents of LiOH and then with 4 equivalents of bromine (88% yield).[1]

Bromination. Ganem *et al.*[2] have used **1** for bromination of 4-methyl-3-nitro-benzoic acid (**2**).

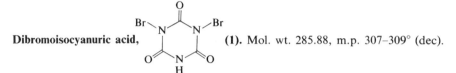

2   **3**

[1] W. Gottardi, *Monats.*, **98**, 507 (1968).
[2] A. R. Leed, S. D. Boettger, and B. Ganem, *J. Org.*, **45**, 1098 (1980).

Di-*t*-butyl dicarbonate, 4, 128; **7,** 91; **8,** 145.

Cleavage of an amide.[1] In the first studies of the structure of the antibiotic nocardicin A (**1**), the acyl side chain was cleaved in low yield by hydrogenation of the C=N grouping followed by Edman degradation with phenylisothiocyanate (**1,** 844–845) to give 3-aminonocardicinic acid (**4**). This degradation has since been accomplished in good yield by treatment of **1** with di-*t*-butyl dicarbonate to give **2** and **3**. Both BOC groups of **2** are easily cleaved by TFA to give **4**.

BOC-Amino acids. Chemists at Fluka[2] have prepared many of these useful derivatives of amino acids by reaction with this reagent in aqueous organic solvents with sodium hydroxide as base. In general yields are 65–95%.

[1] K. Schaffner, B. W. Müller, R. Scartazzini, and H. Wehrli, *Helv.*, **63**, 321 (1980).
[2] O. Keller, W. E. Keller, and G. van Look, and G. Wersin, *Org. Syn.*, submitted (1980).

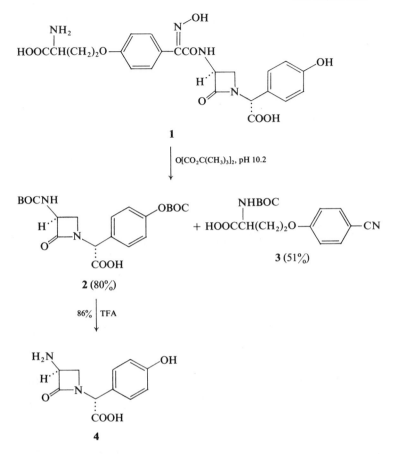

1

$O[CO_2C(CH_3)_3]_2$, pH 10.2

2 (80%)   **3 (51%)**

86% | TFA

4

2,6-Di-*t*-butyl-4-methylpyridine, 8, 145–146; **9**, 141.

Vinyl triflates.[1] Use of this hindered, non-nucleophilic base (**1**) allows direct conversion of aldehydes to vinyl triflates (equation I). The method is also applicable to conversion of ketones to 1-alkenyl triflates.

$$\text{(I)} \quad R_2CHCHO + (CF_3SO_2)_2O \xrightarrow[>80\%]{\text{1, CH}_2\text{Cl}_2, \Delta} R_2C{=}CHOSO_2CF_3$$

[1] P. J. Stang and W. Treptow, *Synthesis*, 283 (1980).

Di-*n*-butyltin oxide, 5, 188, **9**, 141.

Deacetylation.[1] Both N- and O-acetyl groups of the nucleoside **1** are removed by reaction of di-*n*-butyltin oxide in refluxing CH_3OH. Use of CH_3ONa in CH_3OH

under mild conditions results in O-deacetylation only; extensive decomposition occurs under more drastic conditions.

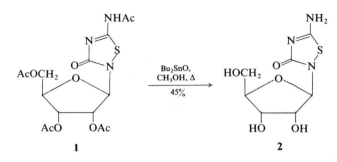

1   **2**

Monoderivatives of symmetrical diols.[2] One possible route to these derivatives is treatment of the diol with di-*n*-butyltin oxide to form a stannoxane, which is then allowed to react with 1 equivalent of an acid chloride or sulfonyl chloride (equation I). Unsymmetrical diesters can be obtained by consecutive reaction of the stannoxane with two different reagents.

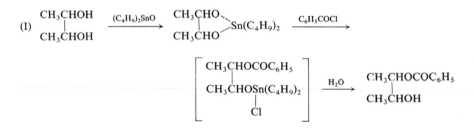

Macrolides. Steliou, Hanessian, and co-workers[3] have prepared lactones and lactams in moderate to excellent yield by treatment of ω-hydroxy or ω-amino carboxylic acids with catalytic amounts of di-*n*-butyltin oxide in refluxing mesitylene or xylene in a Dean-Stark apparatus for 12–24 hours. The method has the advantage of high dilution, since the tin reagent is regenerated continuously. As expected, yields are low for medium-sized rings.

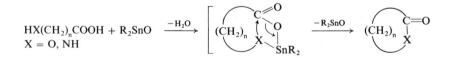

Macrocyclic tetralactones.[4] A method for preparation of these lactones (**2**) uses a stannoxane (**1**) as a covalent template (equation I). Formation of dilactones is not observed.

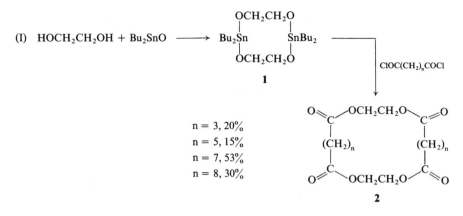

(I) $HOCH_2CH_2OH + Bu_2SnO \longrightarrow$

1

n = 3, 20%
n = 5, 15%
n = 7, 53%
n = 8, 30%

2

[1] T. L. Chwang, J. Nemec, and A. D. Welch, *J. Carbohydrates, Nucleosides-Nucleotides*, **7**, 159 (1980).

[2] A. Shanzer, *Tetrahedron Letters*, **21**, 221 (1980).

[3] K. Steliou, A. Szczygielska-Nowosielska, A. Favre, M. A. Poupart, and S. Hanessian, *Am. Soc.*, **102**, 7578 (1980).

[4] A. Shanzer and N. Mayer-Shocket, *J.C.S. Chem. Comm.*, 176 (1980).

Dicarbonylbis(triphenylphosphine)nickel, 1, 61; **7,** 94. Supplier: Alfa. Preparation.[1]

α-Methylenelactones. α-Methylenelactones can be obtained by intramolecular carbonylation of vinyl bromides such as **1** with this nickel complex.[2] Nickel carbonyl can also be used, but this reagent is probably more toxic and is more difficult to handle. The nickel complex is required in stoichiometric amounts; addition of triethylamine increases the rate of carbonylation. The method is efficient for both five- and six-membered lactones and can also be used for a combination cyclization–carbonylation sequence.

Examples:

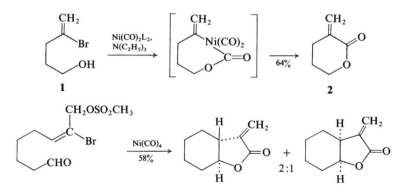

Carbonylation of halides with tetramethyltin.[3] This nickel complex is the most efficient catalyst for the synthesis of methyl ketones by reaction of aryl halides with carbon monoxide and tetramethyltin in HMPT at 120°. No reaction occurs when tetraphenyltin is used. A typical reaction is formulated in equation (I).

$$(I) \quad C_6H_5X + CO + Sn(CH_3)_4 \xrightarrow[\text{73\%}]{\substack{\text{Ni cat., HMPT,} \\ \text{120°, 24 hr}}} C_6H_5COCH_3$$

[1] A. A. Rose and A. A. Stathom, *J. Chem. Soc.*, 69 (1950).
[2] M. F. Semmelhack and J. S. Brickner, *J. Org.*, **46**, 1723 (1981).
[3] M. Tanaka, *Synthesis*, 47 (1981).

Dicarbonylcyclopentadienylcobalt, 5, 172–173; **6,** 153–154; **7,** 94–95; **8,** 146–147; **9,** 142–143.

Annelated benzenes.[1] Vollhardt's cocyclization of bis(trimethylsilyl)ethyne with α,ω-diynes has been extended to several new partners. Thus α,ω-diynes can cocyclize with 1,3-bis(trimethylsilyl)propyne and trimethylsilylalkynes. Examples are shown in equations (I) and (II).

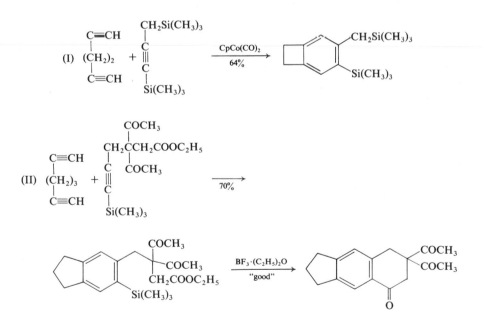

Cyclopentadienones.[2] Cyclopentadienones are generally not useful synthons because of their ready dimerization. A new synthesis of more stable substituted cyclopentadienones involves cyclization of two substituted alkynes with carbon monoxide by the [2 + 2 + 2]cycloadditions shown in equations (I) and (II).

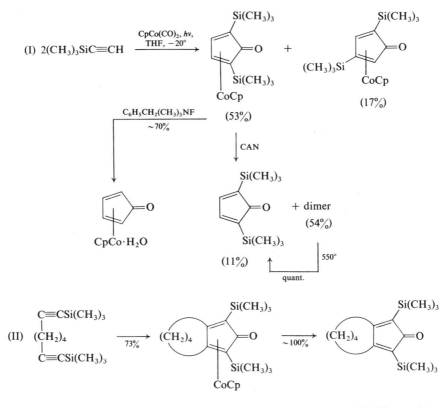

Treatment of the adducts with CAN effects demetallation to give 2,5-di(trimethylsilyl)cyclopentadienones.

[2 + 2 + 2]Cycloadditions.[3] A new approach to tricyclic hydroaromatic systems involves the simultaneous formation of three new C—C bonds in one step using this cobalt catalyst. A typical example is the reaction shown in Scheme (I). An interesting feature is that the reaction is regiospecific and can be stereospecific with respect to the tertiary H and the metal. Complexation with the metal allows a number of interesting transformations as shown.

[1] E. R. F. Gesing, J. A. Sinclair, and K. P. C. Vollhardt, *J.C.S. Chem. Comm.*, 286 (1980).
[2] E. R. F. Gesing, J. P. Tane, and K. P. C. Vollhardt, *Angew. Chem. Int. Ed.*, **19**, 1023 (1980).
[3] E. D. Sternberg and K. P. C. Vollhardt, *Am. Soc.*, **102**, 4839 (1980).

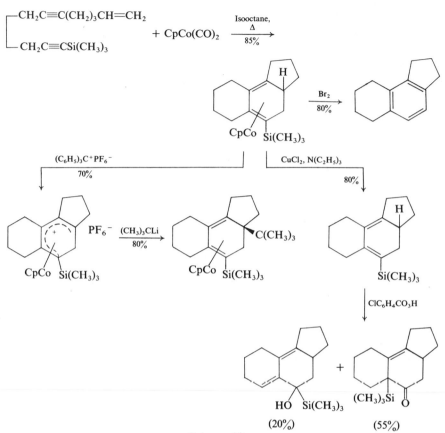

Scheme (I)

Dicarbonylcyclopentadienylironethyl(methylphenyl)sulfonium fluorosulfate (1). Mol. wt. 428.

Preparation:

$$Na[CpFe(CO)_2] + ClCHSC_6H_5 \xrightarrow[78\%]{THF}$$

(with CH_3 on the $ClCHSC_6H_5$)

$$CpFe(CO)_2\overset{CH_3}{CHSC_6H_5} \xrightarrow{FSO_3CH_3} CpFe(CO)_2\overset{CH_3}{CHS^+}\overset{C_6H_5}{\underset{CH_3}{\diagdown}} \quad FSO_3^-$$

2   **1**

Ethylidenation. Helquist et al.[1] have extended the cyclopropanation of olefins with an iron–methylene complex (**9**, 143) to ethylidenation of olefins with the iron–ethylidene complex **1**. Since the sulfide precursor (**2**) is more stable, the reagent is generated in the presence of the olefin. The reagent gives methyl-substituted cyclopropanes in satisfactory yield from monosubstituted, *cis*-disubstituted, and phenylated olefins, but does not react with *trans*-disubstituted olefins.

Examples:

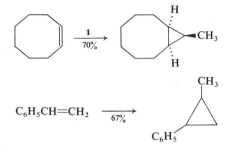

[1] K. A. M. Kremer, P. Helquist, and R. C. Kerber, *Am. Soc.*, **103**, 1862 (1981).

Di-µ-carbonylhexacarbonyldicobalt, $Co_2(CO)_8$, **1,** 224–225; **3,** 89; **4,** 139; **5,** 204–205; **6,** 172; **7,** 99–100; **8,** 148–150; **9,** 144–145.

Propargylation (**8,** 148–149). The propargylation of aromatics and β-dicarbonyl compounds with (propargyl)dicobalt hexacarbonyl cations (**1**) is now used for selective alkylation of ketones and their trimethylsilyl enol ethers and enol acetates. The reaction is regiospecific and involves attack of the more thermodynamically stable enol. In the case of ketones, yields are substantially reduced by use of a solvent. In the case of enol derivatives, CH_2Cl_2 can be used.[1]

Examples:

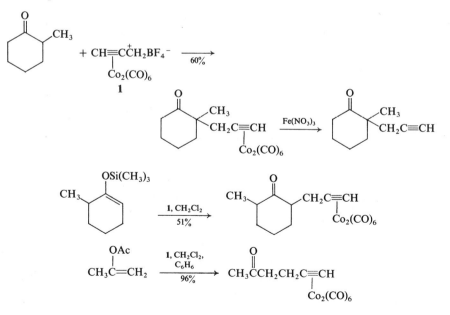

1,5-Enynes.[2] (Propargyl)dicobalt hexacarbonyl cations (**8**, 148–149) couple with allylsilanes to give complexes of 1,5-enynes, generally in satisfactory yields. A typical reaction is formulated in equation (I).

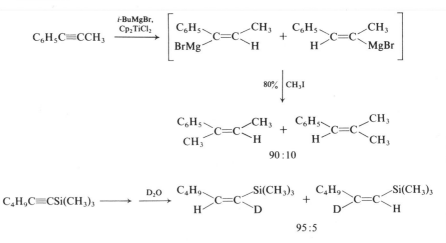

(I)

[1] K. M. Nicholas, M. Mulvaney, and M. Bayer, *Am. Soc.*, **102**, 2508 (1980).
[2] J. E. O'Boyle and K. M. Nicholas, *Tetrahedron Letters*, **21**, 1595 (1980).

Dichlorobis(cyclopentadienyl)titanium (Titanocene dichloride), $(C_5H_5)_2TiCl_2$. Mol. wt. 249.00, m.p. 289–291°. Supplier: Alfa.

Hydromagnesation of acetylenes.[1] Disubstituted acetylenes react with isobutyl-magnesium bromide in the presence of catalytic amounts of Cp_2TiCl_2 to form (E)-alkenyl Grignard reagents in high yield. The actual reagent is probably Cp_2TiH. The reaction is regioselective in the case of unsymmetrical acetylenes.

Examples:

Reductions with Grignard reagents. Aliphatic ketones and aldehydes are reduced to the corresponding alcohols by Grignard reagents with a β-hydrogen (*e.g.*, propyl, isopropyl, 2-methylbutyl Grignard reagents) in the presence of catalytic amounts of

Cp_2TiCl_2. Use of ether as solvent is important for satisfactory yields. Aromatic and α,β-unsaturated ketones are not reduced. The reports suggest that Cp_2TiH is the actual reducing agent. In favorable cases isolated yields of 94–99% can be obtained.[2]

Esters can be reduced under these conditions to either secondary or primary alcohols, depending on the amount of Cp_2TiCl_2 used. In the presence of 2 mole % of catalyst, isobutylmagnesium bromide reduces esters mainly to the corresponding primary alcohols. When less catalyst is used, the major products are secondary alcohols formed by coupling of the ester and the Grignard reagent.[3]

π-Allyltitanium complexes. π-Allyltitanium complexes are formed *in situ* by reaction of Cp_2TiCl_2 with an allyl Grignard reagent or with a 1,3-diene and propylmagnesium bromide (the actual reagent is probably Cp_2TiH). The complexes react with aldehydes or ketones to form, after hydrolysis, homoallylic alcohols (*cf.*, **6**, 361). The reaction is regioselective, with the more substituted carbon of the allyl group becoming attached to the carbonyl carbon.[4]

Examples:

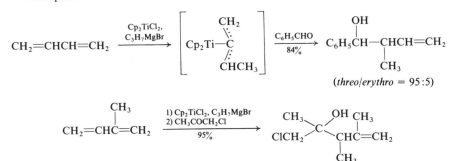

$(threo/erythro = 95:5)$

Hydroalumination. Titanocene dichloride is an effective catalyst for hydroalumination of alkenes and alkynes with bis(dialkylamino)alanes[5] and various complex aluminum hydrides.[6] The adducts can be quenched with water or iodine. The reaction is satisfactory for terminal alkenes and internal alkynes, but is not clean for internal alkenes and terminal alkynes.

[1] F. Sato, H. Ishikawa, and M. Sato, *Tetrahedron Letters*, **22**, 85 (1981).
[2] F. Sato, T. Jinbo, and M. Sato, *Tetrahedron Letters*, **21**, 2171 (1980).
[3] F. Sato, T. Jinbo, and M. Sato, *Tetrahedron Letters*, **21**, 2175 (1980).
[4] F. Sato, S. Iijima, and M. Sato, *Tetrahedron Letters*, **22**, 243 (1981).
[5] E. C. Ashby and S. A. Noding, *J. Org.*, **44**, 4364 (1979).
[6] E. C. Ashby and S. A. Noding, *J. Org.*, **45**, 1035 (1980).

Dichlorobis(cyclopentadienyl)zirconium, $(C_5H_5)_2ZrCl_2$ **(1).** Mol.   wt.   292.32. Supplier: Alfa.

Stereoselective aldol condensation. Studies by Heathcock *et al.*[1] have shown that the stereochemistry of the aldol condensation of lithium enolates can be correlated with the geometry of the enolate; that is, *erythro*-products are formed preferentially from (Z)-enolates and *threo*-products are formed preferentially from (E)-enolates. An example is the reaction of ethyl *t*-butyl ketone with benzaldehyde in which only one of the two possible aldols is formed (equation I). In this case, the enolate has the (Z)-geometry for steric reasons.

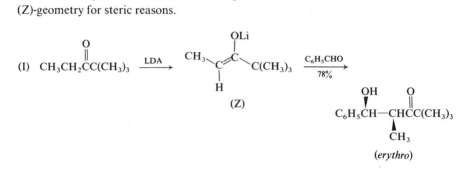

(I)

In contrast, aldol condensation with (Z)- and (E)-chlorobis(cyclopentadienyl)-zirconium enolates results in *erythro*-diastereoselection regardless of the geometry of the enolate.[2,3] These enolates are prepared from lithium enolates by metal exchange with Cp_2ZrCl_2 at $-78°$. The effect is particularly marked with amide enolates (equation II).

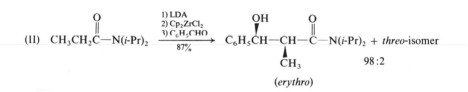

(II)

If the same condensation is carried out on the lithium enolate directly, the *erythro*- and *threo*-aldols are obtained in the ratio 63:37.

Aldol condensation of aldehydes with chiral zirconium enolates. This reaction can exhibit high levels of *erythro*-diastereoselection. Thus the zirconium enolate of the propanamide **1**, reacts with aldehydes to afford predominately a single aldol diastereomer (**2**) in 96–98% ee. The enolate reacts with both (R)- and (S)-aldehydes to form comparable levels of *erythro*-selection. Thus enolate chirality suppresses the influence of chirality of the aldehyde.[4]

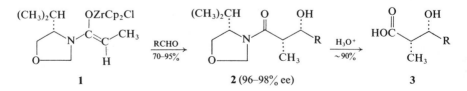

Hydroalumination (*cf.*, **8**, 506–507). This zirconium compound can catalyze hydroalumination of alkenes with trialkylalanes containing β-hydrogens, such as tri-isobutylalane and tri-*t*-butylalane.[5]

[1] C. H. Heathcock, C. T. Buse, W. A. Kleschick, M. C. Pirrung, J. E. Sohn, and J. Lampe, *J. Org.*, **45**, 1066 (1980).
[2] D. A. Evans and L. R. McGee, *Tetrahedron Letters*, **21**, 3975 (1980).
[3] Y. Yamamoto and K. Maruyama, *Tetrahedron Letters*, **21**, 4607 (1980).
[4] D. A. Evans and L. R. McGee, *Am. Soc.*, **103**, 2876 (1981).
[5] E. Negishi and T. Yoshida, *Tetrahedron Letters*, **21**, 1501 (1980).

Dichlorobis(triphenylphosphine)palladium(II), **6**, 60–61; **7**, 95–96; **8**, 151–152. Preparation.[1]

Lactone synthesis.[2] Carbonylation of simple organic halides can be carried out readily with several palladium catalysts such as bis(diphenylphosphinoethane)-bis(triphenylphosphine)palladium(0) and dichlorobis(triphenylphosphine)palladium(II). The latter catalyst is preferred because it is stable and easily converted to Pd(0) *in situ*. Carbonylation of halo alcohols provides a useful synthesis of various lactones.

Examples:

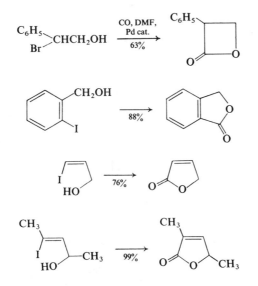

[1] J. F. Fauvarque and A. Jutland, *Bull. Soc.*, 766 (1976).
[2] A. Cowell and J. K. Stille, *Am. Soc.*, **102**, 4193 (1980).

1,3-Dichloro-2-butene, 1, 214–215; **2,** 111. Conrow and Marshall[1] have reported an improved synthesis of the Wichterle reagent from methyl acetoacetate (equation I).

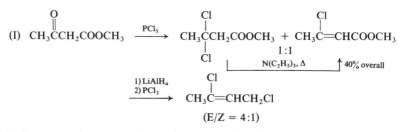

(E/Z = 4:1)

[1] R. E. Conrow and J. A. Marshall, *Syn. Comm.,* **11,** 419 (1981).

1,4-Dichloro-2-butene, $ClCH_2CH{=}CHCH_2Cl$ **(1).** Mol. wt. 125.00, b.p. 72–75°/40 mm. Both the (E)- and (Z)-isomers are available from Aldrich.

Biphenyl synthesis.[1] In the presence of 1,4-dichloro-2-butene, aryl Grignard reagents couple to form biphenyls. The coupling involves electron transfer to the 1,4-dihalide, which is transformed into butadiene (gas). The method serves as an attractive alternative to coupling with thallium(I) bromide, which is toxic. Yields are comparable; both methods fail with *ortho*-substituted Grignard reagents.

Cycloalkylation (**7,** 148).[2] Cycloalkylation of **2** with (Z)-**1** was used as one step in a total synthesis of (±)-sesbanine (**6**), a constituent of *Sesbania drummondii* seeds with antileukemic activity. The hydroxyl group was introduced into the cyclopentene ring of **4** by iodolactonization followed by reduction to give **5**. Final steps included aminolysis of the lactone ring, intramolecular addition of the amide anion to the CN group, and hydrolysis to give **6**.

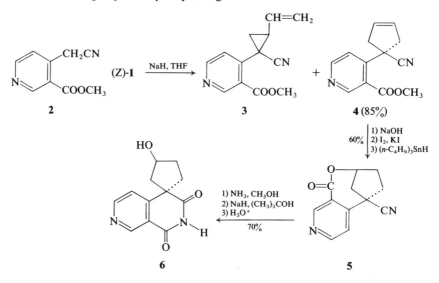

[1] S. K. Taylor, S. G. Bennett, K. J. Heinz, and L. K. Lashley, *J. Org.*, **46**, 2194 (1981).
[2] J. C. Bottaro and G. A. Berchtold, *J. Org.*, **45**, 1176 (1980).

Dichlorodicyanobenzoquinone (DDQ), **1**, 215–219; **2**, 112–117; **3**, 83–84; **4**, 130–134; **5**, 193–194; **6**, 168–170; **7**, 96–97; **8**, 153–156; **9**, 148–151.

Oxidation of aromatic alkyl groups. Aromatic methyl or ethyl groups *para*[1] or *ortho*[2] to an alkoxy function can be oxidized by DDQ in refluxing methanol to aldehydes or methyl ketones, respectively. Simultaneous dehydrogenation can also be effected.

Examples:

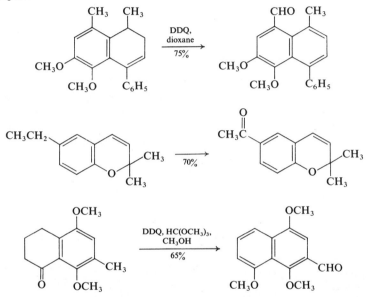

Oxidation of benzylic alcohols.[3] DDQ is a suitable oxidant for alcohols of the type $ArCH_2OH$, $ArCH(OH)R$, and Ar_2CHOH. Electron-donating substituents (OH, OCH_3), particularly in the *para*-position, facilitate oxidation, and electron-attracting substituents render oxidation more difficult. Dioxane is a suitable solvent. Use of methanol can lead to a different reaction as shown in equations (I) and (II).

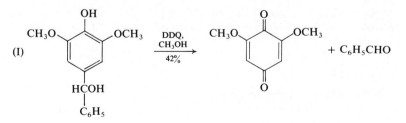

(II)

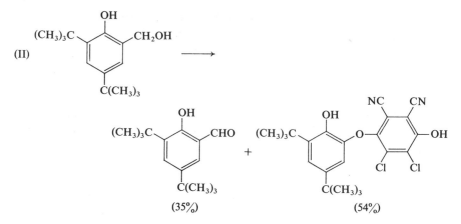

(35%)   (54%)

Azlactone oxidation.[4] Azlactones derived from dipeptides are more readily dehydrogenated than the dipeptides. This route to dehydropeptides has been examined with several reagents. Halogenation–dehydrohalogenation is possible, but yields at best are ~ 50%. Various oxidation procedures are about as effective. The most satisfactory method is oxidation of the corresponding trimethylsilyl enol ether with DDQ. However, this oxidation is limited to aryl azlactones.

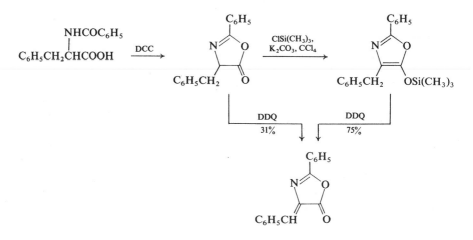

[1] M. V. Naidu and G. S. K. Rao, *Synthesis*, 144 (1979).
[2] A. S. Kende, J.-P. Gesson, and T. P. Demuth, *Tetrahedron Letters*, **22**, 1667 (1981).
[3] H.-D. Becker, A. Björk, and E. Adler, *J. Org.*, **45**, 1596 (1980).
[4] R. S. Latt, E. G. Breitholle, and C. H. Stammer, *J. Org.*, **45**, 1151 (1980).

Di-μ-chlorodimethoxybis(pyridine)dicopper, $(PyCuOCH_3Cl)_2$ **(1).** Mol. wt. 418.26. This deep green crystalline reagent was first prepared by Finkheiner *et al.*[1] in a

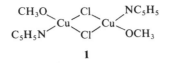

1

study of copper–amine oxidative coupling of 2,6-disubstituted phenols to diphenoquinones or polymeric ethers. Of initial interest was the presence of copper in various oxidases. It can be prepared in various ways from $CuCl + O_2$ or from $CuCl_2$ and the other components, CH_3OH and pyridine.

More extensive studies with **1** have been reported by Rogić and co-workers in connection with new synthetic approaches to caprolactam. They established that **1** can cleave o-quinones, catechols, and phenols under anaerobic conditions. Thus catechol, o-benzoquinone, and phenol are oxidized by **1** to cis,cis-muconic acid methyl ester. The paper presents several lines of evidence that oxygen does not

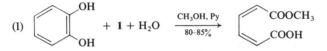

oxidize the organic substrate directly. Rather, the function of oxygen is to reoxidize the Cu(I) generated in the transformation to Cu(II), the actual oxidant.[2]

Reagent **1** also converts anhydrides such as **2**, prepared as indicated, to muconic acid derivatives (**3** and **4**) (equation II).[3]

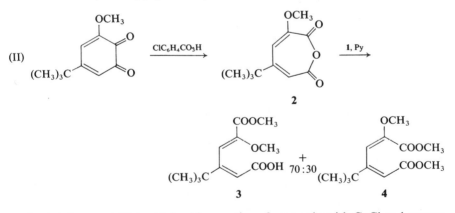

A related reagent (**5**) is obtained by reaction of ammonia with CuCl and oxygen

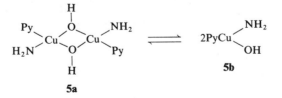

5a

in pyridine. This reagent, in the absence of oxygen, oxidizes *o*-benzoquinones and catechols to muconic acid nitriles (equation III).[4]

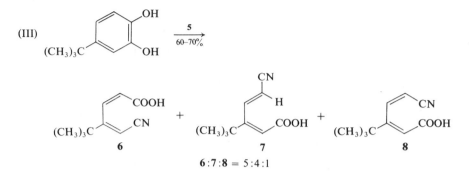

(III)

$$6:7:8 = 5:4:1$$

[1] H. Finkheiner, A. S. Hay, H. S. Blanchard, and G. F. Endres, *J. Org.*, **31**, 549 (1966).

[2] M. M. Rogić, T. R. Demmin, and W. B. Hammond, *Am. Soc.*, **98**, 7440 (1976); M. M. Rogić and T. R. Demmin, *Am. Soc.*, **100**, 5472 (1978).

[3] T. R. Demmin and M. M. Rogić, *J. Org.*, **45**,. 1153 (1980).

[4] T. R. Demmin and M. M. Rogić, *J.Org.*, **45**, 2737 (1980).

Dichlorodimethyltitanium, $(CH_3)_2TiCl_2$. Mol. wt. 158.88. Equivalent amounts of $(CH_3)_2Zn$ and $TiCl_4$ react in CH_2Cl_2 at $-30°$ to form $(CH_3)_2TiCl_2$, which is used without further purification.

Methylation of aldehydes and ketones. The reagent forms 1:1 adducts with aldehydes and ketones at low temperatures; aldehydes undergo addition so much more rapidly that selective reactions are possible in the presence of ketones. Both reactions are diastereoselective. Thus addition to 4-*t*-butylcyclohexanone results in a mixture of axial and equatorial alcohols in the ratio of 82:18.

Of greater interest, *gem*-dimethylation of ketones is possible with reactions conducted in the temperature range $-30°$ to 22° by a two-step sequence involving addition followed by methylation of the intermediate titanium alcoholate. The method is widely applicable, although subject to steric effects, and yields are generally in the range 60–85%. In addition, tertiary alcohols and tertiary alkyl chlorides are methylated by reactions conducted at room temperature in yields of about 75%.[2] *gem*-Dihaloalkanes can also be converted into *gem*-dimethylalkanes.[3] .

The metal plays a crucial role in these reactions. $(CH_3)_3Al$ and $TiCl_4$ (1:1) also induce *gem*-dialkylation of ketones, but $(CH_3)_3Al$ and $AlCl_3$ are ineffective.

[1] M. T. Reetz, R. Steinbach, J. Westermann, and R. Peter, *Angew. Chem. Int. Ed.*, **19**, 900, 901, 1011 (1980).

[2] M. T. Reetz, J. Westermann, and R. Steinbach, *J.C.S. Chem. Comm.*, 237 (1981).

[3] M. T. Reetz, R. Steinbach, and B. Wenderoth, *Syn. Comm.*, **11**, 261 (1981).

Dichloroketene, 1, 221–222; **2**, 118; **3**, 87–88; **4**, 134–135; **8**, 156; **9**, 152–154.

Cyclopentanone annelation. α,α-Dichlorocyclopentanones, obtained by addition of dichloroketene to cycloalkenes followed by ring expansion with diazomethane (**9**, 133–134), are versatile intermediates to a variety of interesting products. Some of the transformations are indicated in Scheme (I) and equation (I).[1]

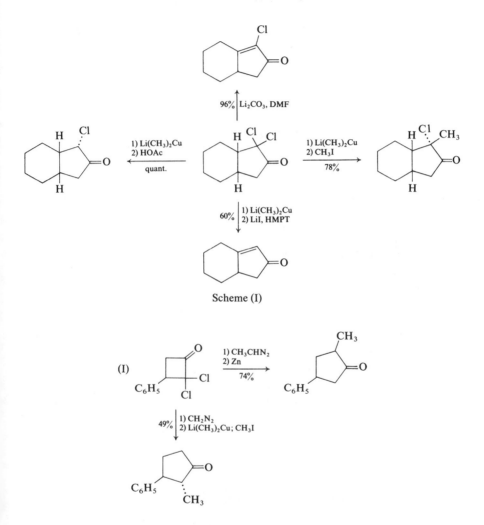

Scheme (I)

This methodology has been used for an iterative synthesis of the tricyclopenta-noid hirsutene (**6**), as outlined in equation (II).[2] Two steps in the synthesis involve conversion of an α-chloro ketone to an alkene. This reaction was accomplished by reduction to a chlorohydrin followed by treatment with chromium(II) perchlorate.[3]

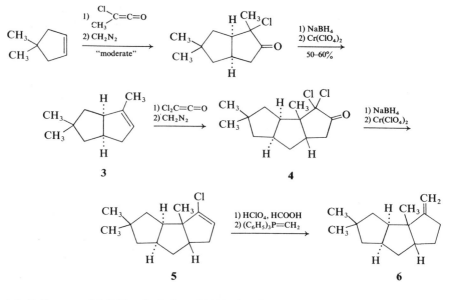

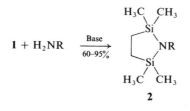

[1] A. E. Greene and J.-P. Deprés, *J. Org.*, **45**, 2036 (1980).
[2] A. E. Greene, *Tetrahedron Letters*, **21**, 3059 (1980).
[3] J. K. Kochi and D. M. Singleton, *Am. Soc.*, **90**, 1582 (1968).

1,4-Dichloro-1,1,4,4-tetramethyldisilylethylene, $(CH_3)_2\overset{\underset{\displaystyle |}{Cl}}{Si}CH_2CH_2\overset{\underset{\displaystyle |}{Cl}}{Si}(CH_3)_2$ **(1).** Mol. wt. 215.3, b.p. 198°/734 mm, m.p. 37°. Supplier: Petrarch Systems.

Protection of primary amines.[1] Primary amines form the cyclic adducts **2** ("Stabase adducts") on reaction with **1** and a base (triethylamine for amines with pK_a 10–11, *n*-butyllithium for less basic amines). The adducts are stable to *n*- and *sec*-BuLi (−25°), LDA, H_2O, CH_3OH, KF, and $NaHCO_3$ but unstable to HOAc, HCl, KOH, and $NaBH_4$.

The adducts are useful for preparation of alkylated amino acids. An example is the preparation of 2-amino-4-pentynoic acid (**4**) from ethyl glycinate (**3**).

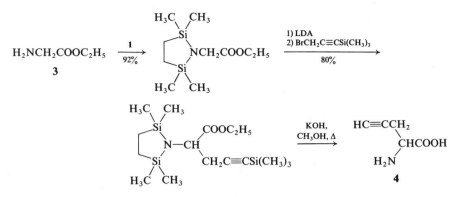

$$H_2NCH_2COOC_2H_5 \xrightarrow[92\%]{1}$$

3

[1] S. Djuric, J. Venit, and P. Magnus, *Tetrahedron Letters*, **22**, 1787 (1981).

Dichlorotris(triphenylphosphine)ruthenium(II), 4, 564; **5**, 740–741; **6**, 654–655; **7**, 99; **8**, 159–161.

Tertiary amines. Primary amines bearing an α-hydrogen atom are converted in high yield to symmetrical secondary amines when heated at 185° in the presence of about 0.2 mole % of this ruthenium complex (equation I).[1] Use of larger amounts of

$$\text{(I)} \quad 2RCH_2NH_2 \xrightarrow{\text{Cat.}} (RCH_2)_2NH + NH_3$$

the catalyst or, more simply, of $RuCl_3 \cdot 3H_2O$ and $P(C_6H_5)_3$ in THF, results in tertiary amines, $(RCH_2)_3N$, in 50–98% yield.[2] This reaction is somewhat sensitive to steric bulk near the nitrogen atom.

α,ω-Diamines are cyclized when heated in diphenyl ether at 180° with $RuCl_2[P(C_6H_5)_3]_3$ (equation II).

$$\text{(II)} \quad H_2N(CH_2)_nNH_2 \xrightarrow[80-90\%]{\text{Cat.}} \underset{NH}{\overset{(CH_2)_n}{\bigcirc}} + NH_3$$

$$n = 4, 5, 6$$

Selective oxidation of alcohols.[3] Primary alcohols are oxidized by this $RuCl_2$ complex about 50 times as rapidly as secondary alcohols. Use of benzene as solvent is critical for this high selectivity. Little or no reaction occurs in CH_3CN, THF, or DMF. Most oxidants, if they show any selectivity, oxidize secondary alcohols more rapidly than primary ones. However, ruthenium-catalyzed oxidations with N-methylmorpholine N-oxide and oxidations with PCC[4] proceed about three times as rapidly with primary alcohols as with secondary ones.

Examples:

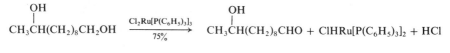

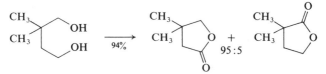

[1] B.-T. Khai, C. Concilio, and G. Porzi, *J. Organometal. Chem.*, **208**, 249 (1981).
[2] B.-T. Khai, C. Concilio, and G. Porzi, *J. Org.*, **46**, 1759 (1981).
[3] H. Tomioka, K. Takai, K. Oshima, and H. Nozaki, *Tetrahedron Letters*, **22**, 1605 (1981).
[4] E. J. Corey and M. Ishiguro, *Tetrahedron Letters*, 2745 (1979).

Dicyclohexylcarbodiimide (DCC), 1, 231–236; **2**, 126; **3**, 91; **4**, 141; **5**, 206–207; **6**, 174; **7**, 100–101; **8**, 162–163; **9**, 156–157.

Tetrapeptides.[1] A novel synthesis of tri- and tetrapeptides is based on the fact that the β-lactam ring can be cleaved by catalytic hydrogenation.[2] Tri- and tetrapeptides can be obtained by cleavage of a substrate containing two β-lactam groups. The method is outlined in Scheme (I). In this case use of DCC does not present the usual difficult separation of dicyclohexylurea from the product because the bis-β-lactam **2** is very soluble in ether, unlike peptides. However this step proceeds in only moderate yield.

[1] N. Hatanake and I. Ojima, *Chem. Letters*, 231 (1981).
[2] I. Ojima, S. Suga, and R. Abe, *Tetrahedron Letters*, **21**, 3907 (1980).

(Diethylamino)sulfur trifluoride (DAST), 6, 183–184; **8**, 166–167.
Caution: The reagent must be handled with suitable precautions.[1]
α,α-Difluoroarylacetic acids.[2] These heretofore rather inaccessible compounds are obtainable by reaction of α-ketoarylacetates with DAST (equation I).

$$\text{(I)} \quad ArCCOOC_2H_5 \quad \xrightarrow[55-90\%]{(C_2H_5)_2NSF_3,\ 20-60°} \quad ArCF_2COOC_2H_5 \quad \xrightarrow[50-95\%]{OH^-\ or\ H^+} \quad ArCF_2COOH$$

[1] W. J. Middleton, *Chem. Eng. News*, May 21, 43 (1979).
[2] W. J. Middleton and E. M. Bingham, *J. Org.*, **45**, 2883 (1980).

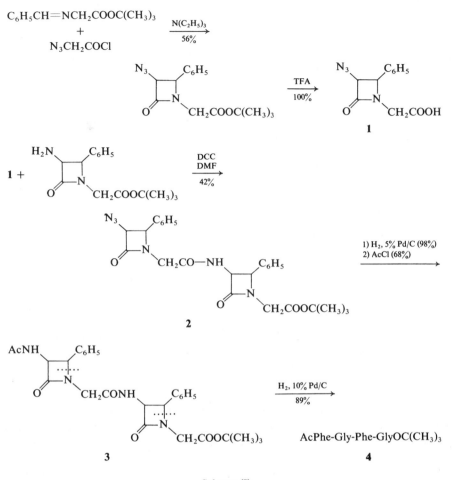

$C_6H_5CH=NCH_2COOC(CH_3)_3$
$+$
N_3CH_2COCl

$\xrightarrow[56\%]{N(C_2H_5)_3}$

Scheme (I)

Diethyl oxomalonate (1), 6, 188.

Allylic carboxylation.[1] Diethyl oxomalonate (**1**) undergoes a thermal ene reaction with mono-, di-, and trisubstituted alkenes at 145–180°. The reaction is also subject to catalysis with Lewis acids, which can lead to a different ene product. The products are α-hydroxymalonic esters. The corresponding malonic acids are converted to carboxylic acids by bisdecarboxylation with $NaIO_4$ and a trace of pyridine[2] or with ceric ammonium nitrate (CAN). Diethyl oxomalonate then functions as an enophilic equivalent of CO_2.

Examples:

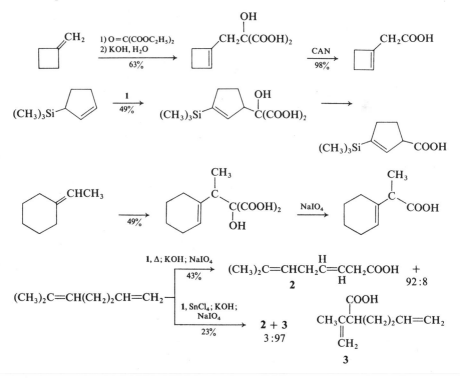

In the course of an examination of isotope effects in the ene reactions of this reagent, Stephenson and Orfanopoulos[3] have observed a potentially useful cyclization. An example is shown in equation (I).

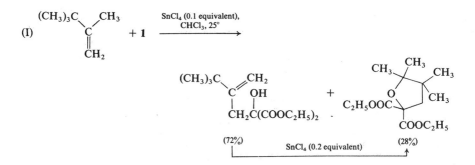

[1] M. F. Salomon, S. N. Pardo, and R. G. Salomon, *Am. Soc.*, **102**, 2473 (1980).
[2] E. J. Corey and R. H. Wollenberg, *Tetrahedron Letters*, 4705 (1976).
[3] L. M. Stephenson and M. Orfanopoulos, *J. Org.*, **46**, 2200 (1981).

Diethyl 1-phenyl-1-trimethylsilyloxymethanephosphonate, $(C_2H_5O)_2PCH{\scriptstyle\begin{array}{c}O\\\parallel\end{array}}\begin{array}{c}C_6H_5\\\diagdown\\ OSi(CH_3)_3\end{array}$

(1). Mol. wt. 316.41, b.p. 95–96°/0.05 mm. The phosphonate is prepared by reaction of $P(OC_2H_5)_3$, C_6H_5CHO, and $ClSi(CH_3)_3$ (94% yield).[1]

α-Hydroxy ketones.[2] The anion **(2)** of **1** reacts with aldehydes and ketones to give a 1,2-adduct **(a)** that rearranges to **b**, which fragments to **c**. Hydrolysis of the

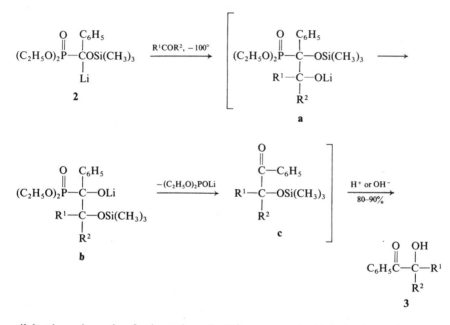

silyl ether gives the final product **3**. When an aryl aldehyde is used as the electrophile the intermediate **a** rearranges more slowly, and the alcohol corresponding to **b** can be isolated.

[1] G. H. Birum and G. A. Richardson, U.S. Patent 3,113,139 (1963) [*C.A.*, **60**, 5551d (1964)].
[2] R. E. Koenigkramer and H. Zimmer, *Tetrahedron Letters*, **21**, 1017 (1980).

Diethyl phosphorocyanidate (DEPC), **5**, 217; **6**, 192–193; **7**, 107; **9**, 163–164.

α-Cyanation of aromatic amine oxides.[1] This reaction is generally carried out with benzoyl chloride and potassium cyanide.[2] The combination of DEPC and triethylamine in acetonitrile is a useful alternative reagent.

[1] S. Harusawa, Y. Mamada, and T. Shiori, *Heterocycles*, **15**, 981 (1981).
[2] M. Henze, *Ber.*, **69**, 1566 (1936).

Examples:

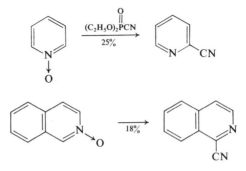

gem-Difluoroallyllithium (1). An improved synthesis is shown in equation (I).

(I) $CH_2=CH_2 + Br_2CF_2 \xrightarrow[76\%]{(C_6H_5COO)_2} F_2CBrCH_2CH_2Br \xrightarrow[74\%]{KOH, H_2O, 120°}$

$F_2CBrCH=CH_2 + CF_2=CHCH_2Br \xrightarrow[-95°]{n-C_4H_9Li,} CH_2\cdots\overset{-}{C}H\cdots CF_2Li^+$

                5:2   **1**

Reactions. The reagent can be used for preparation of 1,1-difluoroallylsilanes by reaction with chlorosilanes (50–90% yield), but the most useful reaction is the difluoroallylation of carbonyl compounds.[1]
Examples:

$(CH_3)_2C=O \xrightarrow[41\%]{1, -95 \to 20°} (CH_3)_2\overset{\displaystyle OH}{\underset{\displaystyle |}{C}}CF_2CH=CH_2$

$n-C_4H_9CHO \xrightarrow[87\%]{} n-C_4H_9\overset{\displaystyle OH}{\underset{\displaystyle |}{C}}HCF_2CH=CH_2$

[1] D. Seyferth, R. M. Simon, D. J. Sepelak, and H. A. Klein, *J. Org.*, **45**, 2273 (1980).

1,3-Dihydrobenzo[c]thiophene 2,2-dioxide (1). Mol. wt. 168.21, m.p. 150–151°.
   Preparation[1]:

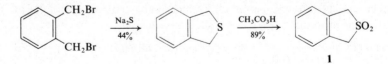

o-*Quinodimethanes*. A recent simple synthesis of $\Delta^{1,3,5(10)}$-estratriene-17-one (5)[2] is based on the fact that on pyrolysis substances such as **1** lose sulfur dioxide with generation of *o*-quinodimethanes.[1] The anion of **1** is generated most satisfactorily with KH (1.1 equivalent) in DME at 0°. It can be converted predominately to monoalkylated products, particularly if an excess of the anion is used. Thus reaction of the anion of **1** with **2** results in the diastereoisomers **3** and **4**. After deketalization, the corresponding ketones are heated at 210° for 8 hours. The *o*-quinodimethane (**a**) is formed and undergoes intramolecular cycloaddition to form **5**.

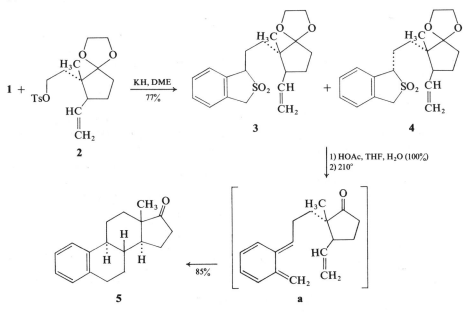

[1] M. P. Cava *et al.*, *Am. Soc.*, **81**, 4266 (1959).
[2] K. C. Nicolaou, W. E. Barnette, and P. Ma, *J. Org.*, **45**, 1463 (1980).

Dihydropyrane, 1, 256–257; **3**, 99; **5**, 220; **7**, 109.

***Reformatsky reaction*.**[1] Tetrahydropyranyl esters are recommended for use in Reformatsky reactions. They are formed from α-bromo acids in quantitative yield in dry benzene without need of an acid catalyst and they are readily hydrolyzed by dilute hydrochloric acid. The Reformatsky reaction is generally conducted in THF at a temperature below 10°. The reaction is more rapid if the zinc is activated with a trace of $HgCl_2$. Yields of β-hydroxy acids are generally 70–90% when aldehydes are used, but usually somewhat lower with ketones.

[1] M. Bogavac, L. Arsenijević, and V. Arsenijević, *Bull. soc.*, II-145 (1980).

Dihydrotetrakis(tri-*n*-butylphosphine)ruthenium(II), $RuH_2[P(n-Bu)_3]_4$ **(1).** The complex is prepared by addition of $NaBH_4$ to a mixture of $RuCl_3 \cdot 3H_2O$ and $P(n-Bu)_3$; 42% yield.

Codimerization of 1,3-butadiene and 1-alkynes. In the presence of this catalyst 1,3-butadiene and terminal alkynes couple to form an (E)-enyne in high yield.[1] The yield of product is low when phenylacetylene is used. Only trialkylphosphine complexes are effective for this coupling.

Example:

$$n\text{-}C_6H_{13}C{\equiv}CH + CH_2{=}CH{-}CH{=}CH_2 \xrightarrow[\substack{88\% \\ \text{isolated}}]{Ru(II)}$$

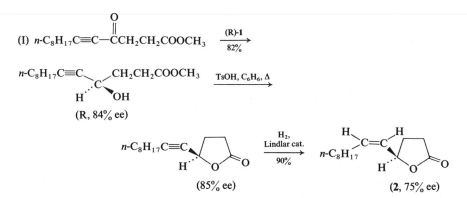

[1] T. Mitsudo, Y. Nakagawa, H. Watanabe, K. Watanabe, H. Misawa, and Y. Watanabe, *J.C.S. Chem. Comm.*, 496 (1981).

(R)- and (S)-2,2′-Dihydroxy-1,1′-binaphthyl–Lithium aluminum hydride, 9, 169–170.

Asymmetric reduction of alkynyl ketones. The (R)-form of the complex **(1)** reduces alkynyl ketones to optically active propargylic alcohols (usually R) in 65–85% chemical yield and in 85–95% optical yield; use of the (S)-form of **1**, as expected, results in the epimeric alcohol. This reduction was used in a synthesis of the natural Japanese beetle pheromone **(2,** equation I).[1]

Asymmetric reduction of Δ^{25}-24-oxosteroids. Reduction of the unsaturated 24-oxosteroid **2** with $LiAlH_4$ and the (R)-(+)-isomer of Noyori's reagent **(1)** gives a mixture of the two diols **3** and **4** in the ratio 95:5. The stereoselectivity is reversed by use of (S)-(−)-**1**. This reaction was used to prepare optically pure (24R)- and (24S)-24-hydroxycholesterol and the epimeric pairs of 24,25-dihydroxycholesterol and 25,26-dihydroxycholesterol.[2]

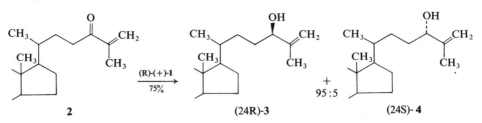

2   (24R)-**3**   (24S)- **4**

[1] M. Nishizawa, M. Yamada, and R. Noyori, *Tetrahedron Letters*, **22**, 247 (1981).
[2] M. Ishiguro, N. Koizumi, M. Yasuda, and N. Ikekawa, *J.C.S. Chem. Comm.*, 115 (1981).

Diiodo(methyl)bis(triphenylphosphine)rhodium, $CH_3RhI_2[P(C_6H_5)_3]_2$ **(1).** Mol. wt. 894.50. Preparation.[1]

Methylation of organomercurials.[2] Alkenyl-, alkynyl-, and arylmercurials are methylated by **1** in HMPT in 90% yield. In theory **1** could serve as catalyst in alkylations with CH_3I, but yields are lower because of interfering side reactions, mainly dimerization. Alkylmercurials are not reactive to **1**.

Example:

$$[(CH_3)_3CC\equiv C]_2Hg \xrightarrow[99\%]{\substack{1, HMPT, \\ LiCl}} 2(CH_3)_3CC\equiv CCH_3$$

[1] D. N. Lawson, J. A. Osborn, and G. Wilkinson, *J. Chem. Soc. A*, 1733 (1966).
[2] R. C. Larock and S. S. Hershberger, *Tetrahedron Letters*, **22**, 2443 (1981).

Diisobutylaluminum hydride (DIBAH), **1**, 260–262; **2**, 140–142; **3**, 101–102; **4**, 158–161; **5**, 224–225; **6**, 198–201; **7**, 111–113; **8**, 173–174; **9**, 171–172.

Reduction of allenes.[1] DIBAH preferentially reduces the more substituted double bond of an allene. Actually the hydride may attack the less substituted double bond and an allylic rearrangement during hydrolysis may be involved. The reduction of 1,1-diphenylallene[2] follows a different course, but may represent a special case.

Examples:

$$n\text{-}C_7H_{15}CH=C=CH_2 \xrightarrow[83\%]{DIBAH} n\text{-}C_8H_{17}-CH=CH_2 + n\text{-}C_{10}H_{22}$$
$$96:4$$

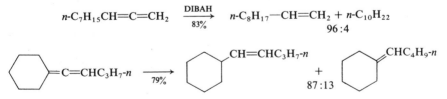

[1] M. Montury and J. Goré, *Tetrahedron Letters*, **21**, 51 (1980).
[2] J. J. Eisch and G. R. Husk, *J. Organometal. Chem.*, **4**, 415 (1965).

Diisobutylaluminum phenoxide, $[(CH_3)_2CHCH_2]_2AlOC_6H_5$ **(1), 8,** 172–173. The reagent is prepared *in situ* from diisobutylaluminum hydride and phenol.

Regioselective aldol condensation of methyl ketones.[1] The first synthesis of muscone (3-methylcyclopentadecanone, **4**)[2] used methylanilinomagnesium bromide, $C_6H_5N(CH_3)MgBr$ **(5,** 440–441), for the intramolecular aldol condensation of 2,15-hexadecanedione **(2).** The reaction presumably generates a magnesium enolate regioselectively, but the desired aldol was obtained in only 17% yield. Orienting experiments with 2-octanone indicated that diisobutylaluminum phenoxide in conjunction with pyridine permits regioselective aldol condensation at the methyl group of methyl ketones, and the method was then used for conversion of **2** to **3** in 65% yield.

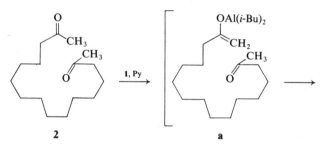

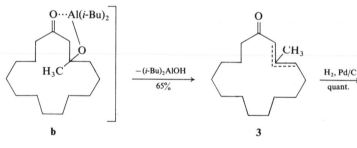

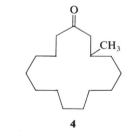

[1] J. Tsuji, T. Yamada, M. Kaito, and T. Mandai, *Bull. Chem. Soc. Japan,* **53,** 1417 (1980); *Tetrahedron Letters,* 2257 (1979).
[2] M. Stoll and A. Rouve, *Helv.,* **30,** 2019 (1947).

Diisopropylethylamine, 1, 371.

α-Substituted α-amino acids. Hünig's base is sufficiently strong to permit alkylation of 5-oxo-3,4-dihydro-1,3-oxazoles (**1**) without catalyzing self-condensation. Thus **1** can be alkylated under mild conditions in DMF or HMPT. Acidic hydrolysis of the 4,4-disubstituted oxazoles (**2**) liberates free amino acids (**3**). The method is not useful if the alkyl halide contains a double or triple bond.[1]

Example:

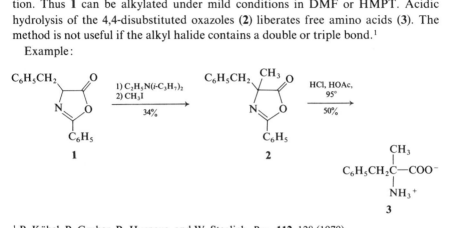

[1] B. Kübel, P. Gruber, R. Hurnaus, and W. Steglich, *Ber.*, **112**, 128 (1979).

Diketene–Iodotrimethylsilane.

N-Acetoacetylation.[1] The N-acetoacetylation of amides with diketene is markedly improved if it is carried out in the presence of iodotrimethylsilane (generated *in situ*). The reactive reagent is probably **a**.

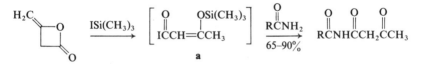

[1] Y. Yamamoto, S. Ohnishi, and Y. Azuma, *Synthesis*, 122 (1981).

(3S,6S)-(+)-2,5-Dimethoxy-3,6-dimethyl-3,6-dihydropyrazine,

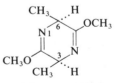

(1). Mol. wt. 170.21, b.p. 75°/8–10 mm, α_D + 82.6°. The reagent is prepared[1] by reaction of cyclo-(L-Ala-L-Ala)[2] with trimethyloxonium tetrafluoroborate.

α-Methylamino acids.[1] The anion (**2**) of **1** (*n*-BuLi or LDA, THF, −70°) is alkylated with high asymmetric induction exclusively at C_3. The products are hydrolyzed by dilute HCl at 25° to methyl alanate and optically active (R)-(−)-α-methyl-α-amino acids esters (**4**) (equation I).

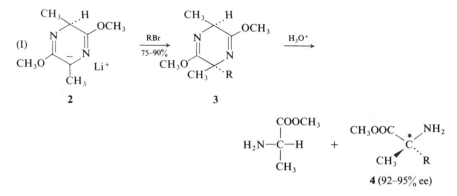

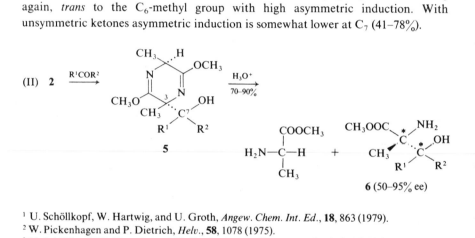

α-*Methylserines*.[3] The synthesis has been extended to asymmetric α-methylserines (**6**), as formulated in equation (II). A carbonyl group adds to **2** at C_3 again, *trans* to the C_6-methyl group with high asymmetric induction. With unsymmetric ketones asymmetric induction is somewhat lower at C_7 (41–78%).

[1] U. Schöllkopf, W. Hartwig, and U. Groth, *Angew. Chem. Int. Ed.*, **18**, 863 (1979).
[2] W. Pickenhagen and P. Dietrich, *Helv.*, **58**, 1078 (1975).
[3] U. Schöllkopf, W. Hartwig, and U. Groth, *Angew. Chem. Int. Ed.*, **19**, 212 (1980).

(S)-1-(Dimethoxymethyl-2-methoxymethyl)pyrrolidine, (**1**). B.p. 48°/0.05 mm, $α_D$ −33°. Preparation from (S)-proline.[1]

Asymmetric synthesis of amines. The reagent **1** has been used as a chiral auxiliary for enantioselective alkylation of DL-amino acids at the α-position as outlined in equation (I).[2]

This method has been used for a general asymmetric synthesis of amines by a similar mono- or dialkylation of propargylamine using the same chiral adjunct

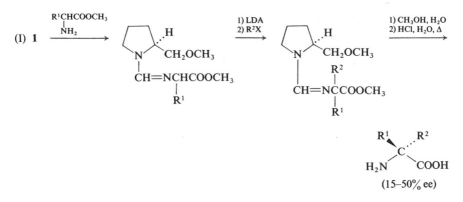

(I) **1**

(equation II). The optically active propargylamines should be useful precursors to various amino compounds.

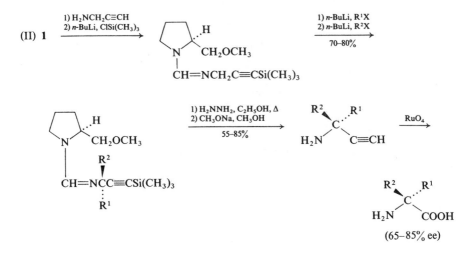

(II) **1**

[1] M. Kolb and J. Barth, *Tetrahedron Letters*, 2999 (1979).
[2] M. Kolb and J. Barth, *Angew. Chem. Int. Ed.*, **19**, 725 (1980).

5,5-Dimethoxy-1,2,3,4-tetrachlorocyclopentadiene, 1, 270; **8**, 178.

Cyclopentenones.[1] One limitation to Jung's cyclopentenone synthesis (**8**, 178) is that tetrasubstituted alkenes react slowly, if at all, with this diene (**1**). An expedient is the use of a dienophile such as isobutenyl acetate (**2**), which does react, even if slowly, to give **3** as the major adduct. The adduct can be converted in several steps to a cyclopentenone such as **4**. The cyclopentenone **4** (R = *p*-tolyl) has been converted to *β*-cuparenone (**5**).[2]

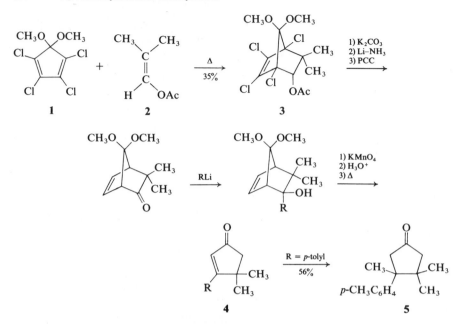

¹ M. E. Jung and C. D. Radcliffe, *Tetrahedron Letters*, **21**, 4397 (1980).
² A. Casaros and L. A. Maldonado, *Syn. Comm.*, **6**, 11 (1976).

N,N-Dimethylacetamide dimethyl acetal, 4, 166–167; **5,** 226–227.
   Preparation¹:

$$CH_3CON(CH_3)_2 + (CH_3O)_2SO_2 \xrightarrow[98\%]{} \underset{\overset{\parallel}{CH_3COCH_3}}{\overset{^+N(CH_3)_2}{}} \ CH_3OSO_3^- \xrightarrow[59\%]{NaOCH_3}$$

$$\underset{CH_3C(OCH_3)_2}{\overset{N(CH_3)_2}{|}}$$

b.p. 100–118°

¹ R. G. Salomon and S. R. Raychaudhuri, *Org. Syn.*, submitted (1981).

(2S,3R)-(+)-4-Dimethylamino-3-methyl-1,2-diphenyl-2-butanol (Darvon alcohol) (1),
5, 231; **8,** 184–186.
   Asymmetric reductions. Hoffmann-LaRoche chemists¹ have examined in detail
the asymmetric reduction of α,β-acetylenic ketones with Darvon alcohol (**1**) and
related alcohols. Use of the enantiomer of **1** resulted predominately in reduction to
the enantiomer (S) of the (R)-carbinols obtained with **1**. Several optically pure 1,3-

amino alcohols related to **1** were also prepared. Of these, ligands **2** and **3** are interesting because they reduce α,β-unsaturated ketones in the same way as the enantiomer of **1**, although the optical yield is less.

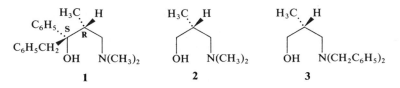

1   **2**   **3**

[1] N. Cohen, R. J. Lopresti, C. Neukom, and G. Saucy, *J. Org.*, **45**, 582 (1980).

2-(Dimethylamino)-3-pentenonitrile, $CH_3CH=CHCH\begin{smallmatrix}N(CH_3)_2\\CN\end{smallmatrix}$ **(1).** Mol. wt. 124.19. The nitrile is prepared by reaction of crotonaldehyde with dimethylamine and HCN in CH_2Cl_2 at 0°.

Cyclopentenones.[1] The anion of **1** (LDA, THF, −78°) is susceptible to attack by electrophiles at either the α- or γ-position. At temperatures of −78° the reaction occurs at the former position; at temperatures near 0° the product is a spirolactone. These lactones are converted into cyclopentenones when heated at 60° with P_2O_5 in methanesulfonic acid.

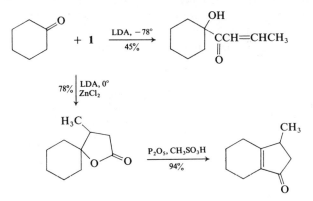

[1] R. M. Jacobson and J. W. Clader, *Tetrahedron Letters*, **21**, 1205 (1980).

4-Dimethylaminopyridine (DMAP), **3**, 118–119; **5**, 260; **9**, 178–180.

Esterification. Acids can be esterified at 25° and in yields of 85–95% by treatment with 2-fluoro-1,3,5-trinitrobenzene (1 equivalent), DMAP (2 equivalents), and an alcohol in acetonitrile for 2–24 hours. The method is successful with hindered acids, but *t*-butyl esters cannot be prepared under these mild conditions. Presumably a trinitrophenyl ester is an intermediate.[1] Thiol esters can be prepared using the same conditions, also in high yield except for reactions with hindered acids.[2]

Tritylation (9, 179).[3] 4-Dimethylamino-N-triphenylmethylpyridinium chloride (1) has been prepared by reaction of trityl chloride with DMAP. It has been presumed to be the effective reagent in tritylations catalyzed by DMAP. It does

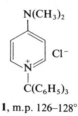

1, m.p. 126–128°

convert primary alcohols into trityl ethers, but is inactive with secondary alcohols. One possible use for **1** is selective protection of primary alcohols and selective protection of an amine in the presence of a hydroxyl group (equation I).

$$\text{(I) HOCH}_2\text{CHCOOCH}_3 \quad \xrightarrow[60\%]{\textbf{1}} \quad \text{HOCH}_2\text{CHCOOCH}_3$$
$$\qquad\quad |\qquad\qquad\qquad\qquad\qquad\qquad |$$
$$\qquad\quad \text{NH}_2 \qquad\qquad\qquad\qquad\qquad \text{NHC(C}_6\text{H}_5)_3$$

4-Chromanones.[4] In the presence of this amine 4-methyl-6-hydroxy-2-pyrone (**1**) undergoes decarboxylative dimerization to form a coumarochromanone (**2**) in 42% yield. The pyrone group in **2** undergoes decarboxylative Diels–Alder reactions (**8**, 326; **9**, 175–176) with acetylenes to form 4-chromanones (**3**).

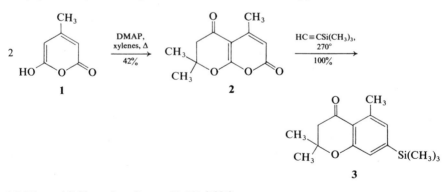

[1] S. Kim and S. Yang, *Syn. Comm.*, **11**, 121 (1981).
[2] S. Kim and S. Yang, *Chem. Letters*, 133 (1981).
[3] O. Hernandez, S. K. Chandhary, R. H. Cox, and J. Porter, *Tetrahedron Letters*, **22**, 1491 (1981).
[4] S. D. Burke, J. O. Saunders, and C. W. Murtiashaw, *J. Org.*, **46**, 2425 (1981).

Dimethyl diazomethylphosphonate (1), 3, 113–114.[1]

Allyl vinyl ethers. Enol ethers of aldehydes can be obtained in moderate to high yield by a Wittig–Horner reaction of aliphatic ketones with **1** and an alcohol in the presence of a base (equation I).[2]

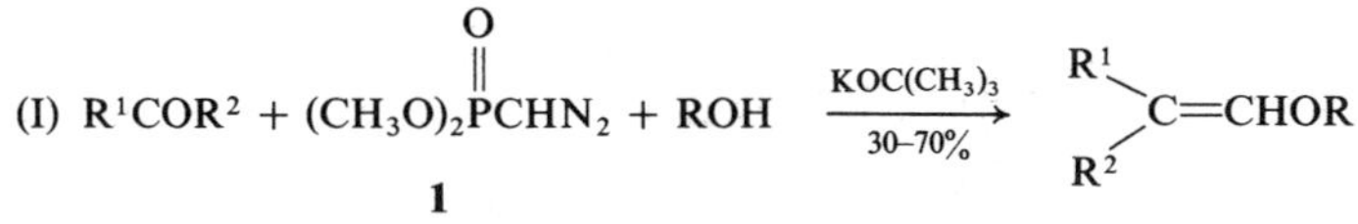

$$\text{(I)} \quad R^1COR^2 + (CH_3O)_2\overset{\overset{O}{\|}}{P}CHN_2 + ROH \xrightarrow[30-70\%]{KOC(CH_3)_3} \quad \overset{R^1}{\underset{R^2}{>}}C{=}CHOR$$

1

This reaction has been extended to preparation of allyl vinyl ethers by use of an allyl alcohol.[3] Yields, however, are only satisfactory with cyclohexanones.

Example:

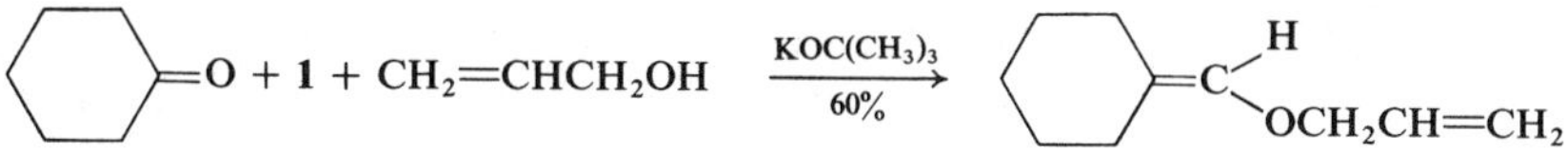

[1] D. Seyferth, R. M. Marmor, and P. H. Hilbert, *J. Org.*, **36**, 1379 (1971).
[2] J. C. Gilbert and U. Weerasooriya, *Tetrahedron Letters*, **21**, 2041 (1980).
[3] J. C. Gilbert, U. Weerasooriya, B. Wiechman, and L. Ho, *Tetrahedron Letters*, **21**, 5003 (1980).

Dimethylformamide diethyl acetal, 1, 281–282; **2**, 154; **3**, 115–116; **4**, 184; **5**, 253–254; **7**, 125; **9**, 182–183.

4-Amino-2-azabutadienes; 1-aminobutadienes.[1] The reaction of dimethylformamide diethyl acetal with azomethines results in 4-amino-2-azabutadienes (**1**, equation I). 1-Aminobutadienes (**2**) are obtained by a similar reaction of α,β- or β,γ-unsaturated esters (equation II).

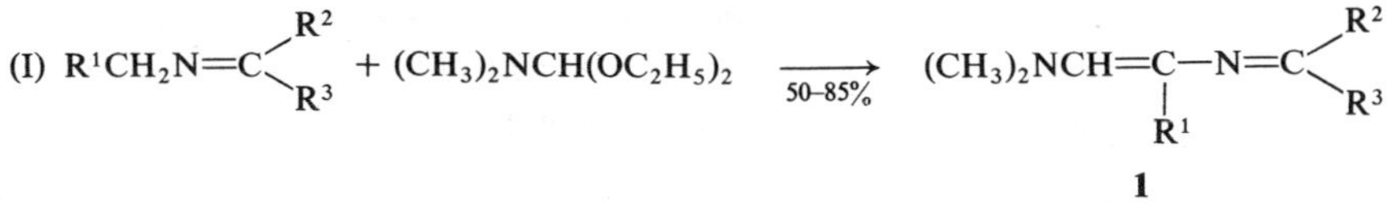

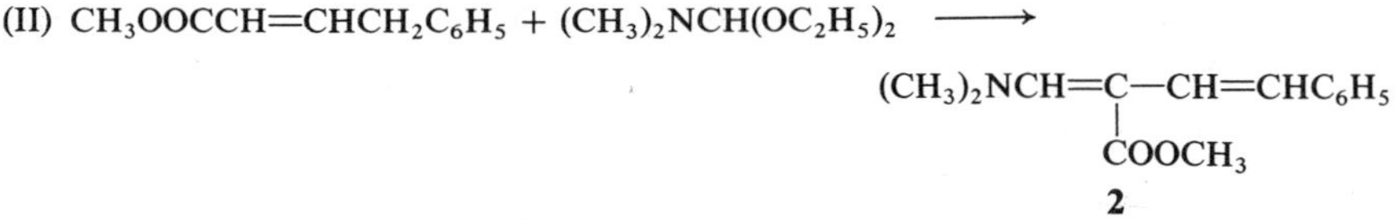

$$\text{(I)} \quad R^1CH_2N{=}C\overset{R^2}{\underset{R^3}{<}} + (CH_3)_2NCH(OC_2H_5)_2 \xrightarrow{50-85\%} (CH_3)_2NCH{=}\underset{R^1}{\overset{|}{C}}{-}N{=}C\overset{R^2}{\underset{R^3}{<}}$$

1

$$\text{(II)} \quad CH_3OOCCH{=}CHCH_2C_6H_5 + (CH_3)_2NCH(OC_2H_5)_2 \longrightarrow$$

$$(CH_3)_2NCH{=}\underset{COOCH_3}{\overset{|}{C}}{-}CH{=}CHC_6H_5$$

2

As expected, **1** and **2** undergo Diels–Alder reactions with dimethyl acetylenedicarboxylate to give pyridines (from **1**) or biphenyl derivatives (from **2**).

Aminopyrimidines can be obtained by a similar sequence formulated in equation (III).

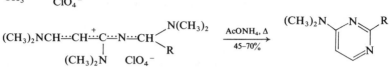

$$(III)$$

$$\xrightarrow[\text{75–90\%}]{(CH_3)_2NCH(OC_2H_5)_2}$$

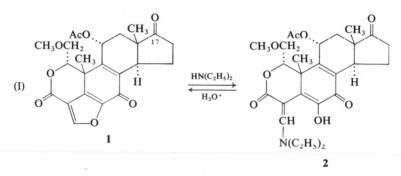

$$\xrightarrow[\text{45–70\%}]{AcONH_4, \Delta}$$

[1] R. Gompper and U. Heinemann, *Angew. Chem. Int. Ed.*, **20**, 296, 297 (1981).

Dimethylformamide dimethyl acetal, 1, 281–282; **2**, 154; **3**, 115–116; **4**, 184–185; **6**, 221–222; **8**, 191–192.

Annelated furanes. Attempts to apply some typical steroid reactions to wortmannin (**1**) are hampered by the high reactivity of the furane unit to nucleophiles. The furane ring can be masked by reaction with an amine to form an

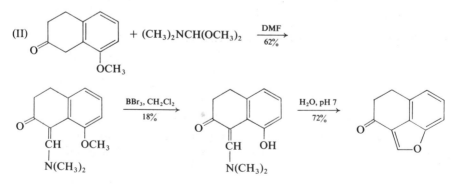

enamine (**2**) (equation I). Grignard and similar reactions with the 17-keto group are then possible. The furane ring can be reformed by treatment with dilute acid.[1]

This method has since been used for synthesis of annelated furanes.[2] An example is formulated in equation (II). The masked furane ring is introduced by reaction of an activated methylene group with dimethylformamide dimethyl acetal.

Naphthalenes; biphenyls.[3] The reaction of 1,3-diphenylacetone (**1**) with DMF dimethyl acetal at 110° affords the bisenamine **2**; when conducted at 180° the reaction results in a naphthalene derivative (**3**) (equation I). This synthesis of naphthalenes is general and yields are usually 50–90%. The same reaction when applied to **4** results in a 4-methoxybiphenyl derivative (**5**) (equation II). When the carbonyl group in **1** is labeled with ^{13}C, it can be identified as the carboxyl group in the carboxamide **3**. Thus an extensive rearrangement is involved.

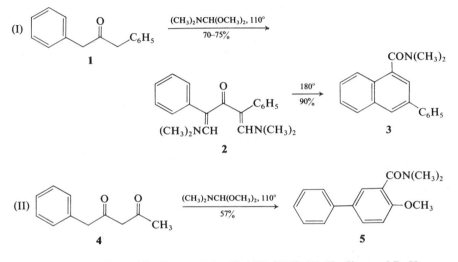

[1] W. Haefliger, Z. Kis, and D. Hauser, *Helv.*, **58**, 1620 (1975); W. Haefliger and D. Hauser, *Helv.*, **58**, 1629 (1975).
[2] W. Haefliger and D. Hauser, *Synthesis*, 236 (1980).
[3] R. F. Abdulla, K. H. Fuhr, R. P. Gajewski, R. G. Suhr, H. M. Taylor, and P. L. Unger, *J. Org.*, **45**, 1724 (1980).

2,2-Dimethyl-3(2*H*)-furanone (1).

Preparation[1]:

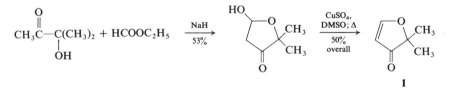

Cyclohexenones.[2] This furanone (**1**) undergoes photochemical [2 + 2]cyclo-additions to alkenes regioselectively (head to tail adducts predominate). The products can be converted into 4-substituted or 4,5-disubstituted cyclohexenones by two methods. The more general route is formulated in equation (I) for the adduct (2) of 1 and tetramethylethylene.

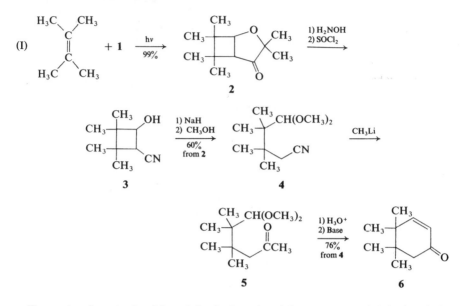

For a related synthesis of 5- and 6-substituted cyclohexenones see 2,2,6-trimethyl-1,3-dioxolenone (this volume).

[1] P. Margartha, *Tetrahedron Letters*, 4891 (1971).
[2] S. W. Baldwin and J. M. Wilkinson, *Tetrahedron Letters*, 2657 (1979).

Dimethyl(methylene)ammonium halides, 3, 114–115; **4**, 186–187; **7**, 130–132; **8**, 194.

Simplified preparation.[1] The reagent (1) can be obtained in high yield by cleavage of N,N,N′,N′-tetramethylmethylenediamine (2, Aldrich) with freshly prepared iodotrimethylsilane.

$$(CH_3)_2NCH_2N(CH_3)_2 \xrightarrow{\text{ISi(CH}_3)_3\text{, ether}} \left[\underset{\underset{I^-}{|}}{(CH_3)_2\overset{Si(CH_3)_3}{\overset{|}{N^+}}CH_2N(CH_3)_2} \right] \xrightarrow{96\%}$$

$$\underset{\mathbf{2}}{} \qquad \qquad \underset{\mathbf{a}}{}$$

$$\overset{+}{CH_2}{=}N(CH_3)_2I^-$$

$$\mathbf{1}$$

Reaction with trimethylsilyl ethers. Regioselective Mannich reactions are possible by reaction of dimethyl(methylene)ammonium iodide with trimethylsilyl enolates and dienolates.[2]

Examples:

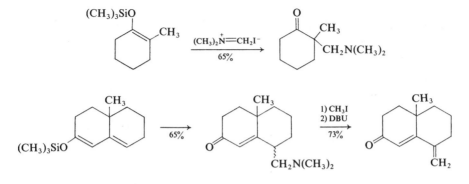

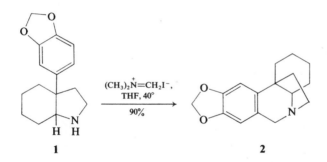

Pictet–Spengler cyclization.[3] The final step in a recent synthesis of the alkaloid crinane (**2**) from **1** involved insertion of the final carbon to form the isoquinoline moiety.[4] Heating **1** with aqueous HCHO and conc. HCl at 60° resulted in **2** in 53% yield. Use of Eschenmoser's salt resulted in **2** in 90% yield.

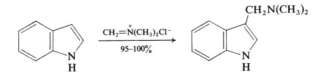

Reaction with indoles.[5] Reaction with N,N-dimethyl(methylene)ammonium chloride is superior to reaction under conventional Mannich conditions for functionalization of various indoles.

[1] T. A. Bryson, G. H. Bonitz, C. J. Reichel, and R. E. Dardis, *J. Org.*, **45**, 524 (1980).
[2] S. Danishefsky, M. Prisbylla, and B. Lipisko, *Tetrahedron Letters*, **21**, 805 (1980).
[3] W. M. Whaley and T. R. Govindachari, *Org. React.*, **6**, 151 (1951).
[4] G. E. Keck and R. R. Webb II, *Am. Soc.*, **103**, 3173 (1981).
[5] A. P. Kozikowski and H. Ishida, *Heterocycles*, **14**, 55 (1980).

2,6-Dimethylphenyl propionate, **(1)**. Mol. wt. 178.22, b.p. 100°/0.7

mm. The ester is prepared by reaction of propionyl chloride with lithium 2,6-dimethylphenoxide in THF at −78°.

Aldol condensation.[1] The lithium enolate of **1** reacts with aldehydes to form aldols **2** and **3**; in all cases the major product is the *threo*-3-hydroxy-2-methyl carboxylic ester (**2**). In fact with α-branched aldehydes no *erythro*-product is formed. The esters can be hydrolyzed with KOH in CH_3OH.

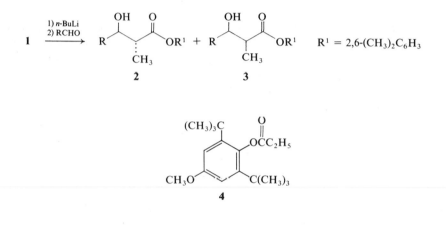

The *threo*-selectivity can be enhanced by use of **4** instead of **1**. However, in this case the aryl group requires oxidation (CAN or AgO) for removal.

[1] M. C. Pirrung and C. H. Heathcock, *J. Org.*, **45**, 1727 (1980).

Dimethylphenylsilyllithium, $C_6H_5Si(CH_3)_2Li$ **(1)**, 7, 133; 8, 196–197.

1 + CuCN reagent (2).[1] A reagent prepared from 2 equivalents of **1** and 1 equivalent of CuCN adds regioselectivity to terminal acetylenes to form an intermediate that is convertible into a wide range of vinylsilanes. Some products obtained from 1-hexyne are formulated in Scheme (I).

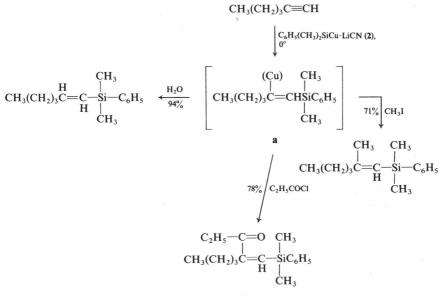

Scheme (I)

[1] I. Fleming and F. Roessler, *J.C.S. Chem. Comm.*, 276 (1980).

Dimethyl(phenylthio)aluminum, $(CH_3)_2AlSC_6H_5$ **(1).** Mol. wt. 166.16. The reagent is prepared from C_6H_5SH and $Al(CH_3)_3$.

Aldol reaction. The reagent (1) adds in a 1,4-fashion to an α,β-unsaturated ketone to give an aluminum enolate, which undergoes aldol condensation with an aldehyde. The adduct is converted into an α-substituted-α,β-unsaturated ketone on sulfoxide elimination.[1]

Examples:

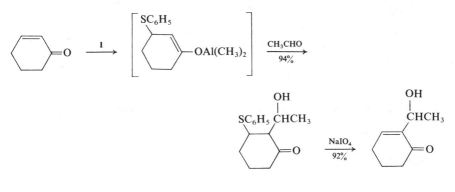

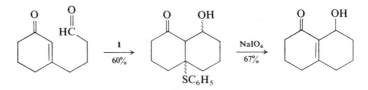

Dimethylaluminum iodide, $(CH_3)_2AlI$, is a useful alternative to **1** because HI is eliminated spontaneously during the aldol reaction. Thus the reaction formulated in the first example occurs in one step in 81% yield.

The ate complex of **1**, $(CH_3)_3Al^-SC_6H_5Li^+$, is necessary for aldol condensations with α,β-unsaturated esters.

[1] A. Itoh, S. Ozawa, K. Oshima, and H. Nozaki, *Tetrahedron Letters*, **21**, 361 (1980); *Bull. Chem. Soc. Japan*, **54**, 274 (1981).

Dimethyl(phenylthiomethyl)amine, $C_6H_5SCH_2N(CH_3)_2$ **(1).** Mol. wt. 167.26, b.p. 112–116°/9–11 mm. The amine is prepared by reaction of thiophenol and dimethylamine with formalin (71% yield).[1]

α-Phenylthiomethylation of carbonyl compounds.[2] This reaction can be effected by reaction of **1** with enamines of both ketones and aldehydes. Morpholine enamines are more satisfactory than pyrrolidine enamines.

Examples:

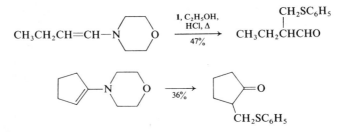

[1] G. F. Grillot and H. G. Thompson, *J. Org.*, **22**, 706 (1957).
[2] K. Suzuki and M. Sekiya, *Synthesis*, 297 (1981).

Dimethylsulfonium ethoxycarbonylmethylide, $(CH_3)_2S=CHCOOC_2H_5$ **(1).**
Preparation:

$$(CH_3)_2S + BrCH_2COOC_2H_5 \xrightarrow[60\%]{} (CH_3)_2\overset{+}{S}CH_2COOC_2H_5Br^- \xrightarrow{NaH} \mathbf{1}$$

Cyclopropanation.[1] The ylide **1** reacts with α,β-unsaturated ketones to form cyclopropanecarboxylic acids in good yield. The reaction with ketones gives glycidic esters (∼ 60% yield).

Tetralone synthesis. Murphy and Wattanasin[2] have devised a simple tetralone synthesis involving cyclization of aryl aroyl cyclopropanes. A chalcone (**2**) is cyclopropanated by **1** to give a 1:1 mixture of **3** and **4**, which can be separated. This step is not necessary since **4** is readily epimerized to **3** under the conditions for

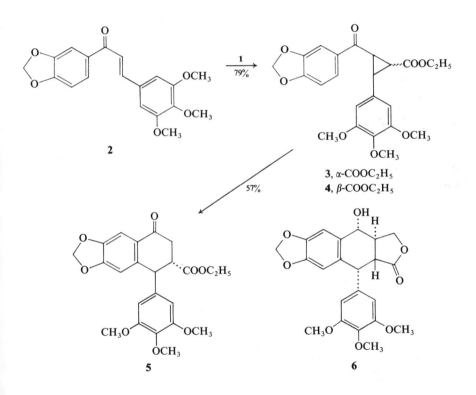

3, α-COOC₂H₅
4, β-COOC₂H₅

2

5

6

cyclization of **3** to **5**. The most effective reagent for this step is BF₃ etherate in nitromethane (yield 57%). Stannic chloride in nitromethane is also satisfactory (yield 53%). Cyclization fails in benzene or methylene chloride. The tetralone **5** has been used for syntheses of podophyllotoxin (**6**), a lignan lactone.

[1] H. Nozaki, D. Tunemoto, S. Maturaba, and K. Kondo, *Tetrahedron*, **23**, 545 (1967); J. Adams, L. Hoffman, Jr., and B. M. Trost, *J. Org.*, **35**, 1600 (1970).
[2] W. S. Murphy and S. Wattanasin, *J.C.S. Chem. Comm.*, 262 (1980).

Dimethyl sulfoxide, 1, 296–310; **2**, 157–158; **3**, 119–123; **4**, 192–194; **5**, 263–266; **6**, 225–229; **7**, 133–135; **8**, 198–199; **9**, 189.

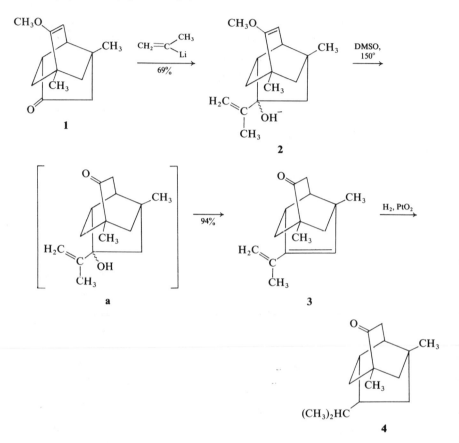

Cleavage of enol ethers.[1] The final steps in a synthesis of (+)-9-pupukeanone (**4**), a sponge and mollusc toxin, from **1** proved unexpectedly difficult. The keto group is too hindered to undergo Wittig reactions or Grignard reactions. It does not react with isopropyllithium, but does react with isopropenyllithium. The product, **2**, was heated in DMSO to effect dehydration (**1**, 301–302). Surprisingly **3** was obtained directly in high yield. Further study revealed that demethylation to give **a** preceded dehydration. This result is apparently the first record of cleavage of enol ethers by DMSO.

Derived reagents. Mancuso and Swern[2] have reviewed activated dimethyl sulfoxide reagents (100 references). Activation of DMSO with either trifluoroacetic anhydride (**7**, 136) or oxalyl chloride (**8**, 200) provides the most generally useful reagents for oxidation of primary and secondary alcohols to carbonyl compounds. The combination of DMSO and sulfur trioxide or trifluoroacetic anhydride is useful for preparation of sulfilimines, $(CH_3)_2\overset{+}{S}\text{-}\overset{-}{N}R$, from amines.[3] Sulfoximines,

$$\overset{\overset{O}{\|}}{(CH_3)_2S}=NAr,$$ can be prepared with DMSO activated by *t*-butyl hypochlorite (**7**, 135–136).

[1] G. A. Schiehser and J. D. White, *J. Org.*, **45**, 1864 (1980).
[2] A. J. Mancuso and D. Swern, *Synthesis*, 165 (1981).
[3] R. W. Heintzelman, R. B. Bailey, and D. Swern, *J. Org.*, **41**, 2207 (1976).

Dimethyl sulfoxide–Chlorosulfonyl isocyanate, $(CH_3)_2S=O$; $ClSO_2N=C=O$.

Oxidation of alcohols. These reagents react at low temperatures ($-78°$) in CH_2Cl_2 to form a zwitterionic complex (**1**), which loses CO_2 at $0°$ to form **2**. The complex **1** in the presence of triethylamine oxidizes primary and secondary alcohols to the corresponding carbonyl compounds in 70–90% yield.

$$\underset{\textbf{1}}{(CH_3)_2\overset{+}{S}O\overset{\overset{O}{\|}}{C}\overset{-}{N}SO_2Cl} \xrightarrow[-CO_2]{0°} \underset{\textbf{2}}{(CH_3)_2S=NSO_2Cl}$$

[1] G. A. Olah, G. D. Vankar, and M. Arvanaghi, *Synthesis*, 141 (1980).

Dimethyl sulfoxide–Oxalyl chloride, 8, 200; 9, 192.

Oxidation.[1] The diol (**1**) is oxidized uniquely to the dialdehyde polygodial (**2**) by

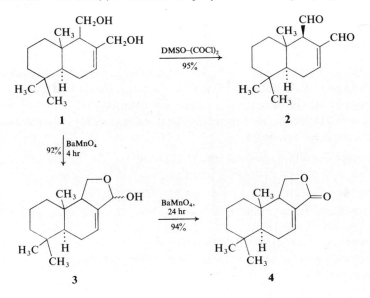

Swern's reagent. It is oxidized to **4** by Fetizon's reagent in quantitative yield and by barium manganate in 92% overall yield.

[1] S. C. Howell, S. V. Ley, M. Mahon, and P. A. Worthington, *J.C.S. Chem. Comm.*, 507 (1981).

Dimethyl sulfoxide–Trifluoroacetic anhydride, 5, 266–267; 7, 136.

cis–trans *Diol conversion.* Conversion of **1** into the *trans*-diol **4** has proved to be unexpectedly difficult because of the tetrahydrofurane oxygen. The two isomeric epoxides derived from **1** can be prepared (93–98% yield with CF_3CO_3H), but neither undergoes normal *trans* diaxial cleavage. In addition, Prévost reactions result mainly in allylic alcohols or acetates rather than *cis*-addition. The desired transformation of **1** to **4** was achieved by hydroxylation with osmium tetroxide to

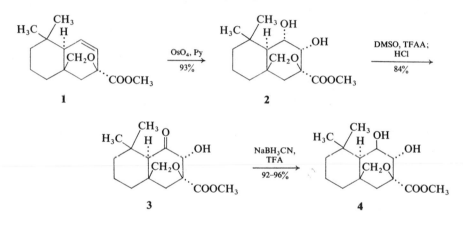

give **2**. Selective oxidation of **2** to **3** also proved difficult, but was best achieved with Swern's reagent. His DMSO–oxalyl chloride reagent (**8**, 200) proved less satisfactory (40% yield of **3**). The last step was carried out with sodium cyanoborohydride in TFA, which was essential for reduction.[1] The diol **4** is a model for the more complex antileukemia agent bruceantin.

[1] O. D. Dailey, Jr., and P. L. Fuchs, *J. Org.*, **45**, 216 (1980).

Dimethylsulfoxonium methylide, 1, 315–318; 2, 171–173; 3, 125–127; 4, 197–199; 7, 133; 8, 194–196; 9, 186–187.

Azetidines.[1] N-Arylsulfonyl-2-phenylazetidines (**2**) can be prepared from N-arylsulfonyl-2-phenylaziridines (**1**) by methylene transfer from dimethyloxosulfonium methylide.

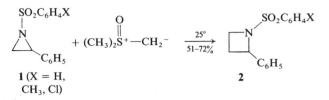

1 (X = H,
CH₃, Cl)

2

¹ U. K. Nadir and V. K. Koul, *J.C.S. Chem. Comm.*, 417 (1981).

Stereoselective reaction with ketones. The reaction of ketone **1** with methyl-lithium, trimethylaluminum, and lithium tetramethylaluminate shows no stereo-specificity. The reaction with methylmagnesium bromide gives the two possible adducts in the ratio 2.4:1. The best stereospecificity is observed with dimethylsulf-oxonium methylide, which converts **1** into **2** and **3** in a ratio about 5:1. Reduction of the epoxides with lithium triethylborohydride gives the desired tertiary alcohols. This reaction was used in a synthesis of (±) stemodin (**4**).²

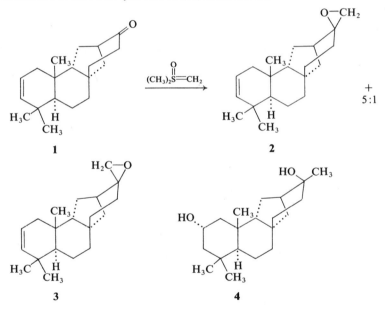

1

2

+
5:1

3

4

² E. J. Corey, M. A. Tius, and J. Das, *Am. Soc.*, **102**, 7612 (1980).

Diperoxo-oxohexamethylphosphoramidomolybdenum(VI), 4, 203–204; **5**, 269–270; **7**, 136; **8**, 206–208; **9**, 197–198.

*Angular hydroxylation.*¹ Vedejs hydroxylation of **1** affords **2** (62% yield), which is a model for the antibiotic bicyclomycin (**3**), obtained from *S. sapporonensis.*

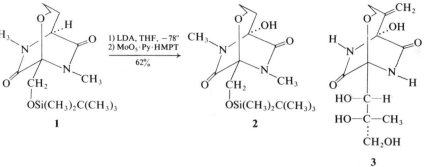

1   2   3

Oxidative desulfonylation.[2] Aryl sulfones are converted into ketones by oxidation of the α-carbanion (LDA) with $MoO_5 \cdot Py \cdot HMPT$ in THF at $-78°$ in 50–97% yield. This conversion permits use of phenyl vinyl sulfone as an equivalent of ketene in Diels–Alder reactions.

Example:

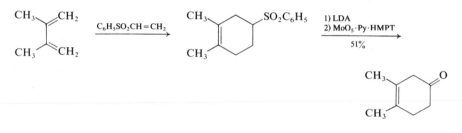

[1] R. M. Williams, *Tetrahedron Letters*, **22**, 2341 (1981).
[2] R. D. Little and S. O. Myong, *Tetrahedron Letters*, **21**, 3339 (1980).

Diphenyl diselenide, 5, 272–276; **6**, 235; **7**, 136–137; **9**, 199. The preparation in *Organic Synthesis* involves the reaction formulated in equation (I), which avoids liberation of (toxic) H_2Se and C_6H_5SeH. The paper includes details for the use of the diselenide for preparation of benzeneselenenyl chloride.[1]

$$(I)\quad C_6H_5Br \xrightarrow[\text{2) Se, }\Delta]{\text{1) Mg, ether}} C_6H_5SeMgBr \xrightarrow[\substack{64-70\% \\ \text{overall}}]{Br_2} C_6H_5Se—SeC_6H_5 \xrightarrow[84-93\%]{Cl_2,\ \text{hexane}} C_6H_5SeCl$$

[1] H. J. Reich, M. L. Cohen, and P. S. Clark, *Org. Syn.*, **59**, 141 (1979).

Diphenyl disulfide, 5, 276–277; **6**, 235–238; **7**, 137; **8**, 210.

Iridoids.[1] The first step in a new route to these terpenes involves regioselective sulfenylation of a β-keto ester (**1**). This product is converted to the dihydropyrane **6** by the series of steps indicated.

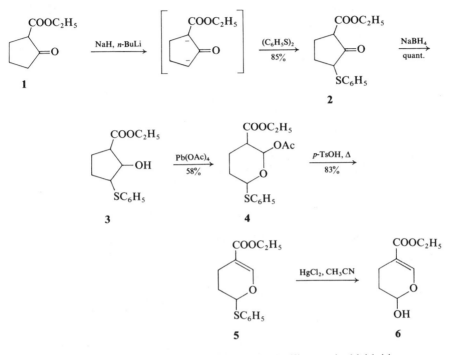

This sequence was used to convert **7** to (±)-loganin (**8**), a typical iridoid.

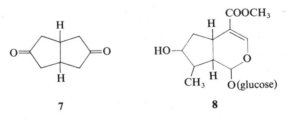

[1] K. Hiroi, H. Miura, K. Kotsuji, and S. Sato, *Chem. Letters*, 559 (1981).

Diphenylphosphinodithioic acid, $(C_6H_5)_2\overset{\text{S}}{\underset{\|}{P}}SH$ **(1).** Mol. wt. 250.32, m.p. 55–56°, stable for months at 25°. The acid is prepared by reaction of benzene and P_4S_{10} in the presence of a large excess of $AlCl_3$ (about 50% yield).[1]

Hydrolysis of nitriles.[2] Nitriles are converted to thioamides by reaction with **1** (2 equivalents) at 40° overnight. Under these conditions most other functional groups are stable. Kinetic studies indicate that the reaction with nitriles is a two-step process, the first of which is analogous to an ene reaction to give **a**. Thioamides are particularly useful precursors to amines by the method of Borch (**2**, 430–431), reaction with triethyloxonium tetrafluoroborate followed by reduction with $NaBH_4$.[3]

$$RCN + 1 \longrightarrow \begin{bmatrix} R \\ \diagdown C{=}NH \\ S \diagup \\ S{=}P(C_6H_5)_2 \end{bmatrix} \xrightarrow{\ \ 1\ \ } \underset{\substack{\\ (75-90\%)}}{\overset{\displaystyle S}{\underset{\displaystyle \|}{R\overset{\|}{C}NH_2}}} + (C_6H_5)_2\overset{\displaystyle S\ S}{\underset{}{P\!\!\overset{\| \ \|}{S}P}}(C_6H_5)_2$$

a

[1] W. Higgins, P. W. Vogel, and W. G. Craig, *Am. Soc.*, **77**, 1864 (1955).
[2] S. A. Benner, *Tetrahedron Letters*, **22**, 1851 (1981).
[3] S. Raucher and P. Klein, *Tetrahedron Letters*, **21**, 4061 (1980).

1-Diphenylphosphonio-1-methoxymethyllithium (1). Mol. wt. 236.16.
Preparation[1]:

$$(C_6H_5)_2PCl \xrightarrow{\text{Li, NH}_3} \left[(C_6H_5)_2PLi \right] \xrightarrow[\text{quant.}]{\text{CH}_3\text{OCH}_2\text{Cl}} (C_6H_5)_2PCH_2OCH_3 \xrightarrow{\substack{\textit{sec}\text{-BuLi,} \\ \text{THF}}}$$

$$\overset{\displaystyle Li}{\underset{\textstyle \mathbf{1}}{\underset{}{(C_6H_5)_2\overset{|}{P}CHOCH_3}}}$$

Aldehyde synthesis.[1] Hindered ketones, which do not react with methoxy-methylenetriphenylphosphorane (**1**, 671), do react with **1** to form an enol ether, which is readily hydrolyzed by acid to the homologous aldehyde. Representative aldehydes (and the yield) available by this method are formulated.

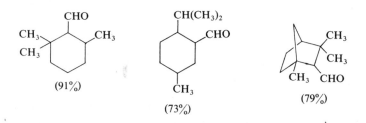

[1] E. J. Corey and M. A. Tius, *Tetrahedron Letters*, **21**, 3535 (1980).

Diphenyl phosphoroazidate, 4, 210–211; **5,** 280; **6,** 193; **7,** 138; **8,** 211–212. Supplier: Aldrich.

The reagent is prepared by reaction of diphenyl phosphorochloridate with sodium azide in acetone (92% yield). It decomposes at 200° and should be protected from light and moisture.[1]

Ring contraction (**7,** 138). Details for the conversion of cyclododecanone (**1**) to cycloundecanecarboxylic acid (**2**) with this reagent are available. With six- to eight-membered cycloalkanones the overall yields are about 75%. In the case of cyclohexadecanone the overall yield is 68%.[2]

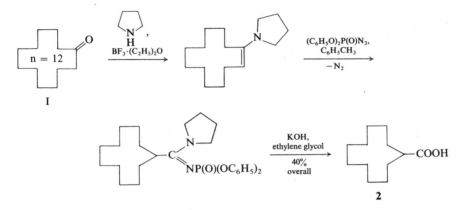

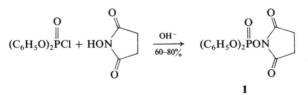

[1] T. Shioiri and S. Yamada, *Org. Syn.,* submitted (1980); *see also Chem. Pharm. Bull.,* **22,** 849 (1974).
[2] Y. Hamada and T. Shioiri, *Org. Syn.,* submitted (1980).

Diphenyl N-succinimidyl phosphate (1). Mol. wt. 347.26, m.p. 88–90°.
Preparation:

$$(C_6H_5O)_2\overset{O}{\overset{\|}{P}}Cl + HON \xrightarrow[60-80\%]{OH^-} (C_6H_5O)_2\overset{O}{\overset{\|}{P}}ON$$

1

Active esters; peptide synthesis.[1] The reagent in combination with $N(C_2H_5)_3$ converts N-protected amino acids into active esters (~85% yield). It can also be used with a tertiary amine to effect peptide formation between an N-protected amino acid and an amino acid ethyl ester. CBZ-Val-Gly-OC_2H_5 was prepared in this way in 89% yield.

[1] H. Ogura, S. Nagai, and K. Takeda, *Tetrahedron Letters,* **21,** 1467 (1980).

S,S-Diphenylsulfilimine, $(C_6H_5)_2S=NH$.   Mol.   wt.   201.28,   m.p.   128.5°.
Preparation.[1] Supplier: Fluka.

Nitriles.[2] This sulfilimine reacts with aldehydes to give the corresponding
nitriles (equation I). The reaction with aliphatic α,β-enals results in 2-cyanoaziri-
dines.

(I) $RCHO + (C_6H_5)_2S=NH \longrightarrow$

$$\left[\begin{array}{c} H \\ | \\ RC-O \\ | \quad | \\ HN-S(C_6H_5)_2 \end{array} \right] \xrightarrow{-(C_6H_5)_2S=O} RCH=NH \xrightarrow[40-85\%]{} RC\equiv N$$

[1] T. Yoshimura, T. Omata, N. Furukawa, and S. Oae, *J. Org.*, **41**, 1728 (1976).
[2] Y. Gelas-Mialhe and R. Vessière, *Synthesis*, 1005 (1980).

Diphosphorus tetraiodide, I_2P-PI_2, **1**, 349–350; **9**, 203–204.

Deoxygenation of pyridine N-oxides.[1] This reaction can be carried out with P_2I_4
in refluxing CH_2Cl_2 in 10–30 minutes with yields of 80–95%.

Deoxygenation of epoxides.[2] Epoxides, particularly α,β-disubstituted ones, are
deoxygenated by reaction with P_2I_4 or PI_3 in CS_2 at room temperature in 70–90%
yield. The reaction occurs with retention (>97%) of configuration. In the case of
terminal, trisubstituted, and tetrasubstituted epoxides, the yields with P_2I_4 or PI_3
are low ($\sim$ 50%) unless pyridine or triethylamine is added.

[1] H. Suzuki, N. Sato, and A. Osuka, *Chem. Letters*, 459 (1980).
[2] J. N. Denis, R. Magnane, M. Van Eenoo, and A. Krief, *Nouv. J. Chem.*, **3**, 745 (1979).

Disodium tetracarbonylferrate, **3**, 267–268; **4**, 461–465; **5**, 624–625; **6**, 550–552; **7**,
341; **8**, 216–217; **9**, 205–207.

Cyclocarbonylation (**9**, 205–206). McMurry and Andrus[1] have examined the
scope of the reaction of Collman's reagent with unsaturated tosylates as a route to
cycloalkanones and have concluded that the reaction is useful for preparation only
of cyclopentanones and cyclohexanones from olefinic tosylates in which the double
bond is monosubstituted. Yields are higher with primary tosylates than with
secondary tosylates.

Examples:

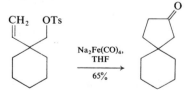

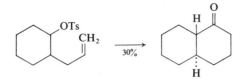

Formylation and acetylation of a primary alkyl bromide.[2] Detailed directions are available for preparation of disodium tetracarbonylferratesesquidioxanate, $Na_2Fe(CO)_4 \cdot (dioxane)_{1.5}$ (1), and for use of the reagent for preparation of aldehydes and ketones from alkyl bromides.

Examples:

$$CH_3OOC(CH_2)_5Br \xrightarrow[\substack{57-63\%}]{\substack{1)\ 1,\ THF,\ 25° \\ 2)\ CO,\ 25° \\ 3)\ CH_3COOH}} CH_3OOC(CH_2)_5CHO$$

$$70-72\% \left\downarrow \substack{1)\ 1 \\ 2)\ CH_3I, \\ N\text{-methylpyrrolidone}} \right.$$

$$CH_3OOC(CH_2)_5COCH_3$$

[1] J. E. McMurry and A. Andrus, *Tetrahedron Letters*, **21**, 4687 (1980).
[2] R. G. Fincke and T. N. Sorell, *Org. Syn.*, **59**, 102 (1979).

Disodium tetrachloropalladate, 3, 134; 9, 207.

N-Vinylation.[1] Imides and lactams are converted to N-vinyl derivatives when heated with vinyl acetate in the presence of Na_2PdCl_4.

Examples:

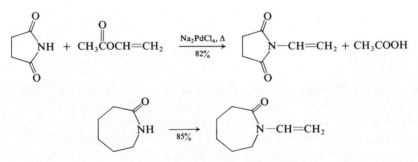

[1] E. Bayer and K. Geckeler, *Agnew. Chem. Int. Ed.*, **18**, 533 (1979).

Disodium tetrachloropalladate–*t*-Butyl hydroperoxide.

β-Keto esters; 1,3-diketones. α,β-Unsaturated esters are oxidized to β-keto esters by Pd(II) (5–20 mole %) and 1–3 equivalents of *t*-butyl hydroperoxide in aqueous acetic acid, isopropyl alcohol, or N-methylpyrrolidone at 50–80°. The same system oxidizes α,β-unsaturated ketones to 1,3-diketones. It is not suitable for oxidation of simple olefins. The function of the hydroperoxide is oxidation of Pd(0) as formed to Pd(II).[1]

Examples:

$$CH_3(CH_2)_2CH=CHCOOCH_3 \xrightarrow[78\%]{\substack{Na_2PdCl_4, \\ (CH_3)_3COOH}} CH_3(CH_2)_2\overset{O}{\overset{\|}{C}}CH_2COOCH_3$$

$$CH_3(CH_2)_5CH=CH\overset{O}{\overset{\|}{C}}CH_3 \xrightarrow[59\%]{} CH_3(CH_2)_5\overset{O}{\overset{\|}{C}}CH_2\overset{O}{\overset{\|}{C}}CH_3$$

[1] J. Tsuji, H. Nagashima, and K. Hori, *Chem. Letters*, 257 (1980).

1,3-Dithiolan-2-yltriphenylphosphonium tetrafluoroborate (1). Mol. wt. 454.3, m.p. 237°.

Preparation:

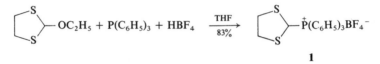

Ketene S,S-acetals.[1] The ylid of **1**, prepared *in situ* with *n*-butyllithium, reacts with aldehydes, but not ketones, to form ketene S,S-acetals (**2**) in high yield.

[1] S. Tanimoto, S. Jo, and T. Sugimoto, *Synthesis*, 53 (1980).

Di(trimethylsilyl)sulfate, $SO_2[OSi(CH_3)_3]_2$ **(1).** Mol. wt. 242.44. The sulfate is prepared by reaction of chlorotrimethylsilane with sulfuric acid.[1]

Sulfonation.[2] This sulfate does not react with benzene, but it has certain advantages for sulfonation of aryl ethers. Thus anisole is converted to trimethylsilyl 4-methoxybenzenesulfonate in 92% yield. The product is hydrolyzed quantitatively to 4-methoxybenzenesulfonic acid without any catalyst. Thus it is more selective than sulfuric acid itself.

The reagent also reacts with acid chlorides and acid anhydrides by substitution of the $SO_3Si(CH_3)_3$ group at the α-position.

[1] N. Duffant, R. Calas, and J. Dunogues, *Bull. Soc.*, 513 (1963).
[2] P. Bourgeois and N. Duffant, *Bull. Soc.*, II-195 (1980).

E

Ethylaluminum dichloride (1), **6,** 251; **7,** 146.

Ene and [2 + 2]cycloaddition reactions of acetylenic esters (**7,** 7–8; **8,** 13). In the early studies on the reaction of acetylenic esters with alkenes, Snider used $AlCl_3$ as the Lewis acid catalyst. The presently preferred catalyst is $C_2H_5AlCl_2$, which is a Lewis acid and also serves as a proton scavenger by reaction with HCl to give ethane and $AlCl_3$. It is generally used in amounts close to 1 equivalent for a neutral alkene.

The catalyzed reaction of acetylenic esters and alkenes can lead to ene products and/or *cis* [2 + 2]cycloaddition. The relative reactivity of alkenes established by reactions with dienes is 1,1-disubstituted > trisubstituted ≫ monosubstituted and 1,2-disubstituted. Ene reactions predominate with alkenes containing two substituents on one carbon.[1]

Ene reaction of aldehydes. Aliphatic and aromatic aldehydes are not reactive enophiles; however, in the presence of dimethylaluminum chloride, which serves as a mild Lewis acid catalyst and proton scavenger, ene reactions occur in reasonable to high yield. Use of $C_2H_5AlCl_2$ results in complex mixtures of products. This ene reaction is a useful route to homoallylic alcohols.[2]

Examples:

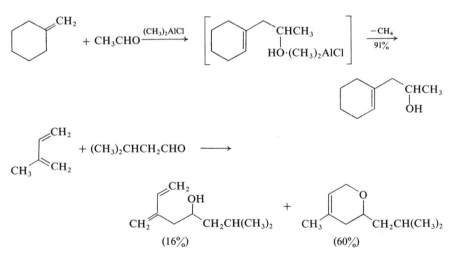

Steroid side chain. Two laboratories[3,4] have reported a stereocontrolled synthesis of the steroid chain with the natural configuration at C_{20} by the catalyzed ene reaction of a (Z)-17-ethylidene steroid (**2**) with methyl propiolate to give **3** in about 90% yield. Catalytic hydrogenation of **3** results in exclusive formation of **4**, with the natural configuration at C_{17}.

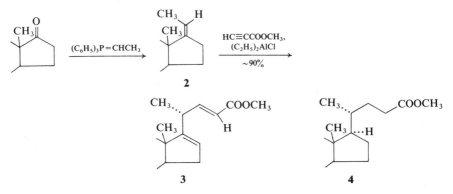

Diastereoselective ene reactions. Thermal cyclization via an intramolecular ene-type reaction of the (Z)-1,6-diene **1** results in a mixture of *trans*-**2** and *cis*-**2** in the ratio 75:25. This reaction is markedly accelerated by addition of $(C_2H_5)_2AlCl$, with only *trans*-**2** being formed. The (E)-isomer of **1** is also cyclized thermally to give

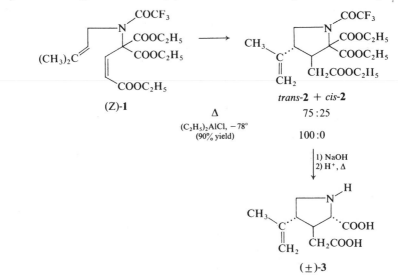

equal amounts of *trans*-**2** and *cis*-**2** but in the presence of $(C_2H_5)_2AlCl$ *trans*-**2** is formed with 89% diastereoselectivity. This reaction was carried out in a synthesis of the natural amino acid α-allokainic acid (**3**) from *Digenea simplex* AG.[5,6]

The reaction was then used to provide an enantioselective synthesis of **3**. Cyclization of the chiral (−)-8-phenylmenthol ester **5** of (Z)-**1** gave the diastereoisomers of *trans*-**6** in the ratio 95:5. The predominate product was converted into pure (+)-**3** in 73% yield. Similar cyclization of this chiral ester of (E)-**5** resulted in the same diastereoisomers, but with reversed optical induction.[7]

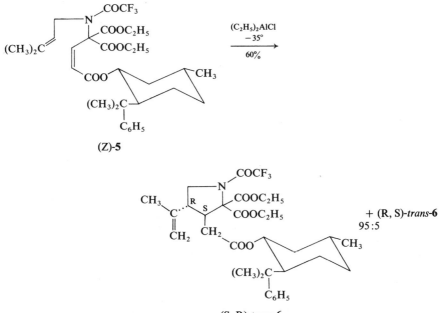

Cyclization of unsaturated carbonyl compounds. Intramolecular ene reactions catalyzed by $(CH_3)_2AlCl$ are useful for cyclization of various unsaturated aldehydes and ketones under mild conditions that avoid side reactions.[8]

Examples:

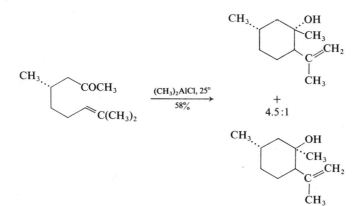

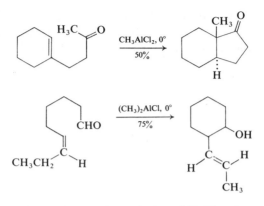

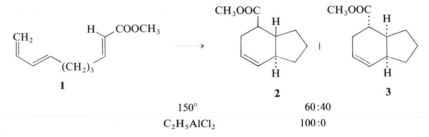

Intramolecular Diels–Alder reactions of trienes.[9,10] The thermal cyclization of the (E,E)-triene **1** affords *trans-* and *cis*-perhydroindenes **2** and **3**, with a slight preference for the former isomer. The reaction is markedly catalyzed by Lewis acids such as $AlCl_3$, $C_2H_5AlCl_2$, and $TiCl_4$, and results in cyclization exclusively to the

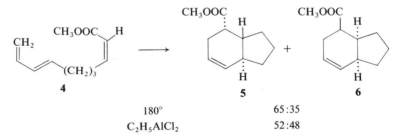

| | 150° | 60:40 |
| | $C_2H_5AlCl_2$ | 100:0 |

trans-fused adduct. In contrast, Lewis acid catalysts exert only a slight effect on the cyclization of the isomeric triene **4**, and both possible adducts are formed in approximately equal amounts.

| | 180° | 65:35 |
| | $C_2H_5AlCl_2$ | 52:48 |

[1] B. B. Snider, D. M. Rousch, D. J. Rodini, D. Gonzalez, and D. Spindell, *J. Org.*, **45**, 2773 (1980).

[2] B. B. Snider and D. J. Rodini, *Tetrahedron Letters*, **21**, 1815 (1980).

[3] W. G. Dauben and T. Brookhart, *Am. Soc.*, **103**, 237 (1981).

[4] A. D. Batcho, D. E. Berger, M. R. Usković, and B. B. Snider, *Am. Soc.*, **103**, 1293 (1981).

[5] W. Oppolzer and C. Robbiani, *Helv.*, **63**, 2010 (1980).

[6] P. D. Kennewell, S. S. Matharu, J. B. Taylor, and P. G. Sammes, *J.C.S. Perkin I*, 2542 (1980).

[7] W. Oppolzer, C. Robbiani, and K. Batig, *Helv.*, **63**, 2015 (1980).

[8] M. Karras and B. B. Snider, *Am. Soc.*, **102**, 7951 (1980).

[9] W. R. Roush, A. I. Ko, and H. R. Gillis, *J. Org.*, **45**, 4264 (1980).

[10] W. R. Roush and H. R. Gillis, *J. Org.*, **45**, 4267 (1980).

Ethyl 4-diethoxyphosphinoyl-3-oxobutanoate, $(C_2H_5O)_2\overset{\overset{O}{\|}}{P}CH_2\overset{\overset{O}{\|}}{C}CH_2COOC_2H_5$ **(1),**

b.p. 120°/0.4 mm[1]; **ethyl 4-diphenylphosphinoyl-3-oxobutanoate,** $(C_6H_5)_2\overset{\overset{O}{\|}}{P}CH_2\overset{\overset{O}{\|}}{C}CH_2COOC_2H_5$ **(2),** m.p. 95°.[2] The reagents **1** and **2** are prepared by reaction of $BrCH_2COCH_2COOC_2H_5$ with $(C_2H_5O)_2PONa$ and $(C_6H_5)_2POC_2H_5$, respectively.

γ,δ-Unsaturated-β-keto esters.[1,2] The dianions (NaH) of these reagents react with aldehydes or ketones to form γ,δ-unsaturated-β-keto esters (Nazarov reagents). In general, the (E)-isomers are formed preferentially, and yields are 60–90% (equation I).

$$(\text{I}) \quad \overset{R^1}{\underset{R^2}{>}}C{=}O + \textbf{1 or 2} \quad \xrightarrow[60-90\%]{2NaH} \quad \overset{R^1}{\underset{R^2}{>}}C{=}CH\overset{\overset{O}{\|}}{C}CH_2COOC_2H_5$$

[1] R. Bodalski, K. M. Pietrusiewicz, J. Monkiewicz, and J. Koszuk, *Tetrahedron Letters*, **21**, 2287 (1980).

[2] J. A. M. van den Goorbergh and A. van der Gen, *Tetrahedron Letters*, **21**, 3621 (1980).

Ethylene glycol–*p*-Toluenesulfonic acid.

Tetrahydrofuranes.[1] An example of a new route to these heterocycles is formulated in equation (I). Cyclization with acid alone is effected in only about 30%

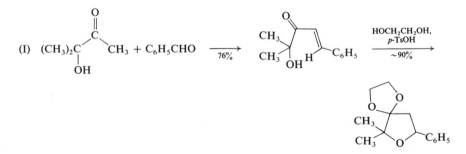

yield. This novel method was used to obtain bullatenone (**1**) and muscarine analogs such as **2**.

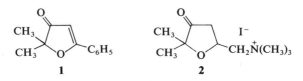

1   **2**

[1] J. E. Semple, A. E. Guthrie, and M. M. Joullié, *Tetrahedron Letters*, **21**, 4561 (1980).

Ethyl malonate, $HOOCCH_2COOC_2H_5$. Mol. wt. 132.11, b.p. 147°/21 mm. Preparation.[1]

β-Keto esters.[2] Ethyl dilithiomalonate, prepared from ethyl malonate with *n*-butyllithium in THF containing a trace of 2,2′-bipyridyl (as indicator), reacts with acid chlorides at about −65° to give, after acidic work-up, β-keto esters. For highest yields at least 1.7 equivalent of the acid chloride is required. Reported isolated yields (nine examples) are 90–99%.

$$C_2H_5O_2C\overset{Li^+}{\underset{}{C}}HCOO^-Li^+ \xrightarrow[90-99\%]{RCOCl} RCOCH_2COOC_2H_5$$

[1] R. E. Strube, *Org. Syn. Coll. Vol.*, **4**, 417 (1963).
[2] W. Wierenga and H. I. Skulnick, *J. Org.*, **44**, 310 (1979); *Org. Syn.*, submitted (1979).

Ethyl nitroacetate–Diethyl azodicarboxylate–Triphenylphosphine.

Oxidation of alcohols. Primary and secondary alcohols are oxidized under neutral conditions by this combination of reagents. An aci-nitro ester is the intermediate, which decomposes slowly when heated in THF (equation I).[1]

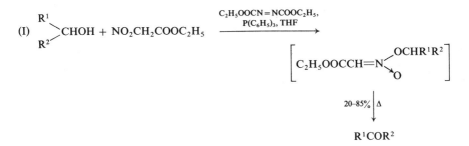

If neutral conditions are not necessary, 2,6-di-*t*-butyl-4-nitrophenol can be used in place of ethyl nitroacetate.[2]

[1] O. Mitsunobo and N. Yoshida, *Tetrahedron Letters* **22**, 2295 (1981).
[2] J. Kimura, A. Kawashima, M. Sugizaki, N. Nemoto, and O. Mitsunobu, *J.C.S. Chem. Comm.*, 303 (1979).

Ethyl 2-phenylsulfinylacetate, 6, 256–257.

Ethyl (E,E)-2,4-dienoates.[1] These diunsaturated esters can be prepared by Knoevenagel condensation (piperidine catalyzed) of ethyl 2-phenylsulfinylacetate with aldehydes and thermolysis of the product in the presence of potassium carbonate (equation I).

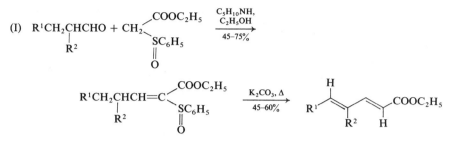

[1] R. Tanikaga, M. Nishida, N. Ono, and A. Kaji, *Chem. Letters*, 781 (1980).

Ethynyl *p*-tolyl sulfone, $p\text{-}CH_3C_6H_4\overset{\displaystyle O}{\underset{\displaystyle O}{\overset{\|}{\underset{\|}{S}}}}C\equiv CH$ (1). Mol. wt. 180.22, m.p. 74–75°.

Preparation[1]:

$$p\text{-}CH_3C_6H_4SO_2Cl + (CH_3)_3SiC\equiv CSi(CH_3)_3 \xrightarrow[75\%]{\substack{AlCl_3, \\ CH_2Cl_2}}$$

$$p\text{-}CH_3C_6H_4SO_2C\equiv CSi(CH_3)_3 \xrightarrow[93\%]{\substack{NaOH, \\ CH_3OH, H_2O}} \mathbf{1}$$

Diels–Alder reactions.[2] This sulfone can be used as an acetylene equivalent in Diels–Alder reactions with 1,3-dienes, since the tosyl group in the adduct is reductively eliminated with sodium amalgam (**7**, 326–327).

Examples:

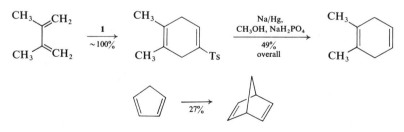

[1] S. N. Bhattacharya, B. M. Josiah, and D. R. M. Walton, *Organometal. Chem. Syn.*, **1**, 145 (1970/1971).
[2] A. P. Davis and G. H. Whitham, *J.C.S. Chem. Comm.*, 639 (1980).

Ethyl vinyl ether, 1, 386–388; **2**, 198; **4**, 234–235.

α,β-Unsaturated aldehydes.[1] In the presence of montmorillonite clay K-10 (**8**, 507–508) ethyl vinyl ether reacts with diethyl acetals or ketals to form 1,1,3-tri-alkoxyalkanes **1** in 50–90% yield. This catalyst is superior to both BF_3 etherate and $FeCl_3$. Yields are higher and the catalyst is removed by simple filtration. Formic acid–sodium formate is recommended for the hydrolysis step to give α,β-enals. When $R^1 = H$, the products have the (E)-configuration.

$$\underset{R^2}{\overset{R^1}{>}}C(OC_2H_5)_2 + CH_2{=}CHOC_2H_5 \quad \xrightarrow[\text{50-95\%}]{\text{Cat.}} \quad \underset{R^2}{\overset{R^1}{>}}C\underset{CH_2CH(OC_2H_5)_2}{\overset{OC_2H_5}{<}} \quad \xrightarrow[\text{70-85\%}]{HCOOH-HCO_2Na}$$

$$\mathbf{1}$$

$$\underset{R^2}{\overset{R^1}{>}}C{=}CHCHO$$

[1] D. Fishman, J. T. Klug, and A. Shani, *Synthesis*, 137 (1981).

F

Ferric chloride, **1**, 390–392; **2**, 199; **3**, 145; **4**, 236; **5**, 307–308; **6**, 260; **7**, 153–155; **8**, 228; **9**, 222.

Ring expansion (**7**, 153–154). Details are available for expansion of cyclohexanone to 2-cycloheptenone via cyclopropanation of enol silyl ethers (equation I). The report includes six other examples of this method; yields range from 80 to 98%. In all cases the more highly substituted C—C bond of the cyclopropane ring is cleaved by $FeCl_3$.[1]

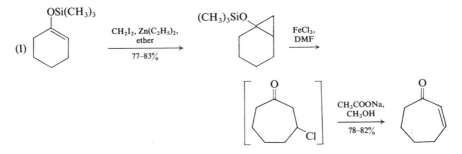

Oxidative coupling of phenols and phenol ethers.[2] This reaction can be conducted with ferric chloride supported on silica gel.

Examples:

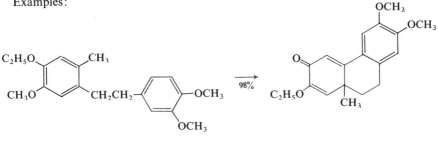

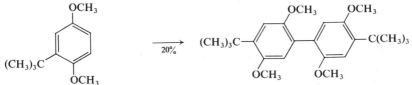

[1] Y. Ito, S. Fujii, M. Nakatsuka, F. Kawamoto, and T. Saegusa, *Org. Syn.*, **59**, 113 (1980).
[2] T. C. Jempty, J. L. Miller, and Y. Mazur, *J. Org.*, **45**, 749 (1980).

Fluorodimethoxyborane, $FB(OCH_3)_2$, **6**, 261–262.

Dienolic primary alcohols. Schlosser's method for allylic hydroxylation of an alkene provides[1] a route to the terpene alcohol anthemol (**2**), briefly described in 1879[2] (equation I). As expected the allylic methyl group of **1** is metallated as desired, but methylene groups are also metallated with about equal ease.[3] The latter reaction gives rise to *p*-cymene.

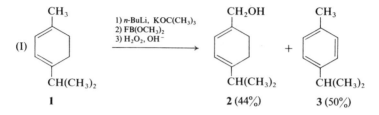

[1] M. Schlosser, H. Bosshardt, A. Walde, and M. Stahle, *Angew. Chem. Int. Ed.*, **19**, 303 (1980).
[2] J. Kobig, *Ann.*, **195**, 92 (1879).
[3] M. Schlosser and G. Ranchschwalbe, *Am. Soc.*, **100**, 3258 (1978).

Formaldehyde, 1, 397–402; **2,** 200–201; **4,** 238–239; **6,** 264–267; **7,** 158–160; **8,** 231–232; **9,** 224–225.

Terminal allenes.[1] Terminal acetylenes can be converted by a one-step reaction into terminal allenes by treatment with formaldehyde, diisopropylamine, and copper(I) bromide in refluxing THF or dioxane. The Mannich base is an intermediate, but, surprisingly, preparation of the quaternary base is not necessary. The source of the introduced hydrogen is not clear. The highest yields are obtained with 2-propynylic alcohols, ethers, and esters.

Examples:

$$\underset{\displaystyle CH_3(CH_2)_4\overset{\displaystyle OH}{\overset{\displaystyle |}{C}}HC{\equiv}CH}{} \xrightarrow[\text{HN[CH(CH}_3)_2]_2]{\text{CH}_2\text{O, CuBr}} \underset{\displaystyle CH_3(CH_2)_4\overset{\displaystyle OH}{\overset{\displaystyle |}{C}}HC{\equiv}CCH_2N[CH(CH_3)_2]_2}{} \xrightarrow[97\%]{}$$

$$CH_3(CH_2)_4\overset{\displaystyle OH}{\overset{\displaystyle |}{C}}HCH{=}C{=}CH_2$$

$$CH_3(CH_2)_5C{\equiv}CH \xrightarrow[26\%]{} CH_3(CH_2)_5CH{=}C{=}CH_2$$

[1] P. Crabbé, H. Fillion, D. André, and J.-L. Luche, *J.C.S. Chem. Comm.*, 859 (1979); P. Crabbé, D. André, and H. Fillion, *Tetrahedron Letters*, 893 (1979).

Formaldehyde–Dimethylaluminum chloride.

α-Allenic alcohols; (Z)-3-chloroallylic alcohols. The 1:1 complex of formaldehyde (trioxane or paraformaldehyde) and dimethylaluminum chloride reacts with terminal alkynes to give a mixture of α-allenic alcohols and (Z)-3-chloroallylic alcohols.[1] An example is formulated in equation (I). The former products may arise by an unusual ene reaction. The latter products involve a stereospecific *syn* Friedel–Crafts addition.

(I) $C_5H_{11}CH_2C{\equiv}CH + CH_2O{\cdot}(CH_3)_2AlCl$ $\xrightarrow[55–70\%]{25°}$

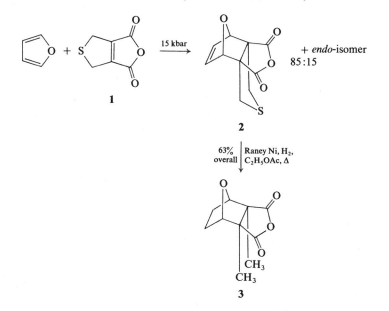

[1] D. J. Rodini and B. B. Snider, *Tetrahedron Letters*, **21**, 3857 (1980).

Furane, 7, 161.

Diels–Alder route to cantharidin. Use of high pressure to facilitate Diels–Alder reactions has led to a simple synthesis of cantharidin.[1] The obvious dienophile component, dimethylmaleic anhydride, is unreactive to furane even under high pressure, but 2,5-dihydrothiophene-3,4-dicarboxylic anhydride (**1**)[2] reacts with furane at 25° under 15 kbar of pressure to give the *exo*-adduct (**2**) and the *endo*-isomer as a 85:15 mixture. The major adduct is converted into cantharidin (**3**) by reduction and desulfurization over Raney nickel in 63% overall yield.

[1] W. C. Dauben, C. R. Vessel, and K. H. Takemura, *Am. Soc.*, **102**, 6893 (1980).
[2] B. R. Baker, M. V. Querry, and A. F. Kadish, *J. Org.*, **13**, 128 (1948).

β-Furyllithium. 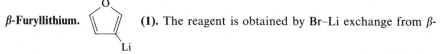 **(1).** The reagent is obtained by Br–Li exchange from β-bromofurane, obtained readily by bromination of the adduct of furane and maleic anhydride followed by pyrolysis.[1]

Cardenolides.[2] Digitoxigenin (**7**) has been synthesized by reaction of a 3β-benzyloxy α,β-unsaturated 17-ketone (**2**) with **1**, followed by further transformations as shown. Allylic rearrangement of the initial product **3** results in **4**. After reduction

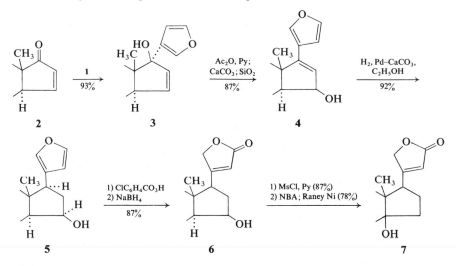

of the 16,17-double bond, the furane group is oxidized by *m*-chloroperbenzoic acid to a hydroxy lactone, which is immediately reduced to the lactone **6**. Remaining steps are dehydration and introduction of a hydroxyl group at C_{14} by a known method.[3]

When **5** is oxidized with NBS, the isomeric lactone **8** is obtained. This product was converted to an isomer (**9**) of the natural cardenolide **7**.

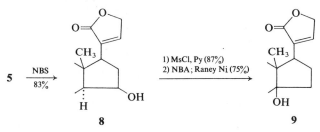

[1] J. Srogl, M. Handa, and I. Stibor, *Coll. Czech. Chem. Comm.*, **35**, 3478 (1970); S. Gronowitz and B. Holm, *Syn. Comm.*, **4**, 63 (1974).

[2] T. Y. R. Tsai, A. Minta, and K. Wiesner, *Heterocycles*, **12**, 1397 (1979).

[3] C. R. Engel and G. Bach, *Steroids*, **3**, 593 (1964).

G

Grignard reagents, 1, 415–424; **2**, 205; **5**, 321; **6**, 269–270; **7**, 163–164; **8**, 235–238; **9**, 229–233.

Ketone synthesis. The synthesis of ketones by reaction of Grignard reagents with esters is usually unsatisfactory because of formation of tertiary alcohols as by-products. However, this secondary reaction is mainly prevented if triethylamine (large excess) is present, and reasonable yields of ketones can be obtained unless the ester is readily enolized.[1]

Synthesis of ketones by reaction of Grignard reagents with acid chlorides bearing substituents susceptible to side reactions ($COOCH_3$, Cl) is possible if the reaction is conducted at $-70°$ in THF.[2]

Reaction with formic acid. The reaction of Grignard reagents with formic acid was originally (1905) reported to give aldehydes in low yield and has seen little use. Actually, the reaction proceeds in fair yields even in ether and in higher yields in THF. Aldehydes can also be prepared by the method outlined in equation (I).[3]

$$\text{(I)} \quad HCOOH + C_2H_5MgBr \xrightarrow[-C_2H_6]{THF} HCOMgBr \xrightarrow[50-80\%]{RMgBr} RCHO$$

Vinyl chlorides.[4] In the presence of $Ni[P(C_6H_5)_3]_4$ (**6**, 570)[5] Grignard reagents reacts stereospecifically with (E)- or (Z)-dichloroethylene to form (E)- or (Z)-1-chloro-alkenes, respectively. The reaction has been used for an efficient synthesis of the 1,3-diene **1**, the sex pheromone of *Lobesia botrane* (equation I).

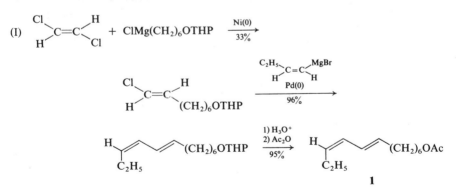

Coupling with silyl enol ethers.[6] Alkenes can be prepared by coupling of Grignard reagents with silyl enol ethers. Nickel acetylacetonate is the most active catalyst, but it favors formation of the most thermodynamically stable isomer. Use of $NiCl_2[P(C_6H_5)_3]_2$ results in regio- and stereospecific coupling (equation I).

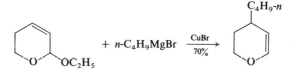

(I) $R^1R^2C=C(R^3)OSi(CH_3)_3$ + R^4MgX $\xrightarrow[\text{55-85\%}]{\text{Ni cat., } C_6H_6, \Delta}$ $R^1R^2C=C(R^3)R^4$

β-Alkylated enol ethers.[7] In the presence of 5% of copper(I) bromide, Grignard reagents react with α,β-unsaturated acetals or dioxolanes (**1**) to form β-substituted enol ethers (**2**), with complete allylic rearrangement.

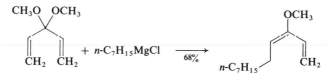

$R^1CH=CH-C(OCH_3)_2$ + $RMgX$ $\xrightarrow[\text{50-85\%}]{\text{CuBr, THF, } P(OC_2H_5)_3}$ $R^1R\,CHCH=C(OCH_3)R^2$
|
R^2

1   (E + Z)-**2**

Examples:

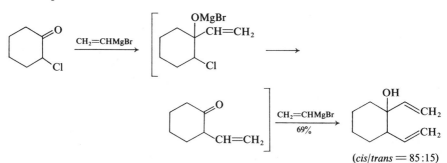

Reaction of vinyl Grignard reagents with α-halo ketones. Vinylmagnesium bromide or vinyllithium (2 equivalents) reacts with α-halocycloalkanones to give 1,2-divinylcycloalkanols, used in oxy-Cope rearrangements for synthesis of large ring ketones (*cf.* **7**, 302–303; **8**, 412–414).[8]

Examples:

(*cis/trans* = 85:15)

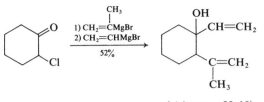

$$(cis/trans = 90:10)$$

Trisubstituted alkenes.[9] A stereoselective synthesis of trisubstituted alkenes uses (E)-alkenyl sulfoxides (**1**)[10] as the starting material. These are reduced to the corresponding sulfides (**2**),[11] which undergo coupling with Grignard reagents in the presence of complexes of nickel chloride and phosphines as catalyst.[12] The products (**3**) are obtained in steroisomeric purity of >99%.

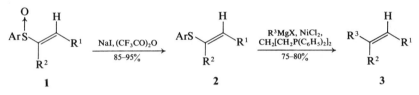

Conjugate addition to α,β-unsaturated sulfoxides. Chiral α-carbonyl-α,β-unsaturated sulfoxides undergo conjugate addition of Grignard reagents with moderate to high asymmetric induction. A copper salt is not required. Organo-copper reagents also undergo this reaction.[13]

Examples:

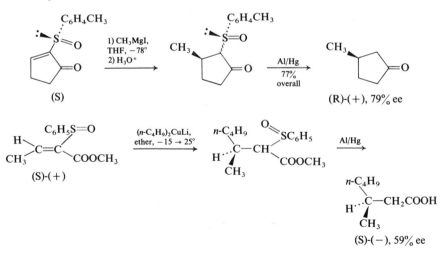

The method was used for a total synthesis of optically active 11-oxoequilenin methyl ether.

Spirolactones.[14] Grignard reagents of the type $BrMgCH_2(CH_2)_nCH_2MgBr$ react with anhydrides of dicarboxylic acids to form spirolactones in one step (equation I).

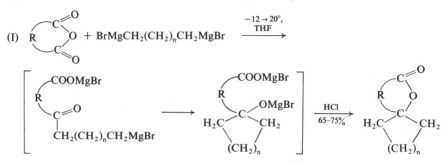

(I)

β-Lactams. During early research on penicillin synthesis, one method examined was cyclization of a β-amino ester with a Grignard reagent. The first detailed report indicated that yields tend to be low (3–45%); even so, the first synthesis of β-propiolactam (2-azetidinone) was achieved in this way (equation I).[15]

(I) $H_2NCH_2CH_2COOC_2H_5$ $\xrightarrow[0.76\%]{C_2H_5MgBr}$

This reaction has since been used in a synthesis[16] of (+)-thienamycin (**4**), a potent antibiotic from *Streptomyces cattlaya*. Thus cyclization of **1** with CH_3MgBr resulted mainly in the desired *trans*-azetidinone **2**. Cyclization with DCC favors formation of the undesired *cis*-isomer (**3**).

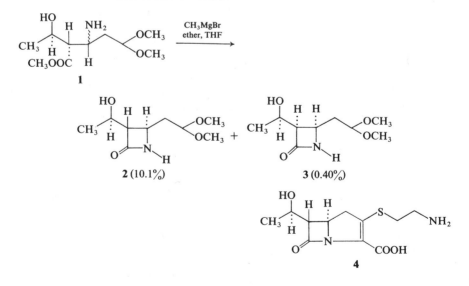

Addition to α-alkoxy carbonyl compounds. n-Butylmagnesium bromide reacts stereoselectively with protected, chiral α-hydroxy ketones. For example, reaction of the Grignard reagent with the MEM ether of 3-hydroxy-2-decanone in THF leads to the *threo*-product [>95% isolated yield (equation I)]. The stereoselectivity is much lower when n-butyllithium is used; in addition the solvent plays an important role. Stereoselectivity is lower in C_5H_{12}, CH_2Cl_2, and ether. A number of protective

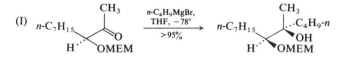

groups can be used; benzyl and benzyloxymethyl groups are preferred when the free diol is desired. Stereoselectivity is much lower with THP ethers. No temperature dependence on stereochemistry is observed. The stereochemistry evidently results from chelation of the two oxygens to magnesium (equation II).[17]

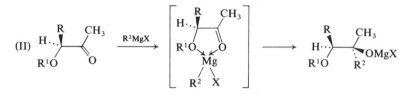

Grignard reagents show some stereoselectivity in reactions with α-alkoxy aldehydes (*threo/erythreo* = 10:1), but only slight stereoselectivity obtains in reactions with β-alkoxy aldehydes. On the other hand, fairly high stereoselectivity is observed in the reaction of lithium dialkylcuprates with β-alkoxy aldehydes, and again formation of the *threo*-product is favored (equation III).[18]

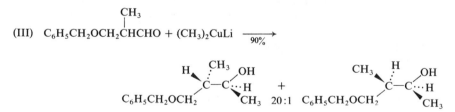

α,β-Unsaturated aldehydes.[19] Grignard reagents react with β-ethoxyacroleins (1) to form 1,2-adducts (2), which are rearranged by acid to (E)-α,β-unsaturated aldehydes (3) in 30–60% overall yield.

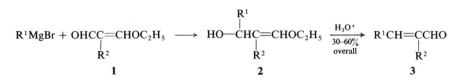

[1] I. Kikkawa and T. Yorifuji, *Synthesis*, 877 (1980).

[2] M. K. Eberle and G. G. Kahle, *Tetrahedron Letters*, **21**, 2303 (1980).

[3] F. Sato, K. Oguro, H. Watanabe, and M. Sato, *Tetrahedron Letters*, **21**, 2869 (1980).

[4] V. Ratovelomanana and G. Linstrumelle, *Tetrahedron Letters,* **22**, 315 (1981).

[5] J. F. Fauvarque and A. Jutand, *J. Organometal. Chem.*, **177**, 273 (1979).

[6] T. Hayashi, Y. Katsuro, and M. Kumada, *Tetrahedron Letters*, **21**, 3915 (1980).

[7] Y. Gendreau and J. F. Normant, *Bull. Soc.*, II-305 (1979).

[8] D. A. Holt, *Tetrahedron Letters*, **22**, 2243 (1981).

[9] H. Takei, H. Sugimura, M. Miura, and H. Okamura, *Chem. Letters*, 1209 (1980).

[10] G. H. Posner, P.-W. Tang, and J. P. Mallamo, *Tetrahedron Letters*, **21**, 3995 (1980).

[11] J. Drabowicz and S. Oae, *Synthesis*, 404 (1977).

[12] H. Okamura, M. Miura, and H. Takei, *Synthesis*, 43 (1979).

[13] G. H. Posner, J. P. Mallamo, and K. Miura, *Am. Soc.*, **103**, 2886 (1981).

[14] P. Canonne and D. Bélanger, *J.C.S. Chem. Comm.*, 125 (1980).

[15] R. W. Holley and A. D. Holley, *Am. Soc.*, **71**, 2124, 2129 (1949).

[16] T. Kametani, S.-P. Huang, Y. Suzuki, S. Yokohama, and M. Ihara, *Heterocycles*, **12**, 1301 (1979); *Am. Soc.*, **102**, 2060 (1980).

[17] W. C. Still and J. H. McDonald III, *Tetrahedron Letters*, **21**, 1031 (1980).

[18] W. C. Still and J. A. Schneider, *Tetrahedron Letters*, **21**, 1035 (1980).

[19] K. Rustemeier and E. Braitmaier, *Angew. Chem. Int. Ed.*, **19**, 816 (1980).

Guaiacylmethyl chloride (GuM-Cl),   **(1).** Mol. wt. 172.61, b.p.

135°/30 mm. The reagent is prepared by reaction of veratrole with PCl_5 in CCl_4 in the presence of dibenzoyl peroxide (80–90% yield).

Protection of hydroxyl groups.[1] The reagent reacts with alkoxides to form ethers (Williamson ether synthesis). The ethers, GuM-OR, are also obtainable by phase-transfer reactions (NaOH, C_6H_6–H_2O, $Bu_4\overset{+}{N}HSO_4{}^-$). This method is particularly useful for selective protection. Thus it is possible to protect selectively a primary hydroxyl group in the presence of a secondary hydroxyl group and to protect a secondary hydroxyl group in the presence of a tertiary hydroxyl group. The ethers are stable to oxidizing and reducing agents, to strong bases, and to $LiAlH_4$. However they are less stable to acids than MEM ethers. They are cleaved by $ZnBr_2$ in CH_2Cl_2 at 25°.

[1] B. Loubinoux, G. Coudert, and G. Guillaumet, *Tetrahedron Letters*, **22**, 1973 (1981).

H

Hexachlorobutadiene, 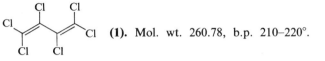 **(1).** Mol. wt. 260.78, b.p. 210–220°.

Supplier: Aldrich.

Condensation with enolates of carbonyl compounds (*cf.* Trichloroethylene, **9**, 479–480). The reagent can react in two ways with carbonyl enolates. The first one evidently is a direct addition–elimination (equation I); the second one is considered to involve perchlorobutenyne (**a**) as an intermediate (equation II).[1]

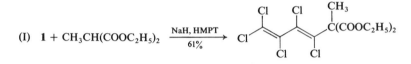

(I) $1 + CH_3CH(COOC_2H_5)_2$ $\xrightarrow[61\%]{NaH, HMPT}$

(II) $1 + (CH_3)_2CHCOOC_2H_5$ $\xrightarrow{LDA, HMPT}$

$$\left[(CH_3)_2CClCOOC_2H_5 + ClC\equiv C-\overset{\overset{\displaystyle Cl}{|}}{C}=CCl_2 \right] \xrightarrow{63\%} Cl_2C=\overset{\overset{\displaystyle Cl}{|}}{C}-C\equiv C-\overset{\overset{\displaystyle CH_3}{|}}{\underset{\underset{\displaystyle CH_3}{|}}{C}}COOC_2H_5$$

$\qquad\qquad\qquad\qquad\qquad$ **a**

[1] A. S. Kende, P. Fludzinski, and J. H. Hill, *Am. Soc.*, **103**, 2904 (1981).

Hexachlorodisilane, Si_2Cl_6. Mol. wt. 268.89, b.p. 146–148°. Supplier: Alfa.

Deoxygenation. Hexachlorodisilane is useful for deoxygenation of nitrones and various N-oxides in good yields under exceptionally mild conditions (25° or below).[1] It is recommended for deoxygenation of 2,3-disubstituted quinoxaline 1,4-dioxides under mild conditions yet at reasonable rates.[2]

[1] A. G. Hortmann, J. Kov, and C.-C. Yu, *J. Org.*, **43**, 2284 (1978).
[2] F. R. Homaidan and C. H. Issidorides, *Heterocycles*, **16**, 411 (1981).

Hexafluoroantimonic acid, 5, 309–310; **6,** 272–273; **7,** 166–167; **8,** 239–240.

m-*Bromination of phenols.*[1] *para*-Alkylated and 2,6-dialkylated phenols (and also the methyl ethers) are brominated in the *meta*-position in reactions conducted in SbF_5–HF at 45°. The unusual orientation is not a result of isomerization of *o*- or *p*-bromo compounds, but is probably due to bromination of the O-protonated phenol or ether.

[1] J.-C. Jacquesy, M.-P. Jouannetaud, and S. Makani, *J.C.S. Chem. Comm.*, 110 (1980).

Hexamethyldisiloxane, 8, 240.

Trimethylsilyl esters. Trimethylsilyl esters are obtained in 65–90% yield by reaction of carboxylic acids with hexamethyldisiloxane in refluxing toluene containing a catalytic amount of sulfuric acid.[1] The actual catalyst is probably bis(trimethylsilyl) sulfate, $[(CH_3)_3Si]_2SO_4$. Pyridinium trifluoromethanesulfonate also can serve as catalyst.

[1] H. Matsumoto, Y. Hoshino, J. Nakabayashi, T. Nakano, and Y. Nagai, *Chem. Letters*, 1475 (1980).

Hexamethylphosphoric triamide (HMPT), **1,** 430–431; **2,** 208–210; **3,** 149–153; **4,** 244–247; **5,** 323–325; **6,** 273–274; **7,** 168–170; **8,** 240–245; **9,** 235.

Reaction of alkynyllithium reagents with epoxides. This reaction, particularly in the case of substituted epoxides, can be sluggish when conducted in THF alone. Doolittle[1] has examined the effect of various cosolvents. Use of THF–HMPT (1:1) results in high yields and reasonable reaction rates at 0°. Of other additives examined, TMEDA is equivalent to HMPT and is more effective than DABCO or ethylenediamine. Note that use of HMPT does not permit reaction of 1-dodecynyllithium with either *cis-* or *trans-*2,3-dimethyloxirane even at 50° for 2 hours.

Conjugate additions (**9,** 235). The reaction of the anion of allyl phenyl sulfone (**1**) with cyclohexene-3-one (**2**) at −78° results in the 1,2 α-adduct (**3**) as the major kinetic product. This product rearranges at 0° to the thermodynamically more stable 1,4 γ-adduct **4**. If the initial reaction is conducted in the presence of HMPT,

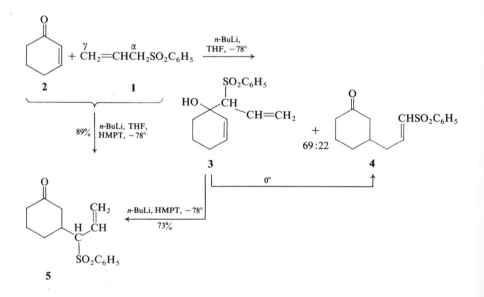

the 1,4 α-adduct **5** is obtained in high yield. This adduct is also formed when **3** is treated with *n*-butyllithium and HMPT at −78°. This selective 1,4 α-addition with HMPT is observed with carbanions of methyl-substituted allyl phenyl sulfones and evidently results from direct conjugate addition, possible in the presence of HMPT.[2]

Haynes *et al.*[3] have examined the effect of HMPT on the reaction of arylthio-allyllithium reagents with cyclopentenone. In THF alone the reaction leads mainly to 1,2-adducts, derived from both α- and γ-substitution. However, when the same

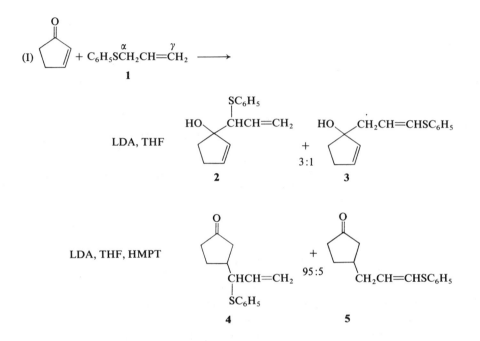

reaction is carried out in the presence of 1 equivalent of HMPT, essentially only products of conjugate addition are formed (*cf.*, **8**, 240–241), and, usefully, the predominant product results from α-1,4-addition (equation I). Similar selectivity is observed on addition of CuCl (1 equivalent). Indeed in the only example cited, CuCl is somewhat more effective than HMPT.

Bicycloannelation. The first step of a new method for this reaction involves conjugate addition of a cyclic dienolate to a nitro olefin followed by an intra-molecular displacement with loss of the nitro group. An example is the reaction of the α′-enolate of isophorone (**1**) with 1-nitropropene. When refluxed with HMPT the initial adduct (**2**), which can be isolated, is converted into the tricyclooctanones **3** and **4** in 63% yield.[4]

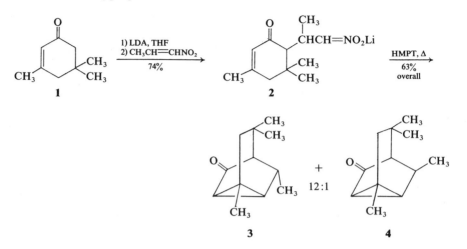

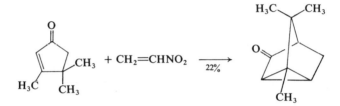

Another example:

Polymer-supported reagent. HMPT supported on a polystyrene-type resin is a catalyst for S_N2 reactions[5-7] and for reduction of ketones by $NaBH_4$.[7] It also has a marked effect on the alkylation of ethyl acetoacetate with diethyl sulfate. In the presence of solid HMPT the enolates undergo 60–70% O-alkylation. In the absence of HMPT, the lithium enolate does not react and the sodium and potassium enolates undergo C-alkylation (90–100%). There is some difference in the effect of solid and liquid HMPT: The solid HMPT increases the reactivity of the K enolate more than the liquid form, whereas the reverse is true with the Li enolate.[8]

[1] R. E. Doolittle, *Org. Prep. Proc. Int.*, **12**, 1 (1980).
[2] M. Hirama, *Tetrahedron Letters*, **22**, 1905 (1981).
[3] M. R. Binns, R. K. Haynes, T. L. Houston, and W. R. Jackson, *Tetrahedron Letters*, **21**, 573 (1980).
[4] R. M. Cory, P. C. Anderson, F. R. McLaren, and B. R. Yamamoto, *J.C.S. Chem. Comm.*, 73 (1981).
[5] Y. Leroux and H. Normant, *Compt. Rend. (C)*, **285**, 241 (1977).
[6] S. L. Regen, A. Nigam, and J. J. Besse, *Tetrahedron Letters*, 2757 (1978).
[7] M. Tomoi, T. Hasegawa, M. Ikeda, and H. Kakiuchi, *Bull. Chem. Soc. Japan*, **52**, 1653 (1979).
[8] G. Nee, Y. Leroux, and J. Seyden-Penne, *Tetrahedron*, **37**, 1541 (1981).

Hexamethylphosphorous triamide, 1, 425; **2,** 207; **3,** 148–149; **6,** 279–280; **9,** 235–236.

Conversion of bromohydrins to alkenes.[1] The 14,15-epoxide (**1**) of arachidonic acid methyl ester has been converted into the 11,12-epoxide (**5**) in two steps. The first is conversion of **1** into a mixture of isomeric bromohydrins (**2** and **3**) with KBr. The mixture is then epoxidized under the conditions of Sharpless (**5,** 75–76) to give, after chromatographic purification, **4,** the epoxide of **2**. Conversion of **4** into **5** presented a problem, but was eventually achieved by treatment of the *vic*-bromo triflate with $P[N(CH_3)_2]_3$ as a Br^- acceptor.

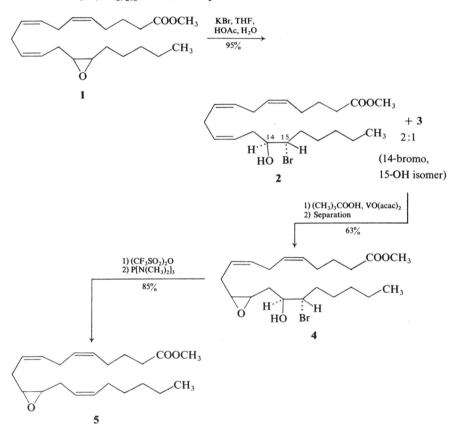

[1] E. J. Corey, A. Marfat, J. R. Falck, and J. O. Albright, *Am. Soc.,* **102,** 1433 (1980).

Hydriodic acid, 1, 449–450; **2,** 213–214; **8,** 246; **9,** 238.

Reduction of quinones.[1] The combination of HI (56% aqueous solution) and phosphorus (to scavenge the I_2 formed) was used to a limited extent in early studies of reduction of quinones to arenes, but had the disadvantage of overreduction. Actually hydriodic acid in refluxing acetic acid is an efficient reagent for reduction not only of polycyclic quinones, but also hydroquinones and phenols to arenes. In resistant cases, HI (neat) or with phosphorus can be used; addition of phosphorus often results in a cleaner reaction. Yields are generally $>95\%$.

Reductive methylation of quinones.[2] Dimethylarenes can be obtained in high yield by reaction of quinones with methyllithium (excess) in benzene to form a dimethyl dihydrodiol, which is reduced without purification with HI in refluxing acetic acid.

Example:

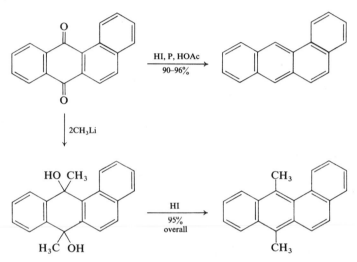

[1] M. Konieczny and R. G. Harvey, *J. Org.*, **44**, 4813 (1979); *Org. Syn.* submitted (1980).
[2] M. Konieczny and R. G. Harvey, *J. Org.*, **45**, 1308 (1980).

Hydrogen bromide, 1, 450–452; **4**, 251.

β-Bromo acetals and ketals.[1] These useful derivatives are generally prepared by addition of an α,β-unsaturated carbonyl compound to a solution of HBr in the diol. The same products can be obtained by addition of HBr to the α,β-enal or enone followed by acetalization. The method is improved if only a stoichiometric amount of HBr is used. Dicinnamalacetone (equation I)[2] is used to determine the end point. Hydrogen bromide is added to the initially yellow solution until a red color persists.

$$\text{(I)}\quad 2C_6H_5CH{=}CHCHO + CH_3COCH_3 \xrightarrow[-2H_2O]{\substack{\text{NaOH,}\\C_2H_5OH}} O{=}C(CH{=}CHCH{=}CHC_6H_5)_2$$

Golden yellow, m.p. 142°

Examples:

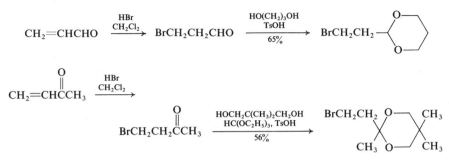

[1] J. C. Stowell, D. R. Keith, and B. T. King, *Org. Syn.*, submitted (1980).
[2] L. Diehl and A. Einhorn, *Ber.*, **18**, 2320 (1885).

Hydrogen chloride–Titanium(IV) chloride.

Intramolecular **ipso-*acylations.*** Grisadienediones such as **2** can be obtained by treatment of a substituted *o*-phenoxybenzoic acid ester (**1**) with dry hydrogen chloride and an excess of $TiCl_4$ at 0–25°. The synthesis involves an intramolecular *ipso*-acylation.[1]

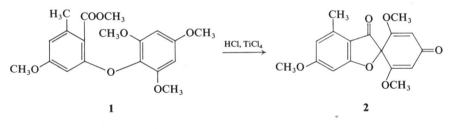

[1] M. V. Sargent, *J.C.S. Chem. Comm.*, 285 (1980).

Hydrogen peroxide, 1, 457–471; 2, 216–217; 3, 154–155; 4, 253–255; 5, 337–339; 6, 286; 7, 174; 8, 247–248; 9, 241–242.

Conversion of nitro compounds to ketones. This reaction has usually involved reductive or oxidative hydrolysis of the corresponding alkali nitronates. A recent method involves direct oxidation with 30% H_2O_2 and K_2CO_3 in methanol at room temperature. Yields are 75–95%.[1]

Bicyclic endoperoxides. The novel bicyclic [3.2.1]endoperoxide **2** has been prepared in high yield by oxidation of the ketone **1** with 90% H_2O_2 in the presence of BF_3 etherate. The mechanism of this oxidation is not clear. Benzophenone-sensitized ultraviolet irradiation converts **2** into the pine beetle pheromone frontalin (**3**) in quantitative yield. In the absence of the sensitizer the epoxide **4** is formed, and then rearranges when heated to **3**. Decomposition mediated by $TiCl_4$ results in cleavage of the one carbon bridge to give the diketone **5**.[2]

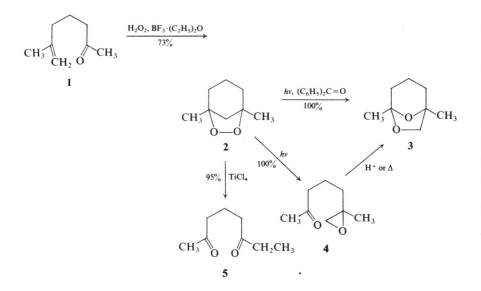

Epoxides of α-methylene ketones.[3] Epoxidation of α-methylene ketones is generally unsatisfactory. The desired epoxides have usually been prepared by epoxidation of allylic alcohols followed by oxidation. Another route involves oxidation of salts of Mannich bases with alkaline hydrogen peroxide (equation I).

$$(I) \quad \underset{O}{\overset{O}{\underset{\|}{R^1C}}} - \underset{R^2}{\overset{R^2}{\underset{|}{CHCH_2\overset{+}{N}(CH_3)_3}}} \, I^- \quad \xrightarrow[40-75\%]{\substack{H_2O_2 \\ NaOH}} \quad R^1C - \overset{R^2}{\underset{O}{C}} - CH_2$$

p-*Hydroxylation of phenols.* A new method for hydroxylation of phenols involves alkylation with cyclopentadiene, isomerization of the allylic double bond to a propenyl double bond, and finally oxidation with H_2O_2 in CH_3NO_2 in the presence of HCl (equation I).[4]

Oxidative deoximation.[5] Ketoximes revert to ketones by reaction with H_2O_2 in an alkaline medium (75–95% yield, five examples).

[1] G. A. Olah, M. Arvanaghi, Y. D. Vankar, and G. K. S. Prakash, *Synthesis*, 662 (1980).
[2] R. M. Wilson and J. W. Rekers, *Am. Soc.*, **103**, 206 (1981).
[3] F. Henin and J. P. Pete, *Synthesis*, 895 (1980).
[4] D. V. Rao and F. A. Stuber, *Tetrahedron Letters*, **22**, 2337 (1981).
[5] T.-L. Ho, *Syn. Comm.*, **10**, 465 (1980).

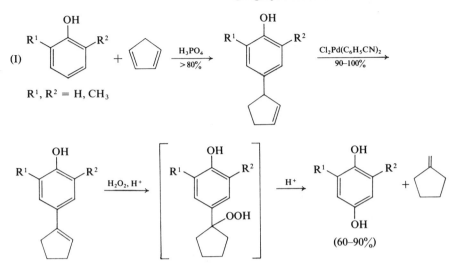

Hydrogen peroxide–Palladium acetate.

Methyl ketones. Hydrogen peroxide in the presence of $Pd(OAc)_2$ as catalyst converts terminal alkenes to methyl ketones in high yield and with high selectivity (equation I).[1] Internal and cycloalkenes are inactive. The actual oxidant is

$$\text{(I)} \quad RCH{=}CH_2 + H_2O_2 \xrightarrow{\quad Pd(OAc)_2 \quad} RCOCH_3 + H_2O$$

considered to be a palladium hydroperoxidic species (PdOOH), and the reaction is thus analogous to that observed with complexes of the type $[RCO_2PdOOC(CH_3)_3]_4$.[2]

[1] M. Roussel and H. Mimoun, *J. Org.*, **45**, 5387 (1980).
[2] H. Mimoun, R. Charpentier, A. Mitschler, J. Fischer, and R. Weiss, *Am. Soc.*, **102**, 1047 (1980).

Hydrogen peroxide–Silver trifluoroacetate, 7, 175; 8, 249–250.

Prostaglandin G_2 (PGG$_2$, **1**). This compound is an early isolable intermediate in the bioconversion of arachidonic acid into prostaglandins; it is the hydroperoxide corresponding to PGH$_2$ (prostaglandin endoperoxide, **8**, 249–250). Porter and co-workers have extended the earlier synthesis of PGH$_2$ to PGG$_2$ using Mukaiyama's alkyl halide synthesis (**8**, 90) and the silver–H$_2$O$_2$ reaction (equation I).[1]

[1] N. A. Porter, J. D. Byers, A. E. Ali, and T. E. Eling, *Am. Soc.*, **102**, 1183 (1980).

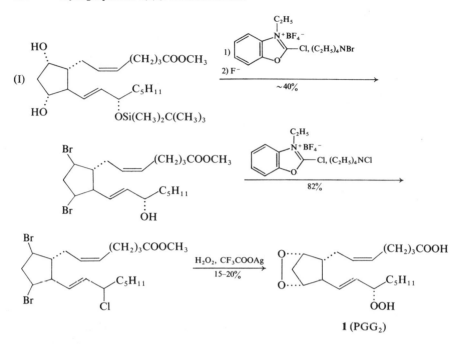

1 (PGG₂)

Hydrogen peroxide–1,1,3,3-Tetrachloroacetone.

Epoxidation. 1,1,3,3-Tetrachloroacetone can mediate the epoxidation of olefins by hydrogen peroxide in the same way as hexafluoroacetone (**9**, 244–245). It has the advantage that it is probably less toxic, and it is inexpensive and commercially available. The actual oxidant presumably is the hydroperoxide **1**, which is converted during the epoxidation into the unstable hydrate (**2**), from which the tetrachloroacetone can be recovered in >70% yield. Monosubstituted alkenes are epoxidized in low yields by this method; more highly substituted alkenes are epoxidized in 65–85% yield (VPC).[1]

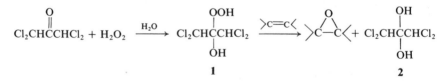

[1] C. I. Stark, *Tetrahedron Letters*, **22**, 2089 (1981).

Hydrogen peroxide–Triethyl orthoacetate.

Epoxidation. Reaction of ortho esters with 90% H_2O_2 generates a species that can epoxidize alkenes. Of several ortho esters tested, triethyl orthoacetate was most satisfactory. An example is shown in equation (I).[1]

(I)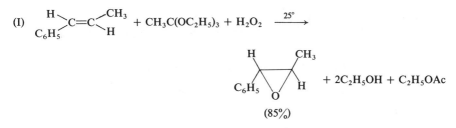

Intramolecular epoxidation is also possible with this system.[2] Thus treatment of the ortho ester (1) of oleic acid with H_2O_2 in CH_2Cl_2 leads to the epoxide 2 in 40% yield, probably via the intermediate **a**.

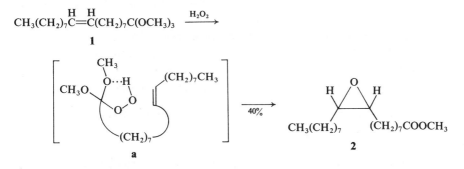

[1] J. Rebek, Jr., and R. McCready, *Tetrahedron Letters*, 1001 (1979).
[2] J. Rebek, Jr., and R. McCready, *Tetrahedron Letters*, **21**, 2491 (1980).

Hydrogen peroxide–Tungstic acid ($WO_3 \cdot H_2O$).

Sulfones. A recent preparation of thietane 1,1-dioxide[1] emphasizes the advantage of tungstic acid catalysis[2] in the hydrogen peroxide oxidation of sulfides to sulfones; an induction period is eliminated and the yield improved.

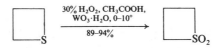

[1] T. C. Sedergran and D. C. Dittmer, *Org. Syn.*, submitted (1981).
[2] H. L. Schultz, H. B. Freyermuth, and R. S. Buc, *J. Org.*, **28**, 1140 (1968).

Hydrogen telluride, H_2Te. Mol. wt. 129.62. unstable gas. The reagent can be generated *in situ* from aluminum telluride, Al_2Te_3 (Alfa), and water.

Reductions. Hydrogen telluride reduces aldehydes and ketones to alcohols in satisfactory yields. The combination Al_2Te_3–D_2O permits deuteration. H_2Te reduces α,β-enones and α,β-enals to the saturated carbonyl compound.[1]

Aromatic nitro compounds are reduced to arylamines in 60–90% yield. Nitrosobenzene, azoxybenzene, and azobenzene are reduced to hydrazobenzene in high yield.[2]

[1] N. Kambe, K. Kondo, S. Morita, S. Murai, and N. Sonoda, *Angew. Chem. Int. Ed.*, **19**, 1009 (1980).

[2] N. Kambe, K. Kondo, and N. Sonoda, *Angew. Chem. Int. Ed.*, **19**, 1008 (1980).

2-Hydroperoxyhexafluoro-2-propanol (1), **9**, 244–245.

—CHO → COOH.[1] This hydroperoxide oxidizes aldehydes to carboxylic acids in generally satisfactory yields. The oxidation is conducted in CH_2Cl_2–Na_2CO_3 or CH_3OH–NaOH. The oxidant does not attack alcohols. The oxidation, as in epoxidations with **1**, can be effected with H_2O_2 and a catalytic amount of **1**.

N-Oxides, sulfoxides, sulfones.[2] The hydroperoxide oxidizes tertiary amines to N-oxides (60–95% yield) and sulfides to sulfoxides (1 equivalent) or to sulfones (2 equivalents).

[1] B. Ganem, R. P. Heggs, A. J. Biloski, and D. R. Schwartz, *Tetrahedron Letters*, **21**, 685 (1980).

[2] B. Ganem, A. J. Biloski, and R. P. Heggs, *Tetrahedron Letters*, **21**, 689 (1980).

(S)-(2-Hydroxy-N,N-dimethylpropanamide-O,O′)oxodiperoxymolybdenum(VI)   (1). Mol. wt. 293.09, m.p. 149°.

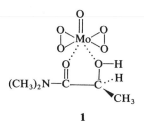

1

Asymmetric epoxidation.[1] Simple, unfunctionalized olefins are epoxidized at 20° in nitrobenzene to (R)-epoxides with 15–35% ee. Chemical yields are 70%. Alkyl substitution increases the rate of epoxidation, but steric hindrance decreases the asymmetric induction.

[1] H. B. Kagan, H. Mimoun, C. Mark, and V. Schurig, *Angew. Chem. Int. Ed.*, **18**, 485 (1979).

Hydroxylamine, 1, 478–481; **7**, 176–177; **9**, 245.

Cleavage of enol esters.[1] Enol esters are converted to oximes of the corresponding ketone by reaction with excess hydroxylamine in THF or pyridine.

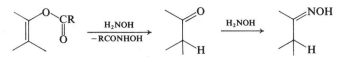

[1] F. W. Lichtenhaler and P. Jarglis, *Tetrahedron Letters*, **21**, 1425 (1980).

Hydroxylamine-O-sulfonic acid, 1, 481–484; **2**, 217–219; **3**, 156–157; **4**, 256; **5**, 343–344; **6**, 290; **8**, 250–251; **9**, 245–246.

Lactams.[1] The reagent is useful for rearrangement of C_5–C_{12} cycloalkanones to lactams.

Example:

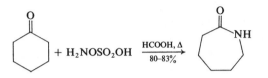

Review.[2] Wallace has reviewed (75 references) uses of this reagent, particularly in synthesis of heterocycles.

[1] G. A. Olah and A. P. Fung, *Synthesis*, 537 (1979); *Org. Syn.*, submitted (1980).
[2] R. G. Wallace, *Aldrichim. Acta*, **13**, 3 (1980).

Hydroxylaminium acetate, $AcO^-\overset{+}{N}H_3OH$ **(1).** The salt is formed from equivalent amounts of hydroxylamine and acetic acid.

2′-O-Deacylation of ribonucleosides.[1] This salt (**1**) is useful for regioselective 2′-O-deacylation of 2′,3′,5′-triacyl derivatives of ribonucleosides. Hydrazinium acetate is somewhat less effective.

[1] Y. Ishido, N. Sakairi, K. Okazaki, and N. Nakazaki, *J.C.S. Chem. Comm.*, 563 (1980).

(2S,2′S)-2-Hydroxymethyl-1-[(1-methylpyrrolidin-2-yl)methyl]pyrrolidine,

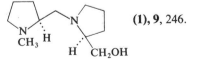

(1), 9, 246.

Asymmetric addition to aldehydes. In the presence of **1**, lithium trimethylsilyl-acetylide reacts with aliphatic aldehydes to form predominately (R)-alkynyl alcohols. One of these products was converted into optically active 5-octyl-2(5H)-furanone (**2**) as shown in equation (I).[1]

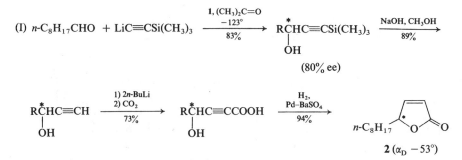

(I) $n\text{-}C_8H_{17}CHO$ + $LiC{\equiv}CSi(CH_3)_3$ $\xrightarrow[\substack{-123° \\ 83\%}]{1,\,(CH_3)_2C{=}O}$ $\overset{*}{R}CHC{\equiv}CSi(CH_3)_3$ $\xrightarrow[89\%]{NaOH,\,CH_3OH}$

$\underset{OH}{|}$

(80% ee)

$\overset{*}{R}CHC{\equiv}CH$ $\xrightarrow[73\%]{\substack{1)\,2n\text{-}BuLi \\ 2)\,CO_2}}$ $\overset{*}{R}CHC{\equiv}CCOOH$ $\xrightarrow[94\%]{\substack{H_2, \\ Pd\text{-}BaSO_4}}$

$\underset{OH}{|}$ $\underset{OH}{|}$

2 $(\alpha_D\ -53°)$

¹ T. Mukaiyama and K. Suzuki, *Chem. Letters*, 255 (1980).

Hypochlorous acid, 1, 245, 487–488.

Chlorohydrins. In a typical procedure[1] for preparation of a chlorohydrin by heterogeneous reaction of an alkene with HOCl, the reagent is generated by reaction of aqueous NaOH with $HgCl_2$ and addition of Cl_2 until the yellow precipitate of HgO disappears. Nitric acid is then added. The chlorohydrin is prepared by reaction of the alkene with this solution with vigorous stirring. In the case of cyclohexene, the yield of 2-chlorocyclohexanol is about 70%.

Marmor and Maroski[2] investigated the reaction of four hindered alkenes with HOCl in acetone–water (homogenous conditions), but the reaction led to a chlorohydrin in only one case: 1-chloro-2,4,4-trimethyl-2-pentanol (34% yield). The products in general are vinyl chlorides.

Allylic chlorides.[3] Actually the reaction of HOCl with highly substituted alkenes is a convenient route to allylic chlorides if CH_2Cl_2 is used as the organic cosolvent. The reagent is prepared by addition of dry ice to calcium hypochlorite (70%) in water. The reaction of 1-methylcyclohexene is typical (equation I). Chlorohydrins are the main products only in the case of 1-alkenes and 1,2-disubstituted alkenes.

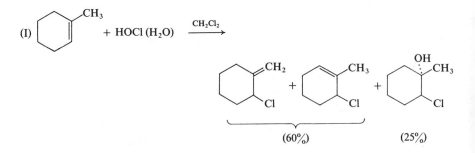

This reaction was used for a convenient synthesis of Rose oxide (equation II).

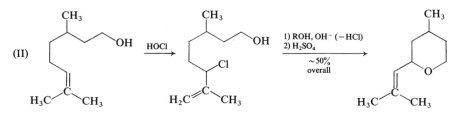

(II)

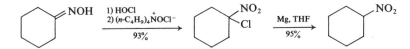

$>C=NOH \rightarrow >CHNO_2$. Corey and Estreicher[4] have developed a mild and efficient conversion of oximes of cyclic ketones to saturated nitro compounds. The first step is usually effected with HOCl in C_6H_6—H_2O (1.2:1) at pH 5.5 (critical). Use of *t*-butyl hypochlorite or of Cl_2–O_2 results in lower yields. The resulting α-chloronitroso compound is oxidized to an α-chloro nitro compound by tetra-*n*-butylammonium hypochlorite (formed *in situ* from tetra-*n*-butylammonium hydrogen sulfate and sodium hypochlorite). These two steps can be conducted as a one-flask operation. A number of reagents can be used for the final step, protodechlorination: Mg, Zn, H_2–Pd/C.

Example:

[1] G. H. Coleman and H. F. Johnstone, *Org. Syn. Coll. Vol.*, **1**, 158 (1947).
[2] S. Marmor and J. G. Maroski, *J. Org.*, **31**, 4278 (1966).
[3] S. G. Hedge, M. K. Vogel, J. Saddler, T. Hrinyo, N. Rockwell, R. Haynes, M. Oliver, and J. Wolinsky, *Tetrahedron Letters*, **21**, 441 (1980).
[4] E. J. Corey and H. Estreicher, *Tetrahedron Letters*, **21**, 1117 (1980).

I

Iododestannylation.[1] This reaction has been used to effect site-specific iodination of tamoxifen (**1**), a drug used for treatment of estrogen-dependent breast carcinoma. Metalation of **1** with *sec*-butyllithium occurs exclusively, even in the absence of TMEDA, at the position *ortho* to the dimethylaminoethoxy group, and reaction of the resulting anion with tri-*n*-butyltin chloride gives **2** in high yield. Iododestannylation of **2** leads to iodotamoxifen (**3**) in practically quantitative yield. Attempts to iodinate **1** with a variety of reagents results in complex mixtures.

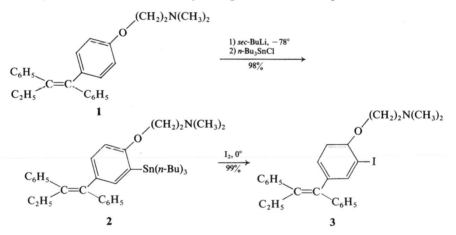

Cyclic iodo carbonates. Allylic and homoallylic alcohols are converted regio- and stereoselectively into five- and six-membered cyclic iodo carbonates, respectively, by carboxylation of the alkoxide followed by reaction with iodine. These products can be converted into *cis*-1,2- or 1,3-diols as illustrated.[2]

Examples:

$CH_3CHOHCH_2CH\!=\!CH_2$ $\xrightarrow{\begin{subarray}{c}\text{1) }n\text{-BuLi}\\\text{2) }CO_2\end{subarray}}$ $\left[\begin{array}{c}CH_3CHCH_2CH\!=\!CH_2\\|\\OCO_2Li\end{array}\right]$ $\xrightarrow[80\%]{I_2}$

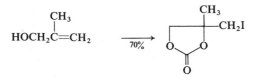

[1] D. E. Seitz, G. L. Tonnesen, S. Hellman, R. N. Hanson, and S. J. Adelstein, *J. Organometal. Chem.*, **186**. C33 (1980).
[2] G. Cardillo, M. Orena, G. Porzi, and S. Sandri, *J.C.S. Chem. Comm.*, 465 (1981).

Iodine–Copper(II) acetate.

α-Iodo ketones.[1] Ketones (or the enol acetate) are converted into α-iodo ketones by reaction with a slight excess of iodine and copper(II) acetate in acetic acid at 60°.

Examples:

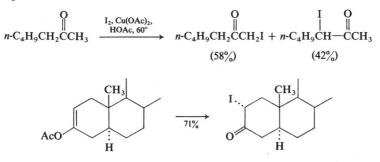

[1] C. A. Horiuchi and J. Y. Satoh, *Synthesis*, 312 (1981).

Iodine azide, 1, 500–501; 2, 222–223; 3, 160–161; 4, 262; 5, 350–351; 6, 297; 8, 260–261.

Amides; lactams.[1] Dimethyl thioketals of cyclic or acyclic ketones react with iodine azide to form an α-azido sulfide in 75–95% yield. The product rearranges to an amide or a lactam in high yield in the presence of trifluoroacetic acid. This sequence is an attractive alternative to the Beckmann rearrangement or Schmidt reaction.

Examples:

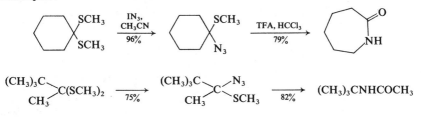

[1] B. M. Trost, M. Vaultier, and M. L. Santiago, *Am. Soc.*, **102**, 7929 (1980).

Iodine monochloride, 1, 502.

Alkyl iodides. Iodine monochloride in the presence of sodium acetate converts trialkylboranes derived from terminal alkenes into alkyl iodides in 75–95% yield (based on conversion of two of the alkyl groups). This reaction has been conducted with iodine, but sodium methoxide is required as base (**6**, 179–180).[1]

Vinyl iodides can be prepared by a similar sequence from terminal alkynes (equation I).[2]

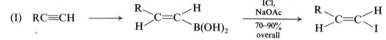

[1] G. W. Kabalka and E. E. Gooch III, *J. Org.*, **45**, 3578 (1980).
[2] G. W. Kabalka, E. E. Gooch, and H. C. Hsue, *Syn. Comm.*, **11**, 247 (1981).

Iodonium di-*sym*-collidine perchlorate, I^+ [structure] ClO_4^- **(1).** Mol. wt. 468.73. Preparation.[1]

α-Linked disaccharides.[2] The reagent has been used in a recent synthesis of sucrose. It functions as a superior source of I^+ probably because of the non-

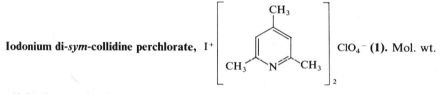

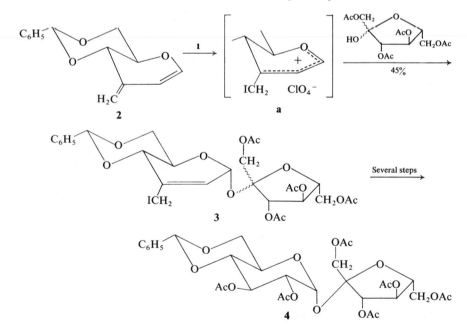

nucleophilic counterion. Thus the pyranoid diene **2** reacts with **1** to give planar ion **a** to which an alcohol adds in a 1,4-sense to give an α-glycoside. Thus tetraacetyl fructose reacts with **a** to give an α-disaccharide **3** in 45% yield. The β-isomer is not detected. The product can be converted in several steps into 4,6-O-benzylidene-hexaacetylsucrose (**4**).[2]

cis,vic-*Hydroxyamino sugars.*[3] The synthesis of methyl α,L-garosamide (**6**), a key component of aminocyclitol antibiotics, is complicated by the presence of a *cis*-hydroxyamino group and by the tertiary character of the hydroxyl group. The problems have been resolved by use of the allylic epoxide **2** as starting material. This epoxide was converted in three steps into **3**. Treatment of **3** with the iodonium salt gave the iodooxazolidinone **4** in 82% yield. This product was reduced and the ethoxyethyl group was removed to give **5**. The final step to **6** involved alkaline hydrolysis at 80° (6 hours).

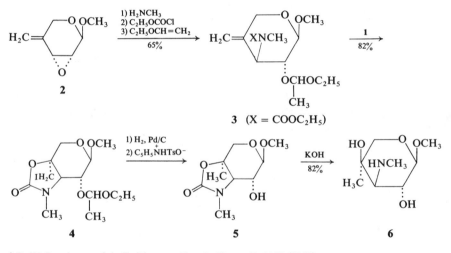

[1] R. W. Lemieux and A. R. Morgan, *Can. J. Chem.*, **43**, 2190 (1965).
[2] B. Fraser-Reid and D. F. Iley, *Can. J. Chem.*, **57**, 645 (1979).
[3] H. W. Pauls and B. Fraser-Reid, *Am. Soc.*, **102**, 3956 (1980).

Iodosobenzene, 1, 507–508.

Oxygen transfer reactions.[1] Iodosobenzene can function as an oxygen transfer reagent. Thus it converts tetracyanoethylene into the oxide in 74% yield. It also effects epoxidation of ketenes to yield, initially, α-lactones that polymerize to polyesters (equation I).

$$\text{(I)} \quad R_2C{=}C{=}O + C_6H_5I{=}O \xrightarrow{25°} C_6H_5I + \left[\begin{array}{c} O \\ R_2C{-}C \\ O \end{array} \right] \longrightarrow \text{polyester}$$

$$(60\text{–}70\%) \qquad\qquad\qquad\qquad (60\text{–}90\%)$$

Iodosobenzene effects decarboxylation of α-keto acids, probably via a mixed anhydride (equation II).

(II) $RCCOOH + C_6H_5I=O \longrightarrow \left[RC-C-OIC_6H_5 \atop OH \right] \longrightarrow C_6H_5I + CO_2 + RCOH$
(75–95%)

Iodosobenzene diacetate in the presence of base converts cyclohexanone into α-hydroxycyclohexanone in 80% yield. A related reaction is the conversion of iso-phorone into the enol of 3,5,5-trimethyl-1,2-cyclohexanedione (equation III). This

(III)

reaction does not involve the epoxy ketone, known to be stable under the conditions used.

α-Hydroxylation of ketones.[2] Aryl methyl ketones are converted by iodoso-benzene and sodium hydroxide (1 equivalent each) in methanol into the acyloin via the dimethyl ketal.

Examples:

$$C_6H_5COCH_3 \xrightarrow[\text{CH}_3\text{OH, NaOH}]{\text{C}_6\text{H}_5\text{IO,}} C_6H_5C-CH_2OH \xrightarrow[\text{60\% overall}]{H^+} C_6H_5CCH_2OH$$

with OCH_3 / OCH_3 on the central carbon.

$$C_6H_5COCH_2CH_3 \xrightarrow[75\%]{} C_6H_5COCHCH_3$$

with OH on the CH.

[1] R. M. Moriarty, S. C. Gupta, H. Hu, D. R. Berenschot, and K. B. White, *Am. Soc.*, **103**, 686 (1981).
[2] R. M. Moriarty, H. Hu, and S. C. Gupta, *Tetrahedron Letters*, **22**, 1283 (1981).

Iodosobenzene diacetate, 1, 508–509; **3**, 166; **4**, 266.

Oxidation of a 1,3,4,6-tetraketone (1).[1] Oxidation of oxalyldiacetone (**1**) with iodosobenzene diacetate results in **2**. Oxidation of **1** with lead tetraacetate gives dehydroacetic acid (**3**), previously obtained by dimerization of diketene.

[1] M. Poje, *Tetrahedron Letters*, **21**, 1575 (1980).

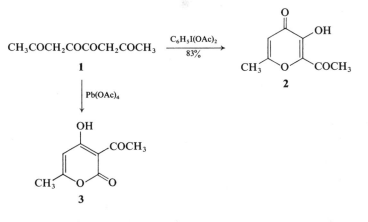

Iodosobenzene dichloride, 1, 505–506; **2**, 225–226; **3**, 164–165; **4**, 264–265; **5**, 352–353; **6**, 298–300; **9**, 249–250.

Chlorination of steroids (**4**, 264–265; **5**, 352–353; **6**, 298–299). Breslow's method for chlorination of steroids at C_9[1] has been developed into a useful method for conversion of 17α-hydroxyprogesterone (**1**) into the triene (**4**), a useful intermediate to highly active corticoids.[2] The hindered 17α-hydroxyl group is esterified with *m*-iodobenzoic anhydride and 4-dimethylaminopyridine. Irradiation of the ester (**2**) with $C_6H_5ICl_2$ in the presence of NaOAc (HCl scavenger) affords **3** in quantitative

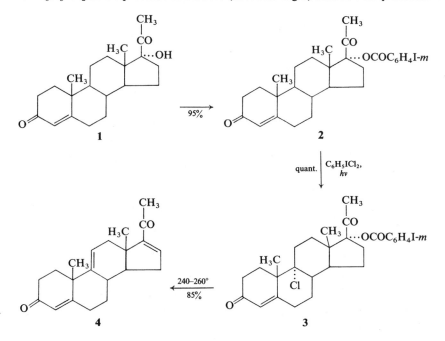

yield. Elimination of HCl and *m*-iodobenzoic acid to give **4** is effected by thermolysis at 240–260°.

[1] R. Breslow, R. J. Corcoran, B. B. Snider, R. J. Doll, P. L. Khanna, and R. Kaleya, *Am. Soc.*, **99**, 905 (1977).
[2] U. Kerb, M. Stahnke, P.-E. Schulze, and R. Wieckert, *Angew. Chem. Int. Ed.*, **20**, 88 (1981).

N-Iodosuccinimide. 1, 510–511.

Glycol cleavage.[1] Glycols are cleaved in high yields by NIS in THF. Irradiation increases the rate.

[1] T. R. Beebe, P. Hii, and P. Reinking, *J. Org.*, **46**, 1927 (1981).

N-Iodosuccinimide–Tetra-*n*-butylammonium iodide.

Oxidation of alcohols.[1] This combination is an efficient oxidant for primary or secondary alcohols in CH_2Cl_2 at 25°. Yields are generally >90%.
    Examples:

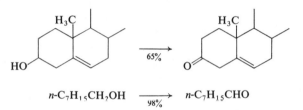

$$n\text{-}C_7H_{15}CH_2OH \xrightarrow{98\%} n\text{-}C_7H_{15}CHO$$

[1] S. Hanessian, D. H. Wong, and M. Therien, *Synthesis*, 394 (1981).

Iodotrimethylsilane, 8, 261–263; 9, 251–256.

Homoallylic ethers.[1] Iodotrimethylsilane catalyzes the allylation of acetals and ketals by allylsilanes, with transposition of the allylic group (equation I). It does not catalyze allylation of aldehydes and ketones. Note that $TiCl_4$ can catalyze both of these reactions (**7**, 370–371). In this respect, $ISi(CH_3)_3$ resembles $(CH_3)_3SiOTf$ (this volume).

$$(I) \quad (CH_3)_3SiCH_2CH{=}CH_2 + \underset{R^2}{\overset{R^1}{{>}}}C(OR^3)_2 \longrightarrow CH_2{=}CHCH_2\underset{OR^3}{\overset{R^1}{\underset{|}{\overset{|}{C}}}}{-}R^2$$

Cleavage of epoxides. Three laboratories have reported cleavage of epoxides with iodo- or bromotrimethylsilane and triethylamine to give (2-haloalkoxy)trimethylsilanes in high yield. Terminal epoxides are cleaved mainly to 1-halo-2-trimethylsiloxyalkanes. The products can be dehydrohalogenated (DBU) to give allylic alcohols[2,3] or oxidized (CrO_3, H_2SO_4)[4] to α-halo ketones. The overall reaction can be conducted as a one-pot operation.

Examples:

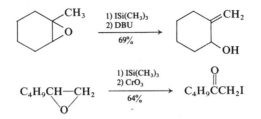

Friedel–Crafts alkylation. Reaction of phenylacetaldehyde (**1a**) with iodotrimethylsilane effects an intramolecular *ortho*-alkylation to give the cyclic ether **2a**.[5]

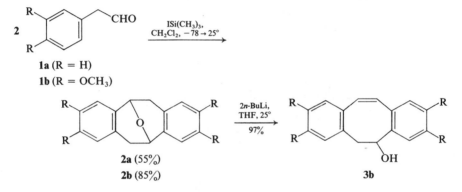

1a (R = H)
1b (R = OCH₃)

2a (55%)
2b (85%)

3b

Even though the reagent usually cleaves methyl ethers, the corresponding ether **2b** can be obtained from **1b** in high yield. The ether is stable to acids, but an E2 elimination to give the alcohol **3b** is possible with base. The alcohol is an obvious precursor to a dibenzocyclooctadiene by dehydration, best effected by thermolysis

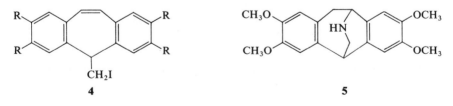

4

5

of the corresponding N-*p*-tosylcarbamate (prepared with TsNCO). Treatment of the alcohol (**3**) with HI results in ring contraction to give an iodomethyldibenzocyclo-heptatriene (**4**) in high yield. This reaction has been used for a total synthesis of isopavine (**5**) in four steps from **1b** in 53% overall yield.[6]

α-Iodo sulfides.[7] These rather inaccessible compounds can be obtained by reaction of $ISi(CH_3)_3$ with O-trimethylsilylhemithioacetals or O-trimethylsilylhemi-ketals (equation I). On treatment with base $[N(C_6H_5)_3$ or $NaOH]$ under phase-transfer conditions α-iodo sulfides are converted into vinyl sulfides by elimination of HI.

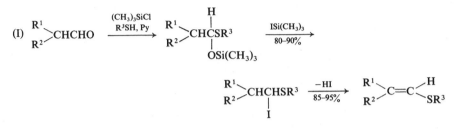

(I)

Cleavage of cyclopropyl ketones.[8] Cyclopropyl ketones are converted by $ISi(CH_3)_3$ under very mild conditions into γ-iodo ketones. The regioselectivity is usually high and is similar to that observed in lithium–ammonia reductions.

Examples:

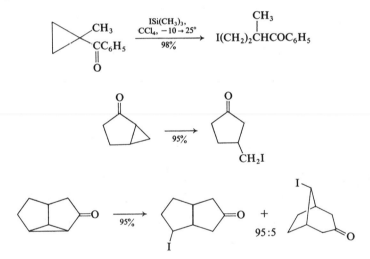

β-Iodo ketones.[9] Cyclobutanones are cleaved by $ISi(CH_3)_3$ in the presence of ZnI_2 regioselectively to β-iodo ketones.

Example:

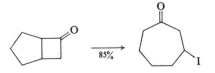

[1] H. Sakurai, K. Sasaki, and A. Hosomi, *Tetrahedron Letters*, **22**, 745 (1981).
[2] G. A. Kraus and K. Frazier, *J. Org.*, **45**, 2579 (1980).
[3] H. Sakurai, K. Sasaki, and A. Hosomi, *Tetrahedron Letters*, **21**, 2329 (1980).
[4] J. N. Denis and A. Krief, *Tetrahedron Letters*, **22**, 1429 (1981).
[5] M. E. Jung, A. B. Mossman, M. A. Lyster, *J. Org.*, **43**, 3698 (1978).
[6] M. E. Jung and S. J. Miller, *Am. Soc.*, **103**, 1984 (1981).
[7] T. Aida, D. N. Harpp, and T. H. Chan, *Tetrahedron Letters*, **21**, 3247 (1980).
[8] R. D. Miller and D. R. McKean, *J. Org.*, **46**, 2412 (1981).
[9] R. D. Miller and D. R. McKean, *Tetrahedron Letters*, **21**, 2639 (1980).

1-Iodo-4-(trimethylstannyl)butane, $I(CH_2)_4Sn(CH_3)_3$ **(1).**   Mol.   wt.   346.81.
Preposition.[1]

Intramolecular conjugate addition; annelation of cyclohexenones.[2] A new method of effecting carbocyclization involves conjugate addition to cyclohexenones mediated by an alkyltin(IV) reagent. An example is formulated in equation (I).

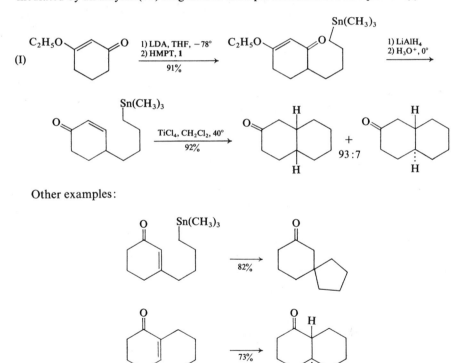

Other examples:

An allylic alcohol unit can also be used to initiate the carbocyclization; this variation can be used for the synthesis of Δ^1-octalins and Δ^4-hydrindenes.[3]

Example:

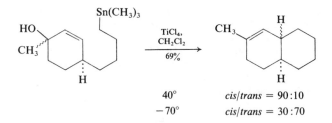

40°   *cis/trans* = 90:10
−70°   *cis/trans* = 30:70

¹ E. J. Bulten, H. F. M. Gruter, H. F. Martens, *J. Organometal. Chem.*, **117**, 329 (1976).
² T. L. Macdonald and S. Mahalingam, *Am. Soc.*, **102**, 2113 (1980).
³ T. L. Macdonald and S. Mahalingam, *Tetrahedron Letters*, **22**, 2077 (1981).

Iodotriphenylphosphonium iodide, $(C_6H_5)_3\overset{+}{P}I\ I^-$. The reagent is formed *in situ* by addition of iodine to triphenylphosphine in benzene.

Reduction of arenesulfonic acids to arenethiols.¹ The reagent reduces arenesulfononic acids to thiols in high yield. In actual practice, iodine can be used in catalytic amounts. The usual stoichiometry employed is $ArSO_3H/I_2/P(C_6H_5)_3 = 2:1:10$.

¹ K. Fujimori, H. Togo, and S. Oae, *Tetrahedron Letters*, **21**, 4921 (1980).

*Mixed acid–base ion-exchange resins.*¹ Rexyn ion-exchange resins (Fisher Scientific Co.) consist of sulfonic acid beads and quaternary ammonium hydroxide beads. Stowell and Hauck¹ have used such a resin to convert the keto acetal **1** in one operation to 2-methyl-2-cyclopentenone-1 (**2**). Two steps are involved: acid-catalyzed hydrolysis of the acetal group and base-catalyzed aldol condensation. The first step is reversible, but the second is irreversible.

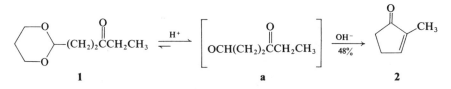

1   **a**   **2**

*Solid phase-transfer reactions.*² A strong base ion-exchange resin (Duolite A-109, Diamond Shamrock) can serve as a solid phase-transfer catalyst for synthesis of β-lactams (**3**) from α-methyl-α,β-dibromopropionyl chloride (**1**) and an amino acid (**2**) (*cf.* **9**, 360).

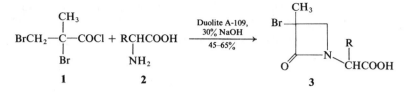

1   **2**   **3**

[1] J. C. Stowell and H. F. Hauck, Jr., *J. Org.*, **46**, 2428 (1981).
[2] T. Okawara, Y. Noguchi, T. Matsuda, and M. Furukawa, *Chem. Letters*, 185 (1981).

Iron carbonyl, 1, 519; **3**, 167; **5**, 357–358; **6**, 304–305; **7**, 183; **8**, 265–267.

Deoxygenation of alcohols. Alcohols, particularly tertiary ones that can form stable carbanions, are deoxygenated by reaction with potassium and iron carbonyl in toluene. The reduction proceeds through the potassium alkoxide.[1]

Examples:

$$(C_6H_5)_3COH \xrightarrow[90\%]{K, Fe(CO)_5} (C_6H_5)_3CH$$

$$CH_3\overset{\underset{\displaystyle C_6H_5}{|}}{\underset{\underset{\displaystyle C_6H_5}{|}}{C}}OH \longrightarrow CH_3CH(C_6H_5)_2 + CH_2\!=\!C(C_6H_5)_2$$
$$(54\%) \qquad\qquad (26\%)$$

Naphthoquinones. Benzocylobutenedione (**1**) when irradiated with iron carbonyl undergoes an insertion reaction to form the iron complex **2** in high yield.[2] This

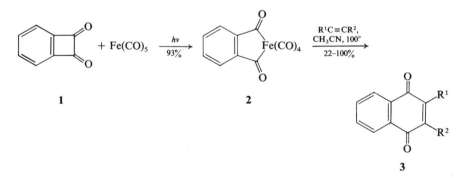

1   **2**   **3**

complex reacts at 100° with a wide variety of alkynes to give naphthoquinones, usually in yields >70%. A cobalt metallacycle similar to **2** also undergoes this reaction, but in this case AgBF$_4$ is required. It is useful for reactions with hindered alkynes, but in general yields are lower than those obtained using **2**.[3]

(E,E)-1,3-Dienes. Recent stereospecific syntheses of insect pheromones with (E)- and (E,E)-1,3-diene skeletons are based on the fact that diene–iron tricarbonyl

complexes (**1**) are locked into the (E)-configuration.[4] An example is the synthesis of (E)-9,11-dodecadienyl acetate **4** from tricarbonyl(butadiene)iron (**1**) by Friedel–Crafts acylation to give **2**. The keto ester is reduced by the mixed hydride to **3**. The

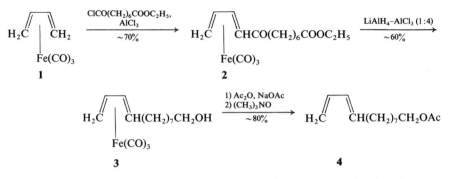

synthesis is completed by acetylation and cleavage of the iron tricarbonyl group with trimethylamine oxide (**6**, 624–625). Cleavage of **3** with this reagent effects oxidation of the primary hydroxyl group to the aldehyde and the carboxylic acid.[5]

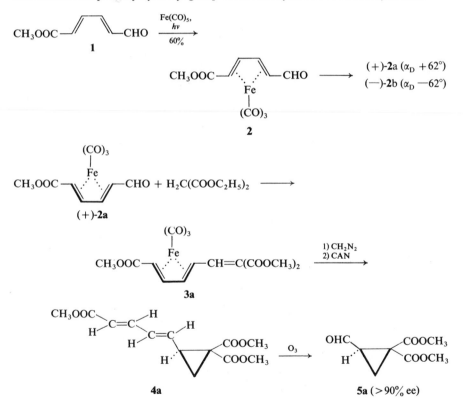

Chiral cyclopropanes. Carrié *et al.*[6,7] have developed a highly enantioselective synthesis of cyclopropanes from the aldehyde **2**, in which the butadiene group is protected as the iron tricarbonyl complex. The complex (**2**) is resolved by the method of Kelly and Van Rheenan (**5**, 289–290), and the two optical isomers are then converted separately into a cyclopropanealdehyde (**5a** and **5b**) as formulated. A sulfur ylide such as $(CH_3)_2S=CHCOOCH_3$ can be used in place of diazomethane for cyclopropanation. Optical yields are $>90\%$.

Allyl to propenyl isomerization.[8] The isomerization of estragole (**1**) to anethole (**2**) is conducted conveniently with iron carbonyl at 140° for 8 hours.

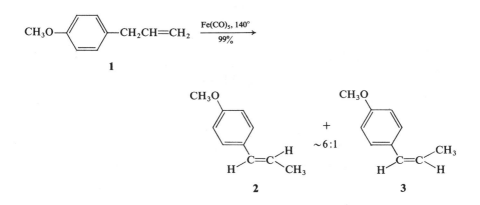

[1] H. Alper and M. Salisova, *Tetrahedron Letters*, **21**, 801 (1980).

[2] L. S. Liebeskind, S. L. Baysdon, M. S. South, and J. F. Blount, *J. Organometal. Chem.*, **202**, C73 (1980).

[3] L. S. Liebeskind, S. L. Baysdon, and M. S. South, *Am. Soc.*, **102**, 7397 (1980).

[4] B. F. Hallam and P. L. Pauson, *J. Chem. Soc.*, 642 (1958).

[5] G. R. Knox and I. G. Thom, *J.C.S. Chem. Comm.*, 373 (1981).

[6] J. Martelli, R. Grée, and R. Carrié, *Tetrahedron Letters*, **21**, 1953 (1980).

[7] A. Monpert, J. Martelli, R. Grée, and R. Carrié, *Tetrahedron Letters*, **22**, 1961 (1981).

[8] R. J. De Pasquale, *Syn. Comm.*, **10**, 225 (1980).

Iron(III) sulfate, $Fe_2(SO_4)_3$.

1,4-Addition of water to quinones.[1] The yields of the known Michael addition of water or an alcohol to a 1,4-naphthoquinone such as 5,8-dimethoxy-1,4-naphthoquinone are improved by addition of an oxidant such as $Fe_2(SO_4)_3$ to convert the initial 2-hydroxynaphthohydroquinone to the corresponding quinone. The presence of a free *peri*-hydroxy group interferes with the reaction.

Example:

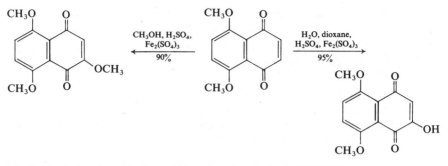

[1] F. Fariña, R. Martinex-Utrilla, and M. C. Paredes, *Synthesis*, 300 (1981).

Isopinocampheylborane, 8, 267.

Asymmetric hydroboration.[1] Hydroboration of phenyl-substituted trisubstituted alkenes, cyclic or acyclic, followed by oxidation results in alcohols with an optical purity of 80–100%, with the (S)-configuration at the hydroxylated carbon predominating with reagent prepared from (+)-α-pinene.

Examples:

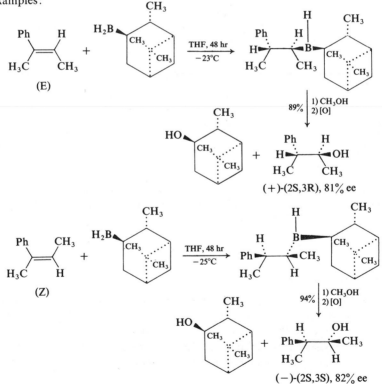

[1] A. K. Mandal, P. K. Jadhav, and H. C. Brown, *J. Org.*, **45**, 3543 (1980).

Isopropenyltriphenylphosphonium bromide, $CH_2{=}C{\Large\langle}\genfrac{}{}{0pt}{}{CH_3}{\overset{+}{P}(C_6H_5)_3}Br^-$ **(1).** Mol. wt. 383.26,

m.p. 197.5°. Preparation.[1]

Bicycloannelation.[2] The α'-enolate of an α,β-cyclohexenone reacts with this phosphonium salt to form a tricyclo[3.2.1.0^{2,7}]octane in low to moderate yield. This reaction was used in a short synthesis of the pentacyclic diterpene trachyloban-19-oic acid (**4**). Reaction of the lithium enolate of **2**, prepared from podocarpic acid, with **1** provided the pentacyclic ketone **3**, which was reduced by the Wolff-Kishner reaction to **4**.

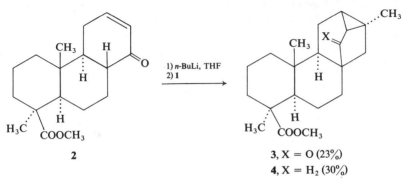

2

3, X = O (23%)
4, X = H$_2$ (30%)

[1] E. E. Schweizer, A. T. Wehman, and D. M. Mycz, *J. Org.*, **38**, 1583 (1973).
[2] R. M. Cory, D. M. T. Chan, Y. M. A. Naquib, M. H. Rastall, and R. M. Renneboog, *J. Org.*, **45**, 1852 (1980).

K

Ketene diethyl acetal, 9, 262–264.

Cyclobutenes.[1] This derivative of ketene undergoes [2 + 2]cycloaddition with ethyl propiolate in refluxing methylene chloride to produce the cyclobutene **1** in 65% yield. The ester group activates **1** sufficiently for Diels–Alder addition with the silyl enol ether **2** to give the 1 : 1 adduct **3** under mild conditions. Hydrolysis of **3** can be effected with base or fluoride ion as usual and also with 3 Å molecular sieves (25°). The product **4** is a key intermediate in a synthesis of (±)-illudol (**5**), a natural sesquiterpene.

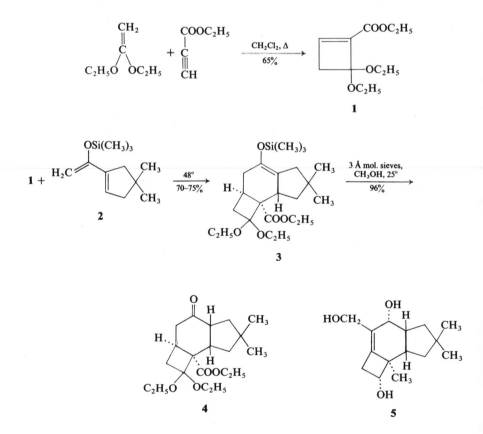

[1] M. F. Semmelhack, S. Tomoda, and K. M. Hurst, *Am. Soc.*, **102**, 7567 (1980).

Ketene dimethyl thioacetal monoxide, 6, 311–312.

Convenient preparation[1]:

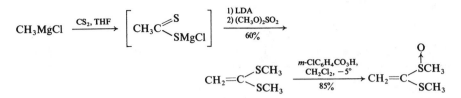

[1] R. Kaya and N. R. Beller, *J. Org.*, **46**, 196 (1981).

L

Lead tetraacetate, 1, 537–563; **2**, 234–238; **3**, 168–171; **4**, 278–282; **5**, 365–370; **6**, 313–317; **7**, 185–188; **8**, 269–272; **9**, 265–269.

Oxidative lactonization of unsaturated diacids **(9,** 265).[1] Further study of this reaction indicates that it is general and can be controlled to result in *cis*-addition of two carboxylic oxygens to the double bond. Two experimental conditions are satisfactory: (1) treatment of the diacid in $CHCl_3$ or CH_3CN with a large excess of $Pb(OAc)_4$ and (2) treatment of the tetra-*n*-butylammonium salt of the diacid in CH_3CN with 6–15 equivalents of $Pb(OAc)_4$. Yields by the latter procedure are generally higher.

Examples:

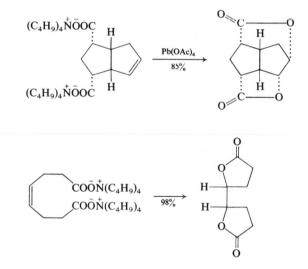

[1] E. J. Corey and A. W. Gross, *Tetrahedron Letters*, **21**, 1819 (1980).

Lead tetraacetate–Boron trifluoride etherate.

Oxidation of enamines.[1] Oxidation of enamines of cyclic ketones in the presence of BF_3 etherate results in a Favorski type rearrangement to esters of contracted cycloalkanoic acids. A related reaction also occurs with enamines of aryl methyl ketones.

Examples:

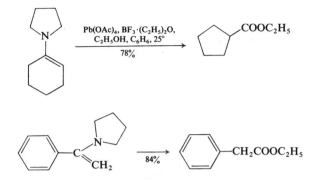

Methyl arylacetates.[2] Aryl methyl ketones are converted into methyl arylacetates by reaction with BF_3 etherate and lead tetraacetate in benzene at room temperature (equation I). Thallium(III) nitrate (**4**, 496) has also been used for this modified Willgerodt-Kindler reaction.

$$(\text{I}) \quad \text{ArCCH}_3 \quad \xrightarrow[\substack{80\text{–}95\%}]{\substack{BF_3 \cdot (C_2H_5)_2O, \\ Pb(OAc)_4, \ CH_3OH}} \quad \text{ArCH}_2\text{COOCH}_3$$

[1] Ž. Čeković, J. Bošnjak, and M. Cvetkovic, *Tetrahedron Letters*, **21**, 2675 (1980).
[2] B. Myrboh, H. Ila, and H. Junjappa, *Synthesis*, 126 (1981).

(S)-(+)-*t*-Leucine *t*-butyl ester, 7, 189; 8, 272–273. B.p. 87–89°/18 mm; α_D + 1.68°. Preparation.[1]

Asymmetric alkylation of α,β-unsaturated cyclic aldehydes.[2] Grignard reagents undergo asymmetric conjugate addition to the chiral aldimines (**2**) of cyclic α,β-unsaturated aldehydes prepared with this α-amino acid, (S)-**1**, to give after hydrolysis the *trans*-2-substituted aldehyde (**4**) in 80–90% optical yield. The intermediate enamine (**a**) can be alkylated also with high stereoselectivity to give the 1,2-disubstituted aldehyde **3**. On the other hand, the aldehyde **4** after metalation is alkylated to give the 1,2-disubstituted aldehyde **5**. The high optical yields of **3** and **4** (80–90%) are believed to result from a fixed conformation owing to chelation in the adduct of **2** with the Grignard reagent and in the magnesioenamine **a**, which directs attack of the nucleophile from the less hindered side. As expected on the basis of this thesis, valine *t*-butyl ester is a less effective chiral auxiliary reagent. The alkylation of the anion of **4** to give **5** is evidently controlled by the steric effect of the R group already present.

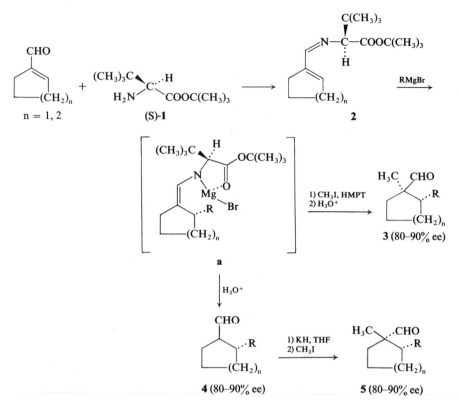

[1] S. Hashimoto, S. Yamada, and K. Koga, *Chem. Pharm. Bull. Japan*, **27**, 771 (1979).
[2] S. Hashimoto, H. Kogen, K. Tomioka, and K. Koga, *Tetrahedron Letters*, 3009 (1979); H. Kogen, K. Tomioka, S. Hashimoto, and K. Koga, *Tetrahedron Letters*, **21**, 4005 (1980); H. Kogen, K. Tomioka, S. Hashimoto, and K. Koga, *Tetrahedron*, **37**, 3951 (1981).

Levulinic acid, $CH_3COCH_2CH_2COOH$, **1**, 564; **2**, 239.

Levulinic esters. Levulinic esters of sugars can also be prepared by Mukaiyama's method for esterification with 2-chloro-1-methylpyridinium iodide (**8**, 95–96). Deblocking of the ester group is possible with hydrazine hydrate (1 equivalent, 20°).[1]

The 4,4-ethylenedithio derivative of levulinic acid selectively esterifies primary hydroxy groups of sugars under the same conditions.[2] Both protective groups are stable to conditions of Koenigs–Knorr glycosidation [$Hg(CN)_2$ and $HgBr_2$].[3] The ethylenedithio group is removed by $HgCl_2$ and HgO in aqueous acetone to give the levulinoyl ester. A branched tetrasaccharide composed of glucose and galactose units has been synthesized by combined use of these two groups.[2]

[1] H. J. Koeners, J. Verhoeven, and J. H. van Boom, *Tetrahedron Letters*, **21**, 381 (1980).
[2] H. J. Koeners, C. H. M. Verdegaal, and J. H. van Boom, *Rec. Trav.*, **100**, 118 (1981).
[3] H. M. Flowers, *Methods Carbohydrate Chem.*, **6**, 474 (1972).

1-Lithiocyclopropyl phenyl sulfide, 5, 372–373; **6**, 319–320; **7**, 190. A new synthesis is outlined in equation (I). The method provides a general route to cyclopropyl phenyl sulfides.[1]

(I) $Br(CH_2)_3Br$ $\xrightarrow[93\%]{NaSC_6H_5}$ $C_6H_5S(CH_2)_3SC_6H_5$ $\xrightarrow[\sim 75\%]{\substack{2n\text{-BuLi,}\\ THF, 0°}}$

[1] K. Tanaka, H. Uneme, and S. Matsui, *Chem. Letters*, 287 (1980).

2-Lithio-1,3-dithianes, 6, 248.

α-Diketones. Aromatic nitrile oxides (**2**) react with 2-lithio-1,3-dithianes (**1**) to give the ketoxime (**3**) of a 1,2-diketone (equation I). The ketoximes are not

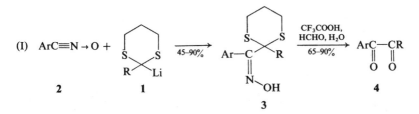

(I) $ArC≡N \rightarrow O$ +

2   **1**   **3**   **4**

hydrolyzed by the usual reagents, but are hydrolyzed by trifluoroacetic acid and aqueous formaldehyde.[1]

[1] T. Yamamori and I. Adachi, *Tetrahedron Letters*, **21**, 1747 (1980).

Lithiomethyl isocyanide, $LiCH_2N≡C$, **4**, 272.

Reaction with lactones.[1] Under controlled conditions and in DMF as solvent lithiomethyl isocyanide reacts with lactones to form alcoholic oxazoles (equation I).

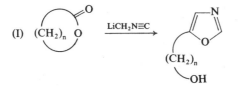

(I) $(CH_2)_n$ $\xrightarrow{LiCH_2N≡C}$

This reaction forms the basis for novel synthesis of natural furane derivatives such as evodone (**1**).

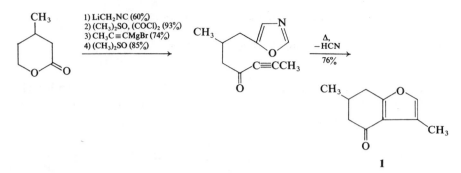

Benzylisoquinolines.[2] A short synthesis of the benzylisoquinoline (**4**) involves reaction of lithiomethyl isocyanide with 2 equivalents of piperonal (**1**) to give, after acetylation, the oxazoline **2**, formed via a 2-lithiooxazoline. The remaining steps are outlined in equation (I).

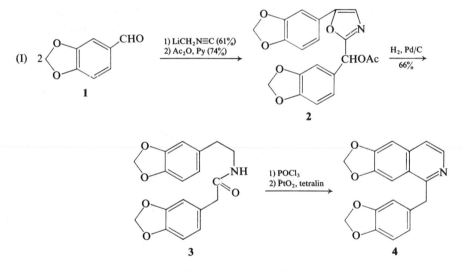

¹ P. A. Jacobi, D. G. Walker, and I. M. A. Odeh, *J. Org.*, **46**, 2065 (1981).
² A. P. Kozikowski and A. Ames, *J. Org.*, **45**, 2548 (1980).

α-Lithiomethylselenocyclobutane (1).

Preparation:

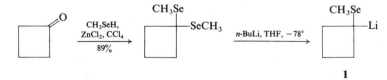

1-Substituted cyclobutenes; 2-substituted 1,3-dienes. The reagent reacts with various electrophiles (alkyl halides, carbonyl compounds, epoxides) to give the corresponding selenides, β-hydroxy- and γ-hydroxyselenides, respectively. Three methods can be used to convert these adducts to cyclobutenes, as shown in equations (I)–(III).[1,2]

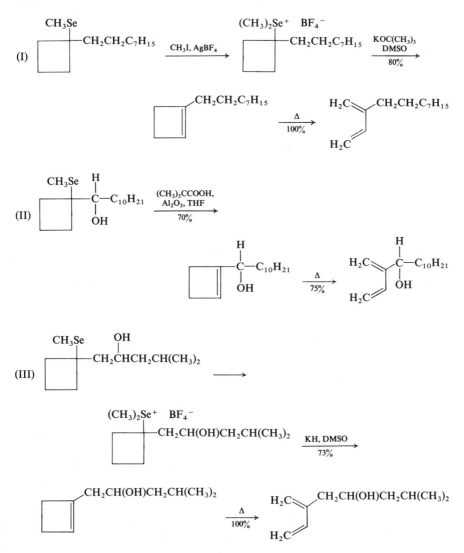

[1] S. Halazy and A. Krief, *Tetrahedron Letters*, **21**, 1997 (1980).
[2] A. Krief, *Tetrahedron*, **36**, 2531 (1980).

1-Lithio-2,4,6-trimethylbenzene 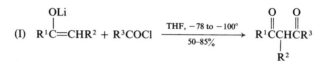 **(1)**. Mol. wt. 126.13. The reagent

is prepared by treatment of 1-bromo-2,4,6-trimethylbenzene with *t*-butyllithium (2 equivalents).

Ketone enolates.[1] The kinetic enolate of unsymmetrical ketones can be generated with **1**. The advantage over LDA is that the resulting enolate solution is free of amine.

Use of **1** was explored in a study of the reaction of enolates with acyl chlorides as a route to 1,3-diketones (equation I).

$$\text{(I)} \quad R^1\overset{\overset{\displaystyle OLi}{|}}{C}{=}CHR^2 + R^3COCl \xrightarrow[50-85\%]{\text{THF, } -78 \text{ to } -100°} R^1\overset{\overset{\displaystyle O}{\|}}{C}\underset{\underset{\displaystyle R^2}{|}}{C}H\overset{\overset{\displaystyle O}{\|}}{C}R^3$$

[1] D. Seebach, T. Weller, G. Protschuk, A. K. Beck, and M. S. Hoekstra, *Helv.*, **64**, 716 (1981).

Lithium, 1, 570–573; **3**, 174–175; **4**, 286–287; **5**, 376–377.

Deoxygenation of epoxides. Finely dispersed lithium in refluxing THF converts epoxides into the corresponding alkenes, usually with retention of configuration.[1]

[1] K. N. Gurudutt and B. Ravindranath, *Tetrahedron Letters*, **21**, 1173 (1980).

Lithium–Ammonia, 1, 601–603; **2**, 205; **3**, 179–182; **4**, 288–290; **5**, 379–381; **6**, 322–323; **7**, 195; **8**, 282–284; **9**, 273–274.

Reductive alkylation. Stork and Logusch[1] have extended the reductive alkylation of α,β-enones (**3**, 179–181) to enediones (equation I). One useful feature is

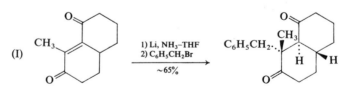

that alkylation leads mainly to the equatorial product, particularly with sterically hindered reagents. Even use of CH_3I results in the two possible products in a ratio of 85:15, again in favor of the equatorial isomer.

This reaction was used for a synthesis of the corticosteroid adrenosterone (**1**), the latter stages of which are shown in equation (II).[2]

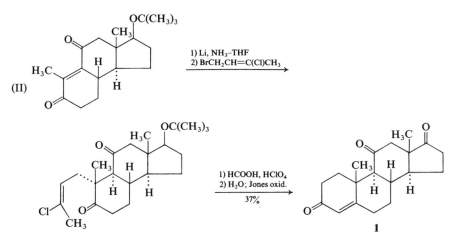

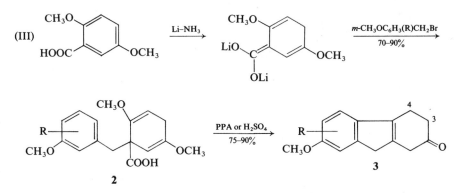

A new synthesis of 3,4-dihydrofluorene-2(1H)-ones (**3**) involves reductive alkylation of 2,5-dimethoxybenzoic acid and cyclization of the resulting acid (**2**) in PPA or 85% H_2SO_4 (equation III). If the carboxyl group is methylated, cyclization

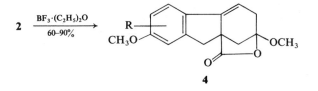

takes place without decarboxylation. Alternatively the acid can be treated with BF_3 etherate in CH_2Cl_2 to give lactones (**4**).[3]

Reductive Michael reactions.[4] The anion obtained on Birch reduction of benzoic acids undergoes Michael reactions with methyl crotonate or acrylate.

Example:

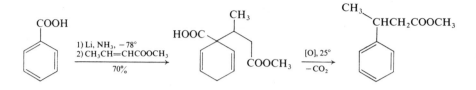

[1] G. Stork and E. W. Logusch, *Am. Soc.*, **102**, 1218 (1980).
[2] G. Stork and E. W. Logusch, *Am. Soc.*, **102**, 1219 (1980).
[3] J. M. Hook and L. N. Mander, *J. Org.*, **45**, 1722 (1980).
[4] G. S. R. Subba Rao, H. Ramanathan, and K. Raj, *J.C.S. Chem. Comm.*, 315 (1980).

Lithium–Ethylamine, 1, 574–581; **2**, 241–242; **3**, 175; **4**, 287–288; **5**, 377–379; **6**, 322; **7**, 194–195; **8**, 284–285.

Reductive cleavage of 5,6-dihydropyranes. The allylic ether bond of 5,6-dihydro-pyranes is cleaved reductively by lithium in ethylamine.[1]

Example:

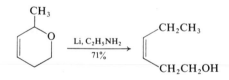

[1] T. Kobayashi and H. Tsurata, *Synthesis*, 492 (1980).

Lithium aluminum hydride, 1, 581–595; **2**, 242; **3**, 176–177; **4**, 291–293; **5**, 382–389; **6**, 325–326; **7**, 196; **8**, 286–289; **9**, 274–277.

Influence of solvents. The effectiveness of $LiAlH_4$ for reduction of alkyl iodides and bromides varies considerably with the solvent. The order of solvent effect is diglyme > monoglyme > THF ≫ ether. In contrast the solvent effect on rate of reduction of tosylates is ether > THF > monoglyme > diglyme. Thus reduction of tosylates by $LiAlH_4$ can be carried out in ether without reduction of a halide substituent.[1]

γ-Amino alcohols.[2,3] Stereoselectivity of reduction of 3,5-disubstituted 2-isoxa-zolines (**1**) to γ-amino alcohols (**2**) is higher with $LiAlH_4$ than with Na(Hg)–C_2H_5OH, Na–C_2H_5OH, or borane–dimethylsulfide. Sodium cyanoborohydride reduces 3 to a 2:1 mixture of *cis*-**4** and *trans*-**4**.

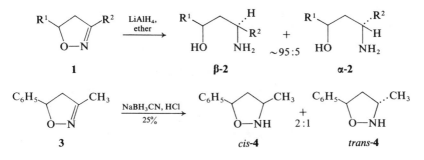

Stereoselectivity is observed in the LiAlH$_4$ reduction of other substituted 2-isoxazolines.

Examples:

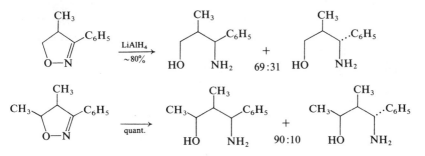

[1] S. Krishnamurthy, *J. Org.*, **45**, 2550 (1980).
[2] V. Jäger and V. Buss, *Ann.*, 101 (1980).
[3] V. Jäger, V. Buss, and W. Schwab, *Ann.*, 122 (1980).

Lithium aluminum hydride–Copper(I) iodide.

Conjugate reduction of enones.[1] α,β-Unsaturated ketones and aldehydes undergo 1,4-reduction in generally high yield with 1 equivalent of lithium aluminum hydride in the presence of 10 mole % of CuI and 1 equivalent of HMPT at −78°. The active agent presumably is LiHCuI. CuI can be replaced by mesitylcopper and copper(I) *t*-butoxide.

[1] T. Tsuda, T. Fujii, K. Kawasaki, and T. Saegusa, *J.C.S. Chem. Comm.*, 1013 (1980).

Lithium aluminum hydride–Dichlorobis(cyclopentadienyl)zirconium.

Hydroalumination of allylic alcohols or ethers.[1] Hydroalumination of these substrates under usual conditions (**9**, 276) proceeds poorly because of deoxygenation associated with titanium catalysts.[2] The desired reaction, however, can be effected in 50–80% yield with ZrCl$_4$ in the case of alcohols and Cp$_2$ZrCl$_2$ in the case of ethers (equations I and II).

(I)   RCHCH=CH$_2$ $\xrightarrow[50–80\%]{\substack{\text{1) LiAlH}_4,\ \text{ZrCl}_4 \\ \text{2) H}_2\text{O}}}$ RCHCH$_2$CH$_3$
      | |
      OH   OH

(II) ROCH$_2$CH=CH$_2$ $\xrightarrow[70–80\%]{\substack{\text{1) LiAlH}_4,\ \text{Cp}_2\text{ZrCl}_2 \\ \text{2) H}_2\text{O}}}$ ROCH$_2$CH$_2$CH$_3$

[1] F. Sato, Y. Tamuro, H. Ishikawa, and M. Sato, *Chem. Letters*, 99 (1980).
[2] F. Sato, Y. Tamuro, H. Ishikawa, and M. Sato, *Chem. Letters*, 103 (1980).

Lithium aluminum hydride–(–)-N-Methylephedrine.

Asymmetric reduction of α,β-enones. This combination of reagents (1:1) in conjunction with N-ethylaniline (2 equivalents) reduces alkyl aryl ketones to alcohols with high stereoselectivity.[1] Under these conditions α,β-unsaturated ketones are reduced to optically active (S)-allylic alcohols. Optical yields of 80–98% have been reported for open-chain enones. Reduction of cyclic enones is somewhat less efficient.[1] The method was used to reduce **1** to **2**, which has been used as an intermediate in an anthracyclinone synthesis.[2]

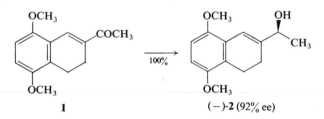

1   (–)-**2** (92% ee)

[1] S. Terashima, N. Tanno, and K. Koga, *J.C.S. Chem. Comm.*, 1026 (1980).
[2] S. Terashima, N. Tanno, and K. Koga, *Chem. Letters*, 981 (1980); *Tetrahedron Letters*, **21**, 2749, 2753 (1980).

Lithium benzeneselenolate, LiSeC$_6$H$_5$.

The selenolate is prepared from LiC$_6$H$_5$ and Se.[1]

Phenyl selenides.[2] In the presence of a ruthenium catalyst, prepared by reaction of RuCl$_3$ with 3 equivalents of potassium, lithium benzeneselenolate reacts with a wide variety of tertiary amines to form phenyl selenides in excellent yield. Anhydrous diglyme is used as solvent.

Examples:

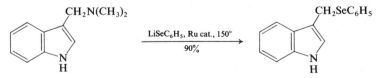

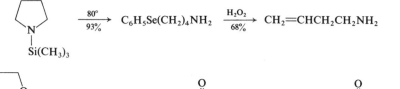

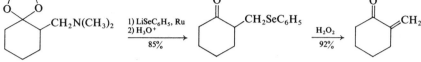

[1] M. Sato and T. Yoshida, *J. Organometal. Chem.*, **67**, 395 (1974).
[2] S.-I. Murahashi and T. Yano, *Am. Soc.*, **102**, 2456 (1980).

Lithium bis(ethylenedioxyboryl)methide (1), 6, 328–329.

Aldehyde homologation (**6**, 328–329). The method utilizing **1** has two drawbacks. One is that the immediate precursor, tris(ethylenedioxyboryl)methane, requires a tedious preparation. The other is the formation of potentially explosive aldehyde peroxides. This risk is decreased by replacing hydrogen peroxide in the oxidation step with sodium perborate, $NaBO_3$, an inexpensive and stable oxidant. Moreover, this substance has less tendency than H_2O_2 to cleave C—C bonds. In fact Matteson and Moody[1] recommend the routine use for peroxide oxidation of organoboranes.

[1] D. S. Matteson and R. J. Moody, *J. Org.*, **45**, 1091 (1980).

Lithium *t*-butylbis(2-methylpropyl)aluminate, $Li[(i-C_4H_9)_2Al(t-C_4H_9)H]$ (1). Mol. wt. 206.26.

The lithium trialkylaluminate is prepared by reaction of diisobutyl-aluminum hydride with *t*-butyllithium. Related hydrides can be prepared with *n*-butyllithium and *sec*-butyllithium.

Stereoselective reduction of ketones.[1] This hydride (**1**) reduces *t*-butylcyclohexanone to about equal amounts of the corresponding axial and equatorial alcohols. Trost *et al.*[2] used **1** to reduce the 3-keto group of **2** to the 3α-alcohol. Other methods were less selective.

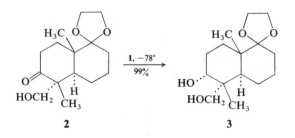

Chemoselective reduction of an ester.[3] The ester group of **2** is reduced in high yield by 3 equivalents of **1** (1 equivalent to neutralize the acid group and 2 equivalents for reduction of the ester).

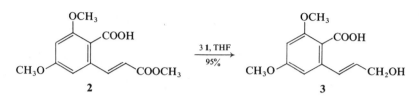

2   **3**

[1] G. Kovács, G. Galambos, and Z. Junanez, *Synthesis*, 171 (1977).
[2] B. M. Trost, Y. Nishimura, and K. Yamamoto, *Am. Soc.*, **101**, 1328 (1979).
[3] B. M. Trost, G. T. Rivers, and J. M. Gold, *J. Org.*, **45**, 1835 (1980).

Lithium chloride–Hexamethylphosphoric triamide.

Spiroannelation. A new method for intramolecular spirocyclization involves decarboxylation of ω-halogeno-β-keto esters with lithium chloride in HMPT at 125–140°. The method appears to be fairly general.[1]

Examples:

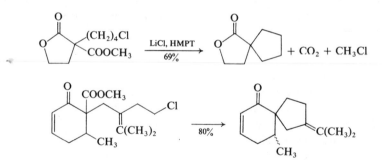

[1] R. G. Eilerman and B. J. Willis, *J.C.S. Chem. Comm.*, 30 (1981).

Lithium 4,4′-di-*t*-butylbiphenyl 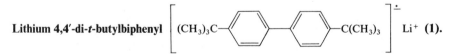 Li⁺ **(1).**

Alkyllithium reagents.[1] This aromatic radical anion (**1**) is particularly effective for conversion of alkyl halides into alkyllithium reagents. These products are formed in 93–95% yield from reactions conducted in THF at −78°.

[1] P. K. Freeman and L. L. Hutchinson, *J. Org.*, **45**, 1924 (1980).

Lithium diisopropylamide (LDA), **1**, 611; **2**, 249; **3**, 184–185; **4**, 298–302; **5**, 400–406; **6**, 334–339; **7**, 204–207; **8**, 292; **9**, 280–283. This base can be prepared in molar quantities by reaction of styrene with 2 equivalents each of granular lithium and diisopropylamine in ether (95% yield).[1]

$$C_6H_5CH{=}CH_2 + 2Li + 2HN[CH(CH_3)_2]_2 \longrightarrow C_6H_5CH_2CH_3 + 2LiN[CH(CH_3)_2]_2$$

***Deprotonation of β-lactones.*[2]** Deprotonation of the 2-oxetanone (**1**) with LDA in THF at −78° generates **2**, which is stable below room temperature. The anion reacts with electrophiles with high stereoselectivity at the *trans*-position at C_3. The paper discusses reasons for the unexpected stability of **2**.

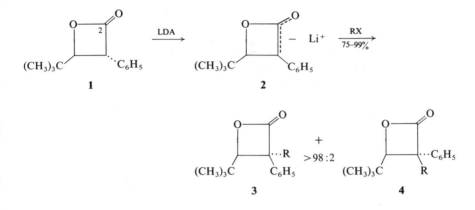

***α-Alkylation of β-hydroxy carboxylic esters* (8,** 258). Dianions of β-hydroxy carboxylic esters (LDA, −50° to −20°) are alkylated stereoselectively to give mainly *threo*-compounds (equation I). The alkylation of dianions of α-mono-

$$(I)\quad CH_3\overset{OH}{\underset{:}{C}}HCH_2COOC_2H_5 \xrightarrow[\substack{80\%}]{\substack{1)\,2LDA,\,THF\\2)\,CH_2{=}CHCH_2Br,\,HMPT}} CH_3\overset{OH}{\underset{:}{C}}H\underset{\underset{CH_2CH{=}CH_2}{\big\uparrow}}{C}HCOOC_2H_5 + \substack{erythro\text{-}\\isomer}$$

$$(96\%)\qquad\qquad (4\%)$$

substituted β-hydroxy esters is also stereoselective (equation II).[3]

This reaction has been extended to an enantioselective synthesis of 4,4- and 6,6-disubstituted cyclohexenones (equation III). Thus (2S,3S)-**2**, prepared by the above procedure, is converted in two steps to the 1,5-diketone **3**. The aldol ring closure of **3** results in either **4** or **5**, depending on the conditions; each product is obtained in 86% optical yield.[4]

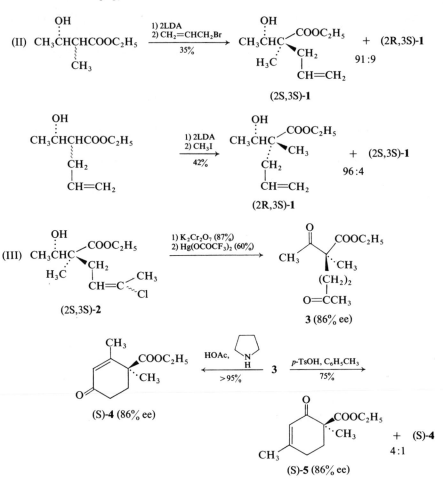

A similar regioselectivity under different conditions for aldol cyclization of 1,5-diketones obtained by the Michael addition of substituted acetoacetic esters to methyl vinyl ketones has been reported by another laboratory (equation IV).[5]

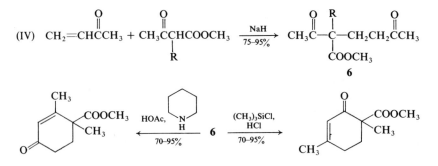

α,β-Enones; acylation of alkenes.[6] 2-Isoxazolines (**1**) are cleaved by LDA in about 60% yield to oximes (**2**) of α,β-enones. Less useful reagents for this reaction

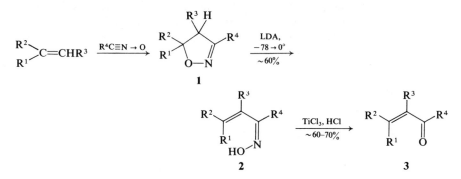

1

2   **3**

are C_2H_5MgBr ($\sim 30\%$ yields) and *n*-butyllithium ($\sim 45\%$ yields). The products (**2**) retain the C=N configuration of **1**. They are cleaved by $TiCl_3$ (**4**, 506–507) to α,β-enones (**3**) in about 60–70% yield. Since the isoxazolines are obtained by dipolar addition of nitrile oxides and alkenes,[7] the sequence constitutes overall acylation of alkenes.

1-Alkynes from methyl ketones.[8] This reaction can be effected by conversion to the enol phosphate followed by β-elimination with LDA (equation I). In the case of a simple ketone such as 2-octanone the yield is low because of formation also of an allene. In such cases lithium tetramethylpiperidide is recommended as base.

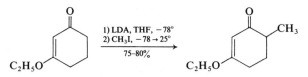

Stork–Danheiser alkylation (**5**, 403–404). An example of this useful kinetic alkylation of 3-alkoxy-2-cyclohexenones to give 6-alkyl derivatives has been submitted to *Organic Synthesis.*[9]

[1] M. T. Reetz and W. F. Maier, *Ann.*, 1471 (1980).
[2] J. Mulzer and T. Kerkmann, *Am. Soc.*, **102**, 3620 (1980).
[3] G. Fráter, *Helv.*, **62**, 2825, 2829 (1979).
[4] G. Fráter, *Tetrahedron Letters*, **22**, 425 (1981).
[5] W. Kreiser and P. Below, *Tetrahedron Letters*, **22**, 429 (1981).
[6] H. Grund and V. Jäger, *Ann.*, 80 (1980).
[7] R. Huisgen, *Angew. Chem. Int. Ed.*, **2**, 565, 573 (1963).
[8] E. Negishi, A. O. King, W. L. Klima, W. Patterson, and A. Silveira, Jr., *J. Org.*, **45**, 2526 (1980).
[9] A. S. Kende and P. Fludzinski, *Org. Syn.*, submitted (1980).

Lithium dimethylaminoethoxide, $(CH_3)_2NCH_2CH_2OLi$ **(1).** The alkoxide is generated from 2-(dimethylamino)ethanol with CH_3Li.

α-Alkylated ketones.[1] α-Alkylated ketones are obtained regiospecifically by alkylation of enol boranes in the presence of this additive (equation I). Unsymmetrical acyclic enol boranes can be synthesized by reaction of a trialkyl-borane with an α-diazoketone.

(I) CH_3COCHN_2 $\xrightarrow{(n-C_4H_9)_3B}$ $n\text{-}C_4H_9CH{=}CCH_3$ (with $OB(C_4H_9)_2$ on the central carbon) $\xrightarrow[50-90\%]{1,\ RX}$ $n\text{-}C_4H_9CHCCH_3$ (with R below and $=O$ above)

[1] J. Hooz and J. Oudenes, *Syn. Comm.*, **10**, 139 (1980).

Lithium 1-(dimethylamino)naphthalenide, 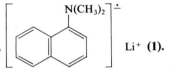 Li^+ **(1).**

Lithium naphthalenide has been used for reductive lithiation of thioketals (**8**, 306; **9**, 284), but has the disadvantage that naphthalene is sometimes difficult to scparatc from final products of alkylation. In such cases, lithium 1-(dimethylamino)-naphthalenide can be used advantageously since dimethylaminonaphthalene is removed from reaction mixtures by extraction with dilute acid.[1]

α-Lithio ethers.[2] These ethers can be obtained by reductive lithiation of α-phenylthio ethers with this reagent (**1**) at -63 to $-78°$. The reaction is considered to involve cleavage of the C—S bond to form a carbon radical, which is then reduced to the carbanion. The second example shows the use of this reaction for synthesis of brevicomin, the aggregation pheromone of the female Western Pine beetle.

Examples:

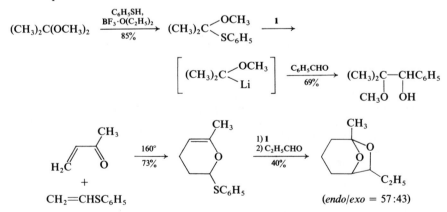

$(endo/exo = 57:43)$

α-Trialkylsilyl ketones.[3] The reagent selectively cleaves the C—Se bond of silyl enol ethers of α-phenylseleno ketones to give, after rearrangement, α-trialkylsilyl ketones. In the case of acyclic ketones, alkynes are obtained as by-products.

Examples:

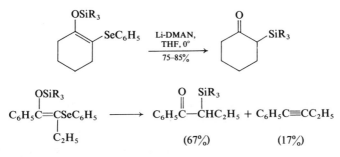

$$C_6H_5C=CSeC_6H_5 \longrightarrow C_6H_5\overset{O}{\overset{\|}{C}}-\overset{SiR_3}{\overset{|}{C}}HC_2H_5 + C_6H_5C\equiv CC_2H_5$$

with $\overset{|}{C_2H_5}$ on the left structure

$$(67\%) \qquad (17\%)$$

[1] T. Cohen and J. R. Matz, *Syn. Comm.*, **10**, 311 (1980).
[2] T. Cohen and J. R. Matz, *Am. Soc.*, **102**, 6900 (1980).
[3] T. Kametani, H. Matsumoto, T. Honda, and K. Fukumoto, *Tetrahedron Letters*, **22**, 2379 (1981).

Lithium L-α,α′-dimethyldibenzylamide, 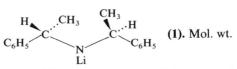 **(1).** Mol. wt. 231.25. The corresponding amine, b.p. 103–105°/0.3 mm., α_D −172.7°, is prepared by catalytic hydrogenation of N-methylbenzylidene-α-methylbenzylamine.[1]

Enantioselective deprotonation.[2] The rearrangement of epoxides to allylic alcohols by lithium dialkylamides involves removal of the proton *syn* to the oxygen.[3] When a chiral lithium amide is used with cyclohexene oxide, the optical yield of the resulting allylic alcohol is 3–31%, the highest yield being obtained with **1**.

[1] C. G. Overberger, N. P. Marullo, and R. G. Hiskey, *Am. Soc.*, **83**, 1374 (1961).
[2] J. K. Whitesell and S. W. Felman *J. Org.*, **45**, 755 (1980).
[3] R. P. Thummel and B. Rickborn, *Am. Soc.*, **92**, 2064 (1970).

Lithium iodide, 5, 410–411; **7**, 208; **9**, 283.

Aldol condensation. Anhydrous lithium iodide (*ca.* 5 equivalents) promotes aldol condensation of ketones with enolizable or nonenolizable aldehydes. The intermediate aldol is usually not isolable, but can be intercepted by addition of $ClSi(CH_3)_3$ and $N(C_2H_5)_3$. In this case LiI can be used in a catalytic amount. The salt cannot be replaced by LiBr or LiCl or NaI.

In this procedure acyclic methyl ketones condense almost exclusively at the methyl group. Enolizable α,β-enals also undergo efficient cross condensation with methyl ketones.[1]

Example:

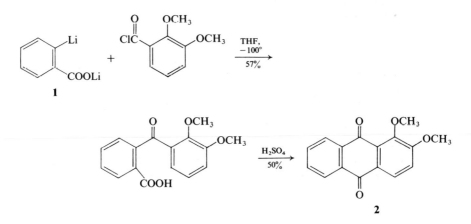

$$C_6H_5CHO + CH_3\overset{O}{\overset{\|}{C}}CH_2CH_3 \xrightarrow[74\%]{\substack{LiI, \\ ether}}$$

[1] R. G. Kelleher, M. A. McKervey, and P. Vibuljan, *J.C.S. Chem. Comm.*, 486 (1980).

Lithium *o*-lithiobenzoate (1). The reagent is prepared by treatment of *o*-bromo-benzoic acid with *n*-butyllithium in THF at $-100°$.[1]

o-Benzoylbenzoic acids.[2] The reaction **1** with aroyl chlorides affords a useful route to *o*-benzoylbenzoic acids, which can be successful in cases where the usual approaches fail. An example is a new synthesis of alizarin dimethyl ether (**2**).

[1] W. E. Parham and Y. A. Sayed, *J. Org.*, **39**, 2051 (1974).
[2] W. E. Parham, C. K. Bradsher, and K. J. Edgar, *J. Org.*, **46**, 1057 (1981).

Lithium methoxy(trimethylsilyl)methylide, $(CH_3)_3Si\overset{Li}{\overset{|}{C}}HOCH_3$ **(1), 9,** 284.

Aldehyde synthesis. The reagent has been used for synthesis of the sesquiterpene aldehydes warburganal (**5**) and isotadeonal (**6**) from the protected keto aldehyde (**2**) via **4**. The ketone reacts only in low yield with a number of other reagents such as $(C_6H_5)_3P{=}CH_2$ and $(C_2H_5O)_2POCH_2OCH_3$.[1]

[1] A. S. Kende and T. J. Blacklock, *Tetrahedron Letters*, **21**, 3119 (1980).

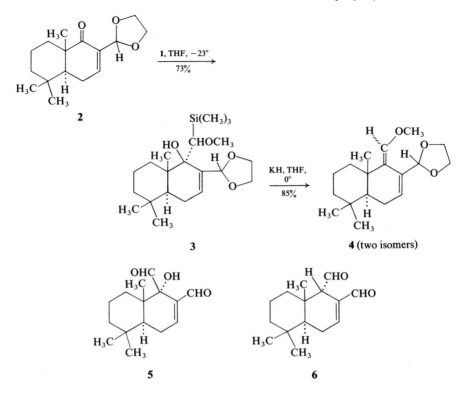

2

3

4 (two isomers)

5

6

Lithium naphthalenide, **2**, 288–289; **3**, 208; **4**, 348–349; **5**, 468; **6**, 415; **8**, 305–306; **9**, 285.

Lithiated nitriles. This base is somewhat more effective than LDA or lithium di-ethylamide for deprotonation of the allylic nitrile **1**, used in a synthesis of **2**, which has the ferulol skeleton.[1]

1

2

[1] K. Takabe, S. Ohkawa, and T. Katagiri, *Synthesis*, 358 (1981).

Lithium phenylethynolate, 6, 343.

β-Lactams.[1] Lithium phenylethynolate (**1**) reacts with imines (**2**) to give β-lactams (**3**) presumably via **a**.

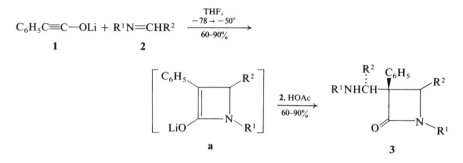

$$C_6H_5C\equiv C-OLi + R^1N=CHR^2 \xrightarrow[\substack{60-90\%}]{\substack{THF, \\ -78 \to -50°}}$$

1   2

[1] R. M. Adlington, A. G. M. Barrett, P. Quayle, A. Walker, and M. J. Betts, *J.C.S. Chem. Comm.*, 404 (1981).

Lithium tetrafluoroborate, $LiBF_4$. Mol. wt. 93.75. Supplier: Alfa.

Cleavage of t-butyldimethylsilyl ethers.[1] The silyl ethers of primary and secondary steroidal alcohols are converted to the alcohols by $LiBF_4$ in $CH_3CN–CH_2Cl_2$ (yields 70–85%).

[1] B. W. Metcalf, J. P. Burkhart, and K. Jund, *Tetrahedron Letters*, **21**, 35 (1980).

Lithium tri-*t*-butoxyaluminum hydride, 1, 620–625; **2**, 251–252; **3**, 188; **4**, 312.

Selective reduction of ketones.[1] Selective reduction of a ketone in the presence of an aldehyde is possible by selective reaction of the aldehyde with *t*-butylamine (4 Å molecular sieves), reduction of the ketone with lithium tri-*t*-butoxyaluminum hydride, and, finally, cleavage of the aldimine to the aldehyde with aqueous HCl. The three-step sequence can be conducted in one pot. The selectivity is high. The method is not applicable to conjugated ketones because of conjugate reduction.

[1] M. P. Paradisi, G. P. Zecchini, and G. Ortar, *Tetrahedron Letters*, **21**, 5085 (1980).

Lithium tri-*sec*-butylborohydride, 4, 312–313; **6**, 348; **7**, 307; **8**, 308–309.

Reduction of uracil derivatives. Uracil itself is not reduced by this hydride in THF, DMF, or NH_3. However N_1,N_3-dialkyluracils are reduced in THF in 70–95% yield, possibly because these derivatives are locked in the enone form. Reaction of these uracils with this hydride and then an electrophile results in 5-alkylated-5,6-dihydro derivatives of uracil (equation I).[1]

Benzyl and benzyloxymethyl ($CH_2OCH_2C_6H_5$) are useful protective groups for the reduction, since both are removed by use of BBr_3.[2]

Demethylation of quaternary ammonium salts.[3] Quaternary ammonium iodides are demethylated by treatment with this borohydride at 25–65° in about 70–90% yield. Demethylation is markedly favored over deethylation.

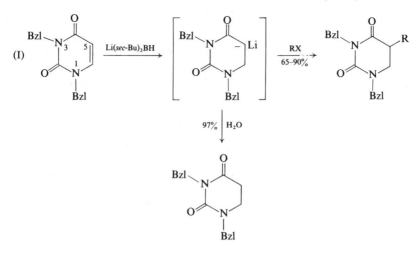

[1] S. J. Hannon, N. G. Kundu, R. P. Hertzberg, R. S. Bhatt, and C. Heidelberger, *Tetrahedron Letters*, **21**, 1105 (1980).

[2] N. G. Kundu, R. P. Hertzberg, and S. J. Hannon, *Tetrahedron Letters*, **21**, 1109 (1980).

[3] G. R. Newkome, V. K. Majestic, and J. D. Sauer, *Org. Prep. Proc. Int.*, **12**, 345 (1980).

Lithium triethylborohydride, 4, 313–314; **6**, 348–349; **7**, 215–216; **8**, 309–310; **9**, 286.

Protodehalogenation.[1] This hydride is the reagent of choice for hydrogenolysis of alkyl halides. Lithium aluminum hydride is somewhat less powerful, particularly for reduction of alkyl chlorides.

Reduction of tosylates. Reduction of the ditosylate **1** with $Li(C_2H_5)_3BH$ gives a mixture of **4** and **5** in the ratio 5.25:1. The reduction proceeds via the epoxides **2** and **3**.[2]

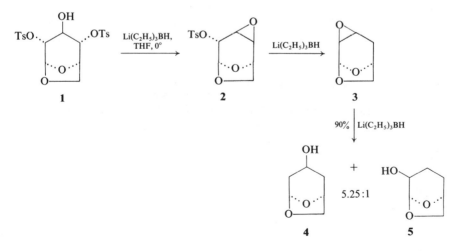

This reduction was useful in a synthesis of a thromboxane B_2 intermediate (**9**) from **6**, prepared in a few steps from a readily available carbohydrate levoglucosan. Hydride reduction of **6** to **7** is regiospecific because of the α-allyl group. RuO_2–$NaIO_4$ oxidation (**2**, 358–359) gives the tricyclic lactone **8**, which is cleaved by Amberlyst 15 in methanol to **9** and **10**. The α-isomer **9** has been converted to thromboxane B_2 (**11**) in two laboratories.[3]

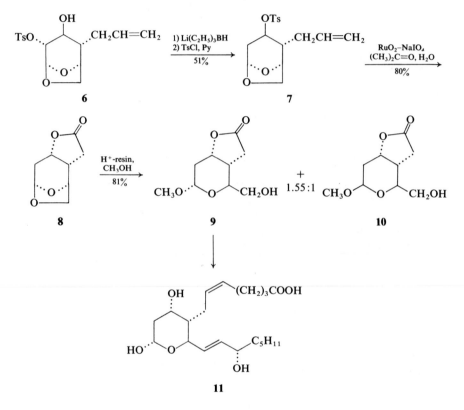

[1] S. Krishnamurthy and H. C. Brown, *J. Org.*, **45**, 849 (1980).
[2] A. G. Kelly and J. S. Roberts, *Carbohydrate Res.*, **77**, 231 (1979).
[3] A. G. Kelly and J. S. Roberts, *J.C.S. Chem. Comm.*, 228 (1980).

M

Magnesium, 1, 627–629; **2**, 254; **3**, 189; **4**, 315; **5**, 419; **6**, 351–352; **7**, 218.

Grignard reagents.[1] Details have been published for preparation of Riecke's highly reactive magnesium. The preparation of 1-norbornylmagnesium chloride is the featured example for use of this magnesium in the preparation of Grignard reagents (equation I). The paper includes examples of preparation of Grignard

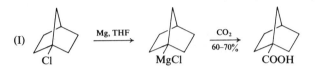

reagents from eight other unreactive halides, including di-Grignard reagents from allyl- and vinylmagnesium halides (*e.g.*, *p*-bromochlorobenzene).

[1] R. D. Reike, S. E. Bales, P. M. Hudnall, and G. S. Poindexter, *Org. Syn.*, **59**, 85 (1979).

Magnesium–Methanol, 6, 351.

Reductions. Magnesium–methanol has been used for selective reduction of α,β-unsaturated nitriles (**6**, 351) and amides[1] and of 1,1- and 1,2-diphenyl-substituted ethylenes.[2]

Addition of palladium on carbon to this system markedly enhances the reactivity and permits reduction of nonactivated carbon–carbon double and triple bonds. One advantage of this system over catalytic hydrogenation is that cyclopropyl groups and benzylic ethers and alcohols are not affected. Yields are generally higher than 80%.[3]

[3] R. Brettle and S. M. Shibib, *Tetrahedron Letters*, **21**, 2915 (1980); R. Brettle and S. M. Shibib, *J.C.S. Perkin* I, 2912 (1981).
[2] J. A. Profitt and H. H. Ong, *J. Org.*, **44**, 3972 (1979).
[3] G. A. Olah, G. K. Prakash, M. Arvanaghi, and M. R. Bruce, *Angew. Chem. Int. Ed.*, **20**, 92 (1981).

Magnesium chloride, $MgCl_2 \cdot 6H_2O$. Mol. wt. 203.30. Supplier: Alfa.

Deethoxycarbonylation of β-keto esters.[1] The β-keto ester **1** undergoes deethoxycarbonylation in only about 30% yield when heated with NaCl in DMSO. Chlorides of bivalent metals were examined and showed the following order of effectiveness for conversion of **1** to **2**: $MgCl_2 \cdot 6H_2O \approx CaCl_2 \cdot 2H_2O > BeCl_2 > BaCl_2 \cdot 2H_2O$, with only the last metal halide being less effective than NaCl.

Evidently the size of the cation is also important. Solvents were shown to be effective in the order HMPT > DMSO > DMF ≫ sulfolane.

The present method is also effective for simpler β-keto esters, but yields are not higher than those obtained by previous methods. Apparently the new method is particularly useful only for hindered β-keto esters.

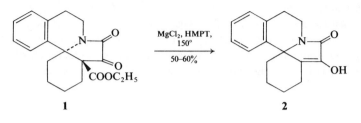

¹ Y. Tsuda and Y. Sakai, *Synthesis*, 119 (1981).

Meldrum's acid, 8, 313–314.

β-Keto esters (8, 313). Complete details are available for use of acyl derivatives of Meldrum's acid as precursors to β-keto esters.[1] A modification of this procedure has been used in a synthesis of thienamycin.[2] The carboxylic acid group is activated by reaction with carbonyldiimidazole, and 4-dimethylaminopyridine is used as catalyst for acylation of Meldrum's acid.

Methyl ketones.[3] Methyl ketones are available by acylation of Meldrum's acid (1) followed by hydrolysis with aqueous acetic acid (equation I). The method fails when R = C_6H_5 because of hydrolysis to benzoic acid.

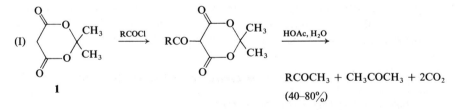

$$RCOCH_3 + CH_3COCH_3 + 2CO_2$$
$$(40–80\%)$$

¹ Y. Oikawa, T. Yoshioka, K. Sugano, and O. Yonemitsu, *Org. Syn.*, submitted (1981).
² D. G. Melillo, I. Shinkai, T. Liu, K. Ryan, and M. Sletzinger, *Tetrahedron Letters*, **21**, 2783 (1980).
³ T. A. Hase and K. Salonen, *Syn. Comm.*, **10**, 221 (1980).

Mercury(II) acetate, 1, 644–652; 2, 264–267; 3, 194–196; 4, 319–323; 5, 424–427; 6, 358–359; 7, 222–223; 8, 315–316; 9, 291.

Hydroxymercuration (2, 265–267; 3, 194; 4, 319–320; 5, 425–427; 6, 359). The reaction of $Hg(OAc)_2$ with simple olefins is somewhat accelerated in the presence of an anionic surfactant such as sodium lauryl sulfate (SLS), but is severely retarded in the presence of a cationic surfactant (hexadecylammonium bromide). Of greater interest, an anionic surfactant can markedly influence selective monohydroxy-mercuration of nonconjugated, unsymmetrical dienes in rigid systems. An example is the reaction of limonene (**1**) with 1 equivalent of Hg(OAc)₂ under the usual

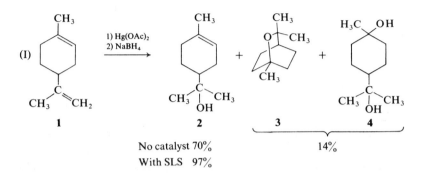

No catalyst 70%
With SLS   97%

conditions and in the presence of SLS (equation I). A similar selectivity can be achieved with the diene **5**. In the catalyzed reaction, the mono-ol **6** is obtained in 90% yield. In the noncatalyzed reaction **6** is obtained in 20–25% yield, together with

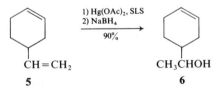

seven other products. However, surfactant catalysis exerts no effect on the mercura-tion of the flexible, linear 1,7-octadiene. Surfactant catalysis can also effect difunctionalization of unsymmetrical dienes. Thus reaction of limonene (**1**) with 2 equivalents of Hg(OAc)₂ gives the ether **3** and the diol **4** in the ratio 1:3. In the presence of SLS, **3** and **4** are formed in the ratio 8:1.[1]

[1] C. M. Link, D. K. Jansen, and C. H. Sukenik, *Am. Soc.*, **102**, 7798 (1980).

Mercury(II) acetate–Sodium trimethoxyborohydride.

Reductive coupling of alkenes and 1,3-dienes. In the presence of Hg(OAc)₂ or Hg(OAc)₂–HgO (1:1) and NaBH(OCH₃)₃ 1,3-dienes undergo reductive coupling in methanol with alkenes substituted with electron-withdrawing groups (CN, COOCH₃, COCH₃) to form the methyl ether of allyl alcohols.[1]

Example:

$$CH_2=CHCH=CH_2 + H_2C=CHCOOCH_3 \xrightarrow[34\%]{\begin{array}{l}1)\ Hg(OAc)_2,\ CH_3OH \\ 2)\ NaBH(OCH_3)_3\end{array}}$$

$$CH_2=CHCHCH_2CH_2CH_2COOCH_3$$
$$\overset{|}{O}CH_3$$

1 B. Giese, K. Heuck, and U. Lüning, *Tetrahedron Letters*, **22**, 2155 (1981).

Mercury(II) nitrate, 7, 223.

Glycosidation.[1] Facile glycosidation can be achieved by use of a heteroatom (*e.g.*, sulfur) in the anomeric position coupled by activation with a metal ion. For example, the pyridinyl l-thio-β-D-glucopyranoside (**1**) reacts within a few minutes with various alcohols in the presence of $AgONO_2$ or $Hg(ONO_2)_2$[2] to form alkyl D-glucopyranosides (**2**). The reaction is slow in the absence of the metal salt even when acid catalyzed. Unfortunately the reaction is not stereoselective, and formation of the alkyl α-D-glucopyranoside by an S_N2 reaction is only slightly favored. Yields of disaccharides prepared in this way are low ($\sim 35\%$).

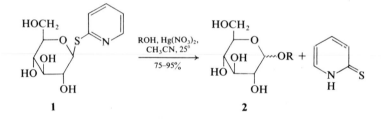

1   **2**

[1] S. Hanessian, C. Bacquet, and N. Lehong, *Carbohydrate Res.*, **80**, C17 (1980).
[2] Commercial $Hg(NO_3)_2 \cdot H_2O$ is heated to 55–60° at 10^{-2} torr.

Mercury(II) oxide–Tetrafluoroboric acid, 9, 293.

Oxidation of allylbenzene.[1] Oxidation of allylbenzene with this reagent and an alcohol results exclusively in a *trans*-cinnamyl ether (equation I).

(I) $C_6H_5CH_2CH=CH_2 + ROH + HgO \cdot 2HBF_4 \xrightarrow{THF}$

$$[C_6H_5CH_2\underset{\underset{OR}{|}}{CH}-CH_2HgBF_4] \xrightarrow[66-80\%]{\begin{array}{l}-HBF_4, \\ -Hg(O)\end{array}} \underset{C_6H_5}{\overset{H}{\diagdown}}C=C\underset{H}{\overset{CH_2OR}{\diagup}}$$

Oxyamination.[2] The reagent has been used to effect hydroxyamination of alkenes with aryl amines (equation I). The rearranged isomer **1** is the major product with terminal alkenes ($R^3 = R^4 = H$).

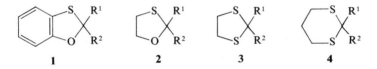

Hydrolysis of cyclic dithioacetals and hemithioacetals.[3] This reaction can be effected in good yield with HgO and 35% aqueous HBF$_4$ in THF in 5–10 minutes. This method is usually more effective than use of HgO–BF$_3$ etherate (**3**, 136). It is useful for hydrolysis of compounds 1–4 to give the corresponding aldehydes or ketones in 80–100% yield.

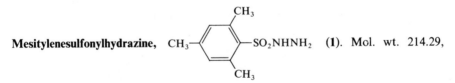

[1] J. Barluenga, L. Alonso-Cires, and G. Asensio, *Tetrahedron Letters*, **22**, 2239 (1981).
[2] J. Barluenga, L. Alonso-Cires, and G. Asensio, *Synthesis*, 376 (1981).
[3] I. Degani, R. Fochi, and V. Regondi, *Synthesis*, 51 (1981).

Mesitylenesulfonylhydrazine, CH$_3$⟨mesityl⟩SO$_2$NHNH$_2$ (**1**). Mol. wt. 214.29,

m.p. 115–116°. Preparation.[1]

Eschenmoser fragmentation (**2**, 419–422). This hydrazine can be more effective than tosylhydrazine for this fragmentation.[2] Thus reaction of **1** with the α-keto epoxide **2** results in formation of 1-cyclononyn-5-one (**3**) in about 43% yield. Another example is the preparation of 2,2-dimethyl-5-hexynal (**5**) from **4**. These reactions proceed in about 15–20% yield with tosylhydrazine.

[1] N. T. Cusack, C. B. Reese, A. C. Risiu, and B. Boozpeikar, *Tetrahedron*, **32**, 2157 (1976).
[2] C. B. Reese and H. P. Sanders, *Synthesis*, 276 (1981).

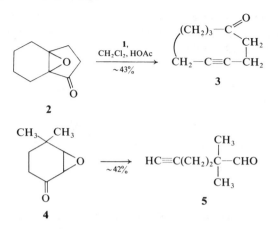

O-Mesitylenesulfonylhydroxylamine (MSH), **5**, 430–433; **6**, 320; **8**, 319.

1H-Aziridines.[1] In the presence of BF_3 etherate, this reagent adds to double bonds substituted with three cyano or carboalkoxy groups to give an adduct that is converted into a $1H$-aziridine by treatment with triethylamine.

Example:

$$\underset{CN}{\overset{H}{}}C=C(COOC_2H_5)_2 \quad \xrightarrow[\substack{2)\ N(C_2H_5)_3 \\ 90\%}]{1)\ MSH,\ BF_3\cdot(C_2H_5)_2O} \quad$$

[structure of aziridine product: CN, COOC₂H₅, COOC₂H₅ on N–H ring]

[1] P. Mitra and J. Hamelin, *J.C.S. Chem. Comm.*, 1038 (1980).

Methanesulfonic acid, 1, 666–667; **2**, 270; **4**, 326.

Cyclopentenones and cyclohexenones. In the presence of CH_3SO_3H or $CH_3SO_3H-P_2O_5$ (**5**, 535) a 2-vinylcyclobutanone such as **1** rearranges by a 1,3-acetyl migration to the fused cyclohexenone **2**. A spiro-2-vinylcyclobutanone such as **3** undergoes 1,2-rearrangement to the fused cyclopentenone (**4**).[1]

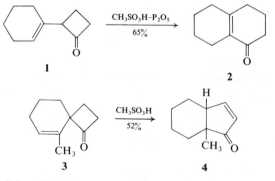

[1] J. R. Matz and T. Cohen, *Tetrahedron Letters*, **22**, 2459 (1981).

Methoxyallene, 7, 225–226; **8**, 320–322.

Methoxycyclopropanes. The reaction of a trialkylborane with 1-lithio-1-methoxyallene forms an ate complex. Reaction of the complex with acetic acid followed by oxidation leads to a methoxycyclopropane as one product (equation I).[1]

(I) $CH_2=C=C$⟨OCH_3 / Li⟩ $\xrightarrow{R_3B}$ $CH_2=C=C$⟨OCH_3 / B^-R_3⟩ Li^+ $\xrightarrow[\text{2) H}_2\text{O}_2,\text{ OH}^-]{\text{1) HOAc}}$ (cyclopropane with R, OCH₃)

In general, yields are low unless a B-alkyl-9-BBN is used. Thus when B-hexyl-9-BBN is used, 1-hexyl-1-methoxycyclopropane is formed in 81% yield.

[1] N. Miyaura, T. Yoshinari, M. Itoh, and A. Suzuki, *Tetrahedron Letters*, **21**, 537 (1980).

(E)-1-Methoxy-1,3-bis(trimethylsilyloxy)-1,3-butadiene, $CH_2=C-C=C$ with $(CH_3)_3SiO$, H, OCH_3, $OSi(CH_3)_3$

(1). Mol. wt. 260.48, b.p. 56–58°/2.0 mm. The diene can be obtained by reaction of the enolate of methyl 3-trimethylsiloxy-2-butenoate (*n*-BuLi, TMEDA, THF) with chlorotrimethylsilane (91–98% yield).[1] An alternate preparation has been published.[2]

Cycloaromatization.[1] The diene can be used as the equivalent of the dianion of methyl acetoacetate for a novel synthesis of methyl salicylates. In the presence of titanium(IV) chloride (1 equivalent) **1** reacts with β-dicarbonyl equivalents to form methyl salicylates. Various reactions of **1** indicate that C_4 is more nucleophilic than C_2. The examples formulated in equations (I) and (II) show that the reaction is

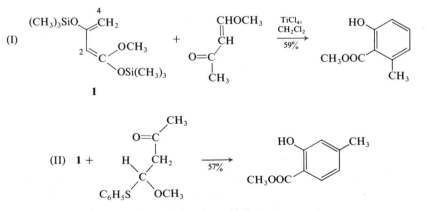

regioselective. In equation (I) the conjugate position of the enone is the more electrophilic site. In equation (II) the ketone group is more reactive as an electrophile than the monothioacetal group. The paper reports seven other examples of this cycloaromatization reaction.

[1] T. H. Chan and P. Brownbridge, *J.C.S. Chem. Comm.*, 578 (1979); *Am. Soc.*, **102**, 3534 (1980).
[2] K. Yamamoto, S. Suzuki, and J. Tsuji, *Chem. Letters*, 649 (1978).

1-Methoxy-1,3-butadiene, CH_2=CHCH=CHOCH$_3$ **(1).** Mol. wt. 84.12, b.p. 91°.

Diels-Alder reactions. Diels-Alder reactions between this diene and carbonyl compounds are possible under high pressure (15–25 kbar) and provide a route to 5,6-dihydro-2H-pyranes (2).[1]

Example:

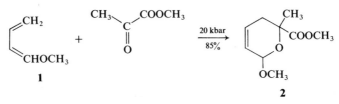

Use of an aldehyde as the dienophile provides a general route to δ-lactones (equation I).[2]

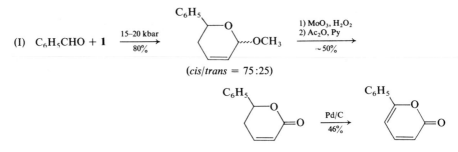

(I) C_6H_5CHO + 1 $\xrightarrow[80\%]{15-20\ kbar}$

(*cis/trans* = 75:25)

[1] J. Jurczak, M. Chmielewski, and S. Filipek, *Synthesis*, 41 (1981).
[2] M. Chmielewski and J. Jurczak, *J. Org.*, **46**, 2230 (1981).

4-Methoxy-3-butene-2-one, $CH_3\overset{\overset{O}{\|}}{C}CH$=CHOCH$_3$. Mol. wt. 100.12, b.p. 200°.

γ-Pyrones. Two laboratories have published a new route to γ-pyrones by condensation of acid chlorides with the enolate of 4-methoxy-3-butene-2-one. One group[1] used the lithium enolate and a temperature of −78°, which resulted in the enol form of the pyrone (equation I). The other group[2] used 2 equivalents of the potassium enolate and obtained the γ-pyrone directly by a hydrolytic step at room temperature (equation II).

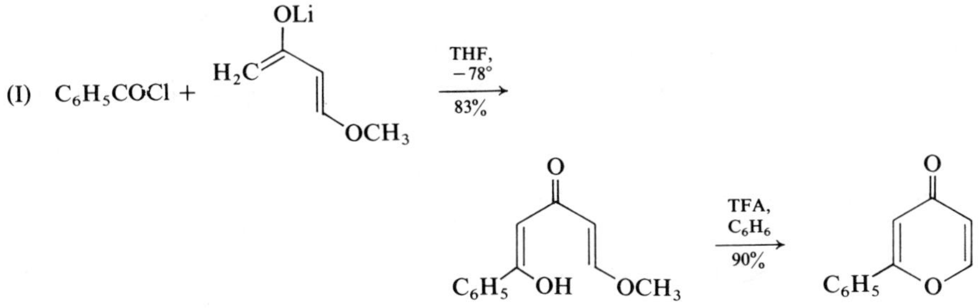

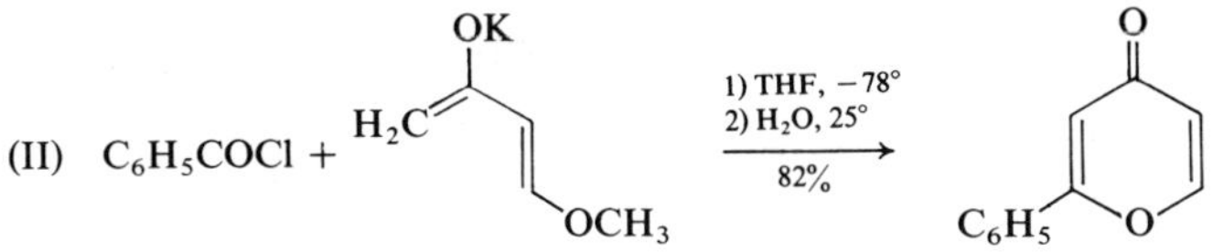

[1] M. Koreeda and H. Akagi, *Tetrahedron Letters*, **21**, 1197 (1980).
[2] T. A. Morgan and B. Ganem, *Tetrahedron Letters*, **21**, 2773 (1980).

(S)-2-Methoxymethylpyrrolidine, 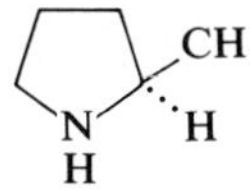 **(1).**

Chiral β-substituted aldehydes.[1] A general route to these aldehydes is formulated in equation (I). The chiral allylamine (S)-**2** is deprotonated to give the ion **a**, which

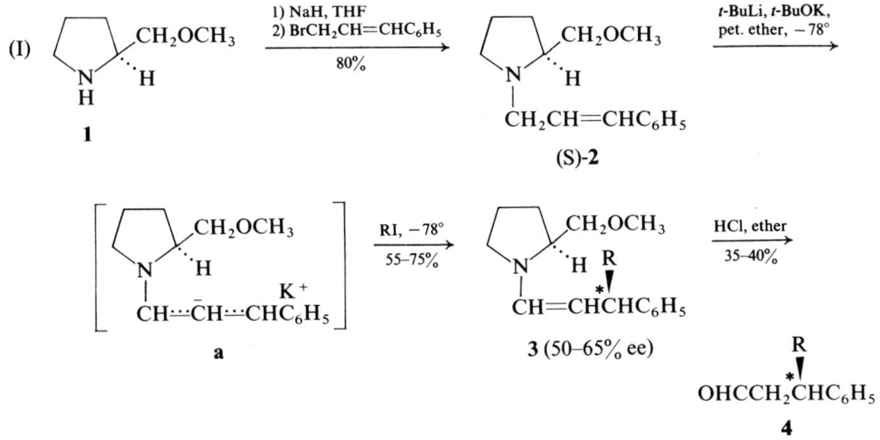

is alkylated to give the enamine **3** in enantiomeric excess of 50–65%. Hydrolysis affords the corresponding aldehyde **4**. Highest asymmetric induction is obtained by use of *t*-butyllithium and potassium *t*-butoxide in petroleum ether at 0°. The enantiomeric excess is almost independent of the size of the alkyl group, but is lower when bromides are used instead of iodides.

[1] H. Ahlbrecht, G. Bonnet, D. Enders, and G. Zimmermann, *Tetrahedron Letters*, **21**, 3175 (1980).

Methoxyphenylthiomethyllithium, 6, 369.

α-Phenylthio aldehydes.[1] The adducts of the reaction of this reagent with ketones are rearranged to α-phenylthio aldehydes by thionyl chloride in pyridine at 0°.

Examples:

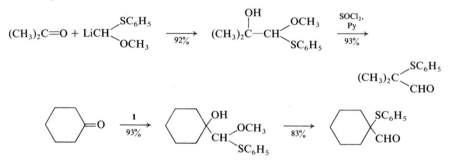

[1] Ae. de Groot and B. J. M. Jansen, *Tetrahedron Letters*, **22**, 887 (1981).

trans-1-Methoxy-3-trimethylsilyloxy-1,3-butadiene, 6, 370; 9, 303–304.

Anthracyclinones.[1] A key step in a stereocontrolled synthesis of 4-demethoxy-daunomycinone (**4**) involves cycloaddition of the epoxide **1** with this diene (**2**). Later steps follow known transformations.

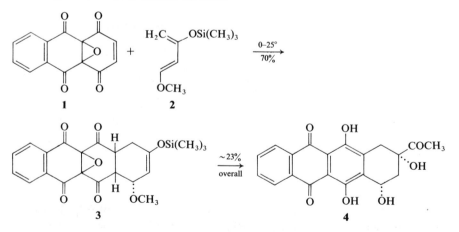

[1] D. A. Jackson and R. J. Stoodley, *J.C.S. Chem. Comm.*, 478 (1981).

Methyl (allylthio)acetate, $CH_2=CHCH_2SCH_2COOCH_3$ **(1)**. Mol. wt. 146.21. The reagent is prepared from allyl bromide and methyl mercaptoacetate (83% yield).

Methyl (2E,4E)-dienoates. The dianion **(2)** of **1** can be prepared by treatment with LDA and then *sec*-butyllithium in THF at $-78°$. The dianion is alkylated selectively at the α-allylic position to give **3** as the major product. Treatment of **3** with LDA induces a [2,3]sigmatropic rearrangement to **4**. Remaining steps to the diunsaturated ester **(6)** are methylation, oxidation, and dehydrosulfenylation.[1]

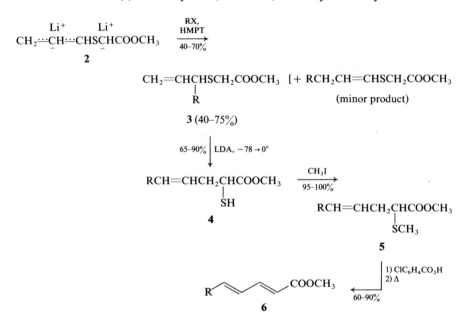

$$CH_2 \overset{Li^+}{\underset{-}{\cdots}} CH \overset{Li^+}{\underset{-}{\cdots}} CHSCHCOOCH_3$$
2
$$\xrightarrow[40-70\%]{RX, HMPT}$$

$$CH_2=CHCHSCH_2COOCH_3 \quad [+ RCH_2CH=CHSCH_2COOCH_3$$
$$\underset{R}{|}$$
(minor product)

3 (40–75%)

$$65-90\% \mid LDA, -78 \to 0°$$

$$RCH=CHCH_2CHCOOCH_3 \xrightarrow[95-100\%]{CH_3I}$$
$$\underset{SH}{|}$$
4

$$RCH=CHCH_2CHCOOCH_3$$
$$\underset{SCH_3}{|}$$
5

$$\underset{R}{\diagup}\diagdown\diagup\diagdown\diagup COOCH_3 \xleftarrow[60-90\%]{\substack{1) ClC_6H_4CO_3H \\ 2) \Delta}}$$
6

[1] K. Tanaka, M. Terauchi, and A. Kaji, *Chem. Letters*, 315 (1981).

O-Methyl-C, O-bis(trimethylsilyl)ketene acetal (1). Mol. wt. 218.44, b.p. 41°/0.5 mm. Preparation:

$$(CH_3)_3SiCH_2COOCH_3 \xrightarrow[81\%]{\substack{1) LDA \\ 2) ClSi(CH_3)_3}} (CH_3)_3SiCH=C\underset{OCH_3}{\overset{OSi(CH_3)_3}{\diagup}}$$
1

| THF | E/Z = 3:1 |
| THF–HMPT | E/Z = 5:95 |

Conjugate addition to enones.[1] In the presence of 1 equivalent of $TiCl_4$, **1** undergoes 1,4-addition to cycloalkenones. An example is formulated for a recent synthesis of methyl jasmonate **(2)** in equation (I).

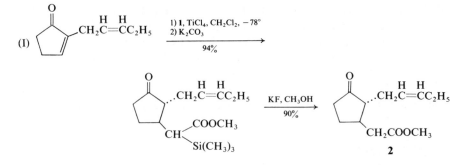

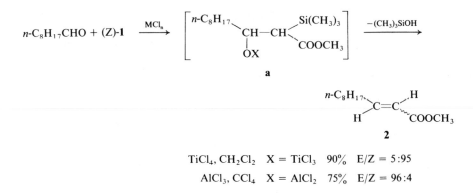

α,β-Unsaturated carboxylic esters.[2] The aldol reaction of (Z)-**1** with aldehydes provides a route to α,β-unsaturated esters. The method is attractive because the configuration of the product can be controlled by the Lewis acid used. When condensation of nonanal with (Z)-**1** is conducted with $TiCl_4$ in CH_2Cl_2 the product (**2**) is predominately (Z)-**2** formed by *anti*-elimination of $(CH_3)_3SiOH$ from the adduct **a**. When $AlCl_3$ in CCl_4 is used, (E)-**2** is formed stereoselectively.

$$n\text{-}C_8H_{17}CHO + (Z)\text{-}1 \xrightarrow{MCl_n}$$

$$\begin{bmatrix} n\text{-}C_8H_{17} & & Si(CH_3)_3 \\ & CH\text{-}CH & \\ & | & COOCH_3 \\ & OX & \end{bmatrix} \xrightarrow{-(CH_3)_3SiOH}$$

a

$$\begin{array}{cc} n\text{-}C_8H_{17} & H \\ & C=C \\ H & COOCH_3 \end{array}$$

2

$TiCl_4, CH_2Cl_2$ X = $TiCl_3$ 90%   E/Z = 5:95
$AlCl_3, CCl_4$ X = $AlCl_2$ 75%   E/Z = 96:4

[1] I. Matsuda, S. Murata, and Y. Izumi, *J. Org.*, **45**, 237 (1980).
[2] I. Matsuda and Y. Izumi, *Tetrahedron Letters*, **22**, 1805 (1981).

Methyl bromide–Lithium bromide.

   (E)-Trisubstituted allylic bromides. The stereoselectivity in the [2,3]sigmatropic rearrangement of the sulfoxide of **1** by the Evans technique (**6**, 30–31) is low (E/Z alcohols = 64:36). A new stereoselective method involves treatment of **1** with CH_3Br (excess) and LiBr in DMF (20°, 10 hours). The desired (E)-allylic bromide **2** is obtained as the major product in 60% yield. A similar reaction of **1** with CH_3I and NaI leads to the iodide corresponding to **2** (80–85% yield). A possible pathway proceeds via **a**.

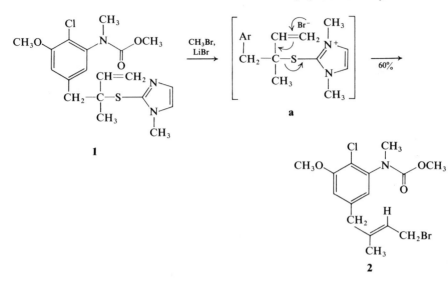

1

2

[1] P.-T. Ho, *Can. J. Chem.*, **58**, 861 (1980).

Methyl cyclobutenecarboxylate, 8, 335–336.

trans-*Decalins.* The [2 + 2] photoadduct (**3**) of **1** and 3-methylcyclohexene-2-one-1 when heated undergoes transformation to the *trans*-decalin **4**. The reaction

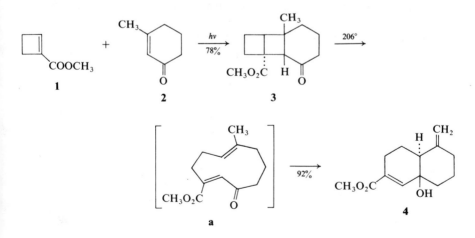

1   2   3

a   4

probably involves fragmentation to **a** followed by an ene closure. This scheme has been used for a synthesis of (±)-calameon **5** (equation I)[1] and of several eudesmanes, including (±)-isoatlantolactone **6** (equation II).[2]

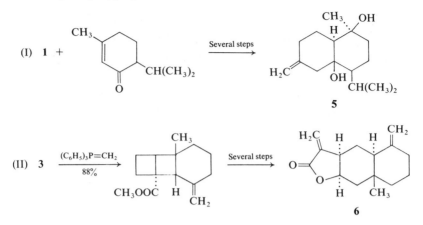

(I) **1** +

Several steps

5

(II) **3**

$(C_6H_5)_3P=CH_2$

88%

Several steps

6

¹ P. A. Wender and J. C. Hubbs, *J. Org.*, **45**, 365 (1980).
² P. A. Wender and L. J. Letendre, *J. Org.*, **45**, 367 (1980).

Methylenetriphenylphosphorane, 1, 678; **6,** 380–381; **8,** 339–340; **9,** 307.

Wittig reactions. The low yields that are encountered in some Wittig reactions with ketones result from formation of the lithium enolate as a side reaction. Regeneration of the ketone by addition of a stoichiometric amount of water can improve the yield of olefin. In a typical example, the yield of olefin was increased in this way from 38 to 89%.¹

Some time ago Conia and Limasset² reported that hindered, enolizable ketones, which do not undergo Wittig reactions readily, do react, although slowly, at elevated temperatures (refluxing benzene or xylene).

The final step in a synthesis of modhephene (**3**) involved addition of a methyl group to the hindered ketone **1**.³ The ketone does not react, however, with

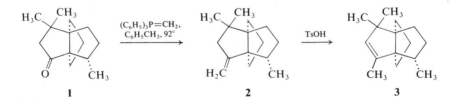

$(C_6H_5)_3P=CH_2$,
$C_6H_5CH_3$, 92°

TsOH

1   **2**   **3**

CH_3MgBr, CH_3Li, or related methyl nucleophiles. It also does not react with the Wittig reagent in DMSO at 70°. Addition does take place if the ketone is added to an excess of reagent generated in a minimum of solvent (toluene) preheated to about 92°.

Related successful Wittig reactions at 68° have been reported in another route to **3**.⁴

[1] P. Adlercreutz and G. Magnusson, *Acta Chem. Scand.*, **B34**, 647 (1980).

[2] J. M. Conia and J.-C. Limasset, *Bull. Soc.*, 1936 (1967); *see also* S. R. Schow and T. C. McMorris, *J. Org.*, **44**, 3760 (1979).

[3] A. B. Smith III, and P. J. Jerris, *Am. Soc.*, **103**, 194 (1981).

[4] H. Schostarez and L. A. Paquette, *Am. Soc.*, **103**, 722 (1981).

(−)-N-Methylephedrine–Lithium aluminum hydride–3,5-Dimethylphenol (1). The complex is prepared from the three components in the ratio 1:1:2 (**9**, 308).

4-Alkyl-γ-butyrolactones.[1] α-Acetylenic ketones (**2**) are reduced by this complex to the corresponding (R)-alcohols (**3**) with an optical purity of 75–90% (**9**, 308). These alcohols can be used for synthesis of 4-alkyl-γ-butyrolactones with an optical purity of about 95% (equation I).

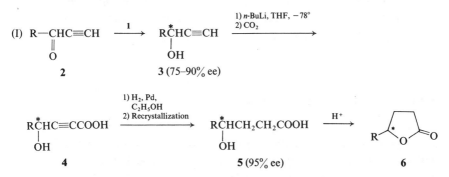

The synthesis shown above can be modified for preparation of 4-alkylbutenolides (**8**) by reduction of the acetylenic acid with quinoline-poisoned catalyst (equation II). The optically active butenolides have been used for a synthesis of optically pure enantiomers of *threo-* and *erythro-*4-methyl-3-heptanols.[2]

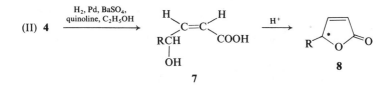

[1] J. P. Vigneron and V. Bloy, *Tetrahedron Letters*, **21**, 1735 (1980); J. P. Vigneron and J. M. Blanchard, *Tetrahedron Letters*, **21**, 1739 (1980).

[2] J. P. Vigneron, R. Méric, and M. Dhaenens, *Tetrahedron Letters*, **21**, 2057 (1980).

2-(N-Methyl-N-formyl)-aminopyridine (1), 8, 341.

Unsymmetrical secondary alcohols.[1] The formylation of Grignard reagents with **1** has been extended to a one-pot synthesis of secondary alcohols by reaction of the intermediate chelate with a second Grignard reagent (equation I).

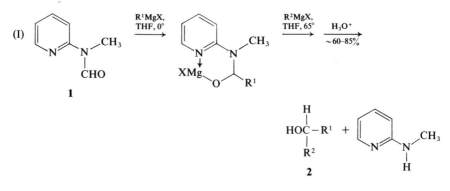

(I)

1

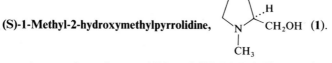

HOC—R¹ +

2

Yields are low if both Grignard reagents are bulky and contain β-hydrogens. One Grignard reagent with β-hydrogens can be tolerated if it is added first to **1**.

[1] D. L. Comins and W. Dernell, *Tetrahedron Letters*, **22**, 1085 (1981).

(S)-1-Methyl-2-hydroxymethylpyrrolidine, (1).

Asymmetric conjugate addition of CH_3MgBr.[1] The reaction of the α,β-enone **2** with CH_3MgBr in the presence of copper(I) bromide and **1** results in asymmetric conjugate addition to give **3** (equation I). The optical yield (68%) is the highest that has been reported for such a reaction.

(I) $C_6H_5CH=CHCOC_6H_5$ + CH_3MgBr $\xrightarrow[71\%]{CuBr, 1, THF}$ (S)-$C_6H_5\overset{\bullet}{C}HCH_2COC_6H_5$

2 CH_3

3 (68% ee)

[1] T. Imamoto and T. Mukaiyama, *Chem. Letters*, 45 (1980).

N-Methyl-N-phenylaminoacetylene, $HC\equiv CN\underset{C_6H_5}{\overset{CH_3}{<}}$ (1). Mol. wt. 131.17, b.p. 83–84°/10 mm.

Preparation[1]:

$CCl_2=CClN\underset{C_6H_5}{\overset{CH_3}{<}}$ $\xrightarrow[-2C_4H_9Cl, -LiCl]{2C_4H_9Li}$ $LiC\equiv CN\underset{C_6H_5}{\overset{CH_3}{<}}$ $\xrightarrow[70\%]{H^+}$ **1**

Spiroannelation.[2] Acylation of **1** by the enol lactone **2** results in formation of only one (**3**) of the two possible spiro[4.5]decenes. The bulky isopropyl group is responsible for the stereoselectivity; when it is replaced by a methyl group, the two possible spiro[4.5]decenes are formed in the ratio 4:1. The reaction thus is useful because of control of the relative stereochemistry at two chiral centers. The product **3** was converted in six steps to the sesquiterpene acoradiene (**4**).

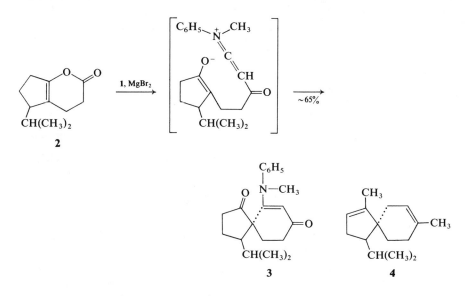

[1] J. Ficini and C. Barbara, *Bull. Soc.*, 2787 (1965).
[2] J. Ficini, G. Revial, and J. P. Genèt, *Tetrahedron Letters*, **22**, 629, 633 (1981).

N-Methyl-N-phenylaminoethynyllithium (1). Mol. wt. 137.10.
  Preparation[1]:

$$CCl_3COOH \rightarrow CCl_3CON\overset{C_6H_5}{\underset{CH_3}{\diagdown}} \xrightarrow[\text{overall}]{(C_2H_5)_3P \atop 75\%} CCl_2=CClN\overset{C_6H_5}{\underset{CH_3}{\diagdown}} \xrightarrow{2n\text{-BuLi}}$$

$$LiC\equiv CN\overset{C_6H_5}{\underset{CH_3}{\diagdown}}$$

1

Macrolides.[2] A new route to macrolides uses an ynamine group as an activated $-CH_2COOH$ group. The method was used to obtain 14-tetradecanolide (**6**). The hydroxy group was protected as the bromopropenyl ether, which is stable to base, but removable by β-elimination with *t*-butyllithium.[3] Ring closure is effected with BF_3 etherate; the lactone (**6**) is liberated by acid hydrolysis of **5**.

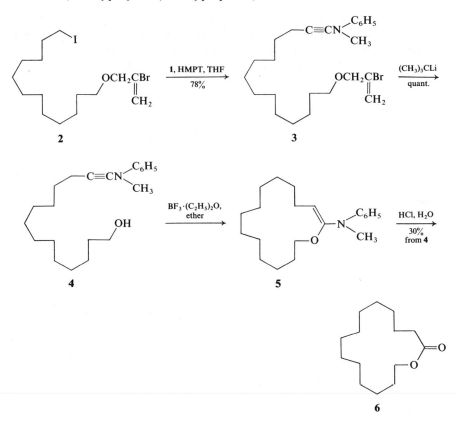

2   **3**

4   **5**

6

[1] J. Ficini and C. Barbara, *Bull Soc.*, 2787 (1965).
[2] J. P. Genet and P. Kahn, *Tetrahedron Letters*, **21**, 1521 (1980).
[3] J. F. W. McOmie, *Chem. Ind.*, 603 (1979).

N,N-Methylphenylamino(tri-*n*-butylphosphonium) iodide, 8, 345–346.

Allenes; enynes.[1] The reagent can be used to effect γ-alkylation of propargyl alcohols by diorganocuprates (equation I). The actual nucleophile is believed to be

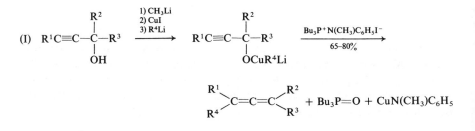

$R^4Cu^- N(CH_3)C_6H_5$. The reaction follows a different course when applied to an enyne alcohol; conjugated (Z)-enynes are the major products (equation II).

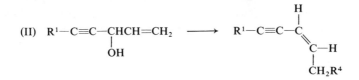

(II) $R^1—C≡C—CHCH=CH_2$ $\longrightarrow$

[1] Y. Tanigawa and S.-I. Murahashi, *J. Org.*, **45**, 4536 (1980).

N-Methyl-N-phenylbenzohydrazonyl bromide (1). Mol. wt. 289.18.
   Preparation[1]:

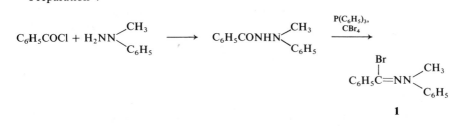

Peptide synthesis.[2] In aqueous acetone **1** ionizes to a nitrilium ion (**2**), which reacts rapidly and stereospecifically at carbon with a carboxylate ion to form a

$$1 \xrightarrow{(CH_3)_2CO,\ H_2O} C_6H_5C≡\overset{+}{N}N\begin{smallmatrix}CH_3\\C_6H_5\end{smallmatrix}\quad Br^- \xrightarrow{CBONHCH_2COO^-}$$

2

(Z)-**3** $\xrightarrow[95\%]{H_2NCH_2COOC_2H_5}$ $CBONHCH_2\overset{O}{\overset{\|}{C}}NHCH_2COOC_2H_5$

4

$+ C_6H_5CONHN\begin{smallmatrix}CH_3\\C_6H_5\end{smallmatrix}$

relatively stable (Z)-O-acylisoimide (**3**). This isoimide is an activated ester and reacts with an amine to form an amide (**4**) in high yield. Peptide synthesis is carried out by adding **1** to a solution of the acid and amine at pH 6; the amide is formed when the pH is adjusted to ~8. Racemization is slight with chiral amino acids.

[1] M. T. McCormack and A. F. Hegarty, *J.C.S. Perkin II*, 1701 (1976).
[2] A. F. Hegarty and D. G. McCarthy, *Am. Soc.*, **102**, 4537 (1980).

3-Methyl-2-selenoxo-1,3-benzothiazole, 7, 245.

Deoxygenation of epoxides.[1] The reagent is also useful for deoxygenation of epoxide groups of carbohydrates and cyclitols. The reaction is compatible with 1,6-anhydro linkages and various functional groups (acetal, O-benzyl, O-tosyl, azido, keto), but anomalous results are obtained if the epoxide bears a free vicinal hydroxyl group.

[1] H. Paulsen, F. R. Heiker, J. Feldmann, and K. Heyns, *Synthesis*, 636 (1980).

3-Methylsulfonyl-2,5-dihydrofurane (1). Mol. wt. 148.2, m.p. 42–43°, b.p. 100–105°/0.2 mm.

Preparation[1]:

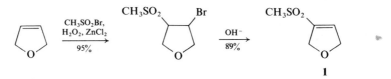

Diels-Alder reactions.[2] This vinyl sulfone has been used as an acetylene equivalent in a new synthesis of vitamin B_6. Thus 1 undergoes a Diels-Alder reaction with 4-methyloxazole (2) to give a pyridoxine derivative (3) with elimination of methanesulfinic acid. The product is converted into the vitamin (4) in three simple steps.

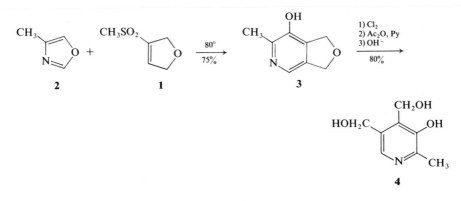

[1] W. Böll, *Ann.*, 1665 (1979).
[2] W. Böll and H. König, *Ann.*, 1657 (1979).

Methyltitanium trichloride, CH_3TiCl_3; Dimethyltitanium dichloride, $(CH_3)_2TiCl_2$.

These alkyltitanium compounds are prepared from $(CH_3)_2Zn$ and $TiCl_4$.

Addition to carbonyl groups. These alkyltitanium(IV) compounds (and related reagents) add readily to both aldehydes and ketones to form alcohols in high yield. Addition to aldehydes is so much faster than addition to ketones that selective addition to an aldehyde is possible. Diastereoselective addition to ketones is a useful feature. Thus the reaction affords an excellent route to the axial alcohol (2).

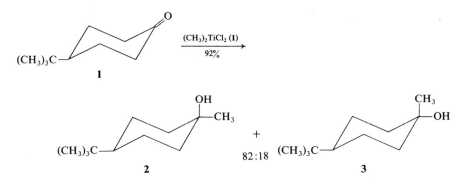

High asymmetric induction is observed in the addition of either CH_3TiCl_3 or $(CH_3)_2TiCl_2$ to 2-phenylpropionaldehyde (equation I).[1]

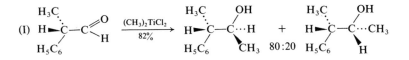

[1] M. T. Reetz, R. Steinbach, J. Westermann, and R. Peter, *Angew. Chem. Int. Ed.*, **19**, 1011 (1980).

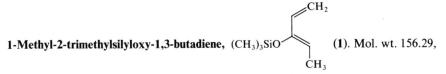

1-Methyl-2-trimethylsilyloxy-1,3-butadiene, $(CH_3)_3SiO$ **(1). Mol. wt. 156.29,**

b.p. 46–48°/12 mm; 82–85°/75 mm. This trimethylsilyl enol ether of ethyl vinyl ketone is prepared by reaction of the ketone with $(CH_3)_3SiCl$ and $N(C_2H_5)_3$ in DMF (55–60% yield).[1,2]

Diels-Alder reaction. The Diels-Alder reaction of **1** with **2** provides a key step in a total synthesis of *dl*-coriolin (**5**), a hirsutane sesquiterpene.[3] Only one product (**3**) is formed, because if **1** added in the alternative mode, the less stable *trans*-fusion of the bicyclopentane system is formed. The product is converted in two steps into

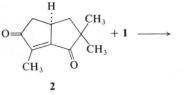

2

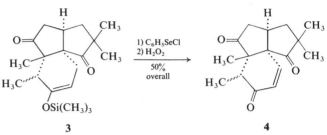

3   **4**

the enone **4** in 50% overall conversion. Conversion of **4** into **5** requires numerous steps.

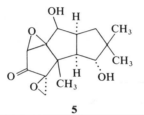

5

[1] G. A. Mock, A. B. Holmes, and R. A. Raphael, *Tetrahedron Letters*, 4539 (1977).
[2] S. Danishefsky and C. F. Yan, *Syn. Comm.*, **8**, 211 (1978).
[3] S. Danishefsky, R. Zambone, M. Kahn, and S. J. Etheridge, *Am. Soc.*, **102**, 2097 (1980).

Methyl vinyl ketone, 1, 697–703; **2**, 283–285; **5**, 464; **6**, 407–409; **7**, 247.

Phenol annelation.[1] Methyl vinyl ketone (and substituted vinyl ketones) undergo Robinson annelation with the β-keto sulfoxide (**1**) to afford the 5,6,7,8-tetrahydro-2-naphthol (**2**) with loss of benzenesulfenic acid. Sodium methoxide is used as base, and the reaction proceeds at 0 → 25°.

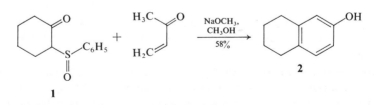

1

[1] D. L. Boger and M. D. Mullican, *J. Org.*, **45**, 5002 (1980).

Molecular sieves, 1, 703–705; **2**, 286–287; **3**, 206; **4**, 345; **5**, 465; **6**, 411–412; **9**, 316–317.

1,4-Dihydro-4-oxonicotinic acid derivatives.[1] Thermal cyclization of β-keto esters (**1**) with β-aminocrotonates (**2**) in the presence of molecular sieves (3–5 Å) provides a simple route to 5,6-disubstituted-1,4-dihydro-4-oxonicotinic acid derivatives (**3**).

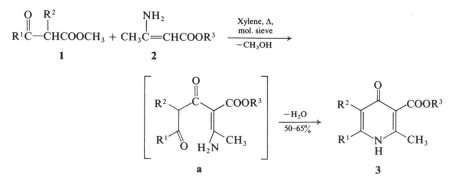

Presumably the first step is elimination of methanol, since use of ethyl esters of β-keto acids results in lower yields. No reaction occurs even at elevated temperatures in the absence of a molecular sieve.

Acetalization. β-Chloropropionaldehyde diethyl acetal can be prepared from acrolein in 88% yield by reaction of ethanol in the presence of hydrogen chloride dissolved in ether and with a 4 Å molecular sieve as water scavenger.[2] In the absence of a molecular sieve the yield drops to 65%. If ethanolic hydrogen chloride is used the yield is 34%.[3]

Oxidations with pyridinium chlorochromate [PCC] and pyridinium dichromate [PDC].[4] Oxidations with PCC and PDC of secondary hydroxyl groups of sugars and nucleosides is slow and incomplete. The reaction is markedly catalyzed by 3 Å molecular sieves. Celite, alumina, and silica are not effective. CH_2Cl_2 is the most satisfactory solvent; oxidations are slower in $ClCH_2CH_2Cl$ and C_6H_6. The rate of oxidation increases in the order 5 Å < 10 Å < 4 Å < 3 Å.

[1] O. Makabe, Y. Murai, and S. Fukatsu, *Heterocycles*, **13**, 239 (1979).
[2] Y. Sato, M. Miyazaki, and N. Nade, *Org. Syn.*, submitted (1980).
[3] E. J. Witzmann, W. L. Evans, H. Hess, and E. F. Schroeder, *Org. Syn. Coll. Vol.*, **2**, 137 (1943).
[4] J. Herscovici and K. Antonakis, *J.C.S. Chem. Comm.*, 561 (1980).

Molybdenum carbonyl, 2, 287; **3**, 206–207; **4**, 346; **7**, 247–248; **9**, 317.

Cyclopropanation.[1] Molybdenum carbonyl catalyses cyclopropanation of α,β-unsaturated esters and nitriles by diazo ketones and esters, even at 25°. Molybdenum carbenes may be intermediates, and indeed carbene dimers are obtained in the absence of excess substrate.

Examples:

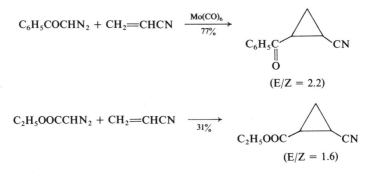

$$C_6H_5COCHN_2 + CH_2{=}CHCN \xrightarrow[77\%]{Mo(CO)_6}$$

(E/Z = 2.2)

$$C_2H_5OOCCHN_2 + CH_2{=}CHCN \xrightarrow[31\%]{}$$

(E/Z = 1.6)

Desulfuration of thiols.[2] Aromatic, benzylic, and aliphatic thiols are converted into hydrocarbons when heated in acetic acid with molybdenum carbonyl. The bridgehead adamantane-1-thiol is an exception and gives instead a thioester, $RSCOCH_3$. Yields are 50–90%. The actual reagent is probably tetrakis(acetato)-dimolybdenum.[3] Molybdenum carbonyl deposited on silica can also be used.

[1] M. P. Doyle and J. G. Davidson, *J. Org.*, **45**, 1538 (1980).
[2] H. Alper and C. Blais, *J.C.S. Chem. Comm.*, 169 (1980).
[3] A. B. Brignole and F. A. Cotton, *Inorg. Syn.*, **13**, 87 (1972).

Molybdenum(V) chloride–Triphenylphosphine.

Cyclic carbonates. Reaction with $MoCl_5–(C_6H_5)_3P$ (1:6) allows the insertion of CO_2 into methyloxirane to proceed at 20° and 1 atm.[1] This reaction had been

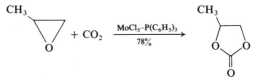

carried out previously in lower yield with catalysis by $RuCl_2[P(C_6H_5)_3]_3$ or $CoCl[P(C_6H_5)_3]_3$ at 100–200° and 5–200 atm.[2]

[1] M. Ratzenhofer and H. Kisch, *Angew. Chem. Int. Ed.*, **19**, 317 (1980).
[2] H. Koinuma, H. Kato, and H. Hirai, *Chem. Letters*, 517 (1977).

Molybdenum(V) chloride–Zinc.

Deoxygenation of N-oxides; reduction of $ArNO_2$.[1] An Mo(III) species obtained by reduction of $MoCl_5$ with zinc in aqueous THF reduces aromatic N-oxides in 45–65% yield. The reagent also reduces aromatic and heteroaromatic nitro compounds to amines in 30–70% yield.

[1] S. Polanc, B. Stanovnik, and M. Tišler, *Synthesis*, 129 (1980).

N

Nafion-H, 9, 320.

Hydration of epoxides.[1] Nafion-H can serve as a catalyst for hydration or methanolysis of epoxides.

Examples:

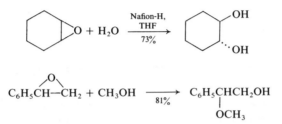

Dimethyl acetals.[2] Nafion-H is recommended as catalyst for the synthesis of dimethyl acetals and ethylene thioacetals. It is also useful for hydrolysis of dimethyl acetals.

Examples:

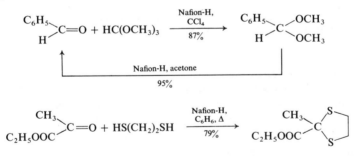

[1] G. A. Olah, A. P. Fung, and D. Meidar, *Synthesis*, 280 (1981).
[2] G. A. Olah, S. C. Narang, and D. Meidar, *Synthesis*, 282 (1981).

Nafion-H–Palladium on carbon.

Sequential aldol condensation and hydrogenation.[1] Since the aldol condensation of ketones is an equilibrium reaction, there are some advantages in hydrogenation of the α,β-unsaturated ketone as formed to shift the equilibrium. Thus it is possible to convert acetone into methyl isobutyl ketone in one step using two different heterogeneous catalysts (equation I).

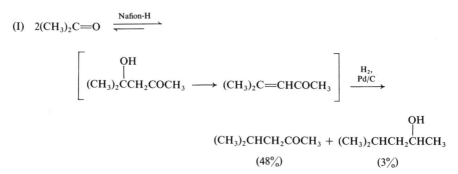

(I) $2(CH_3)_2C=O$ $\xrightleftharpoons{\text{Nafion-H}}$

$$\left[\begin{array}{c} \underset{\displaystyle (CH_3)_2\overset{\textstyle OH}{\underset{|}{C}}CH_2COCH_3}{} \longrightarrow (CH_3)_2C=CHCOCH_3 \end{array} \right] \xrightarrow{\begin{array}{c} H_2, \\ Pd/C \end{array}}$$

$$\underset{\text{(48\%)}}{(CH_3)_2CHCH_2COCH_3} + \underset{\text{(3\%)}}{(CH_3)_2CHCH_2\overset{\textstyle OH}{\underset{|}{C}}HCH_3}$$

[1] C. U. Pittman, Jr., and Y. F. Liang, *J. Org.*, **45**, 5048 (1980).

Nickel carbonyl, 1, 720–723; **2**, 290–293; **3**, 210–212; **4**, 353–355; **5**, 472–474; **6**, 417–419; **7**, 250.

α,β-Unsaturated ketones.[1] 2-Iodomercuric ketones, prepared as shown,[2] react with $Ni(CO)_4$ and an aldehyde in DMF at 50° to form α,β-unsaturated ketones (equation I). Ketones can also be used, but are less reactive.

(I) $R^1\overset{\textstyle OSi(CH_3)_3}{\underset{|}{C}}=CH_2$ $\xrightarrow[\text{80\%}]{\begin{array}{c} \text{1) HgO, Hg(OAc)}_2 \\ \text{2) HgI}_2 \end{array}}$ $R^1\overset{\textstyle O}{\underset{||}{C}}CH_2HgI$ $\xrightarrow[\text{75–90\%}]{\begin{array}{c} R^2CHO, \\ Ni(CO)_4, DMF \end{array}}$ $R^1\overset{\textstyle O}{\underset{||}{C}}CH=CHR^2$

[1] I. Rhee, I. Ryu, H. Omura, S. Murai, and N. Sonoda, *Chem. Letters*, 1435 (1979).
[2] H. O. House, R. Auerbach, M. Gall, and N. P. Peet, *J. Org.*, **38**, 514 (1973).

Nickel(II) chloride–Ethylene.

β-Hydride abstraction from alkyllithiums.[1] β-Hydride abstraction from alkyllithiums or Grignard reagents occurs in the presence of $NiCl_2$ and ethylene (hydride acceptor) at −20 to 0°. The thermodynamically less stable (Hofmann) alkene is favored slightly. Olefin isomerization is not a problem with this nickel salt, but is with complex organometallic nickel reagents.

Examples:

$$CH_3(CH_2)_8CH_2MgCl \xrightarrow[\text{71\%}]{\begin{array}{c} NiCl_2, CH_2=CH_2, \\ \text{ether, } -20° \end{array}} CH_3(CH_2)_7CH=CH_2$$

$$CH_3(CH_2)_3\overset{\textstyle CH_3}{\underset{|}{C}}HMgCl \xrightarrow[\text{62\%}]{0°} \underset{\text{56\%}}{CH_3(CH_2)_3CH=CH_2} + \underset{\text{44\%   (E/Z = 3:1)}}{CH_3(CH_2)_2CH=CHCH_3}$$

[1] M. T. Reetz and W. Stephan, *Ann.*, 171 (1980).

Nickel chloride–Zinc.

Ketone synthesis.[1] The reaction of an alkyl iodide with the carboxylic esters **1** in the presence of 10 mole % of $NiCl_2$ and zinc dust (3 equivalents) provides a one-pot synthesis of unsymmetrical ketones, usually in 70–90% yield (equation I). The reaction apparently involves an alkylnickel formed from a low-valence nickel compound.

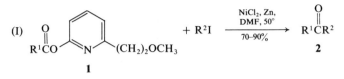

(I)

$$\text{1} + R^2I \xrightarrow[\text{70–90\%}]{\substack{\text{NiCl}_2, \text{ Zn,} \\ \text{DMF, 50°}}} R^1CR^2 \quad \text{2}$$

[1] M. Onaka, Y. Matsuoka, and T. Mukaiyama, *Chem. Letters*, 531 (1981).

Nickel peroxide, 1, 731–732; 5, 474; 7, 250–251; 8, 357–358; 9, 322–323.

Disulfides.[1] Sulfides are oxidized to disulfides usually in 85–95% yield by 1 equivalent of nickel peroxide.

[1] K. Nakagawa, S. Shiba, M. Horikawa, K. Sato, H. Nakamura, N. Harada, and F. Harada, *Syn. Comm.*, **10**, 305 (1980).

o-Nitrobenzenesulfenyl chloride, 1, 745; 4, 359–360.

Protection of thiols.[1] The *t*-butyl group has been of limited service for protection of thiols because removal has involved rather drastic treatment with liquid HF. *t*-Butyl thioethers can be cleaved by treatment with *o*-nitrobenzene-sulfenyl chloride in acetic acid at 25° to give *S*-*o*-nitrophenylsulfenyl derivatives. These products are reduced to thiols with $NaBH_4$, $HSCH_2CO_2H$, or $HOCH_2CH_2SH$.

J. P. Pastuszak and A. Chimiak, *J. Org.*, **46**, 1868 (1981).

3-Nitro-2-cyclohexenone (1). Mol. wt., 142.14, m.p. 35°.

Preparation:

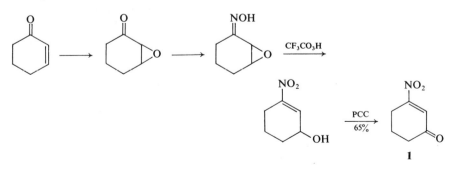

Diels-Alder reactions.[1] This reagent is more reactive in Diels-Alder reactions than cyclohexenone. Treatment of the β-nitro ketone with DBN effects elimination of HNO_2 to form the α,β-enone.[2] Thus the final products correspond to adducts of a "cyclohexynone," but with reversed regioselectivity. 3-Nitro-2-cyclopentenone, m.p. 69–71°, is also a useful dienophile (second example).

Examples:

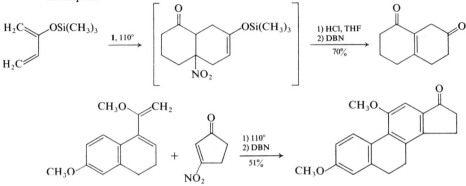

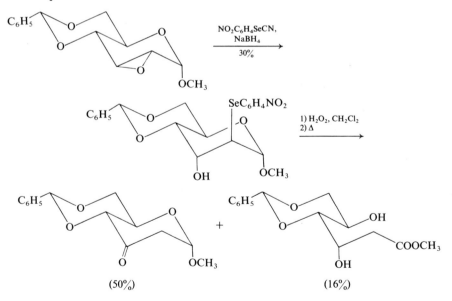

[1] E. J. Corey and H. Estreicher, *Tetrahedron Letters*, **22**, 603 (1981).
[2] J. W. Patterson and J. E. McMurry, *Chem. Comm.*, 488 (1971).

o-Nitrophenyl selenocyanate, 6, 420; **7,** 252; **9,** 325.

Ketone synthesis.[1] The preparation of ketones via β-hydroxy *o*-nitrophenyl selenoxides has been extended to deoxy keto sugars.

Example:

[1] K. Furuichi, S. Yogai, and T. Miwa, *J.C.S. Chem. Comm.*, 66 (1980).

1-Nitro-1-(phenylthio)propene (1). Mol. wt. 195.24, b.p. 70–75°/1 mm.
  Preparation:

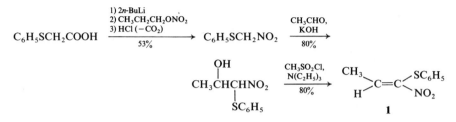

3-Methylfuranes.[1] Potassium fluoride catalyzes the reaction of 1,3-diketones with nitroalkenes to give diastereomeric dihydrofuranes. An example is shown in equation (I) for synthesis of evodone (**2**). 3-Methylfuranes are not obtained by use of 1-nitropropene itself because of tautomerization to give hydroxyiminofuranes.

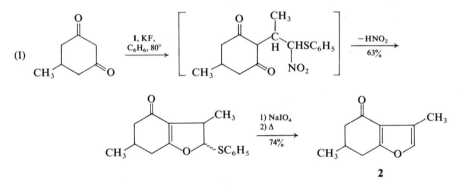

(I)

[1] M. Miyashita, T. Kumazawa, and A. Yoshikoshi, *J. Org.*, **45**, 2945 (1980).

O

1,7-Octadiene-3-one (1). Mol. wt. 124.18, b.p. 31°/4 mm.
Preparation:

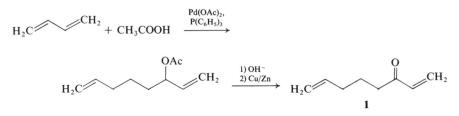

Bisannelation. The reagent **1** can be used for addition of two six-membered cyclohexane rings to a 1,3-cycloalkanedione. The first step involves Michael addition to the enone group of **1**. After formation of the first six-membered ring by aldol condensation, the terminal methylene group is then oxidized to a methyl ketone (**7**, 278). This product is then cyclized to a tricyclic ketone.[1]
Example:

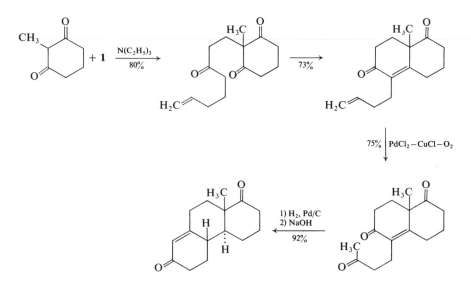

The same sequence has been used for synthesis of (+)-19-nortestosterone (**3**) from the optically active keto ester **2**.[2]

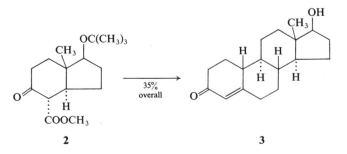

2   3

[1] J. Tsuji, I. Shimizu, H. Suzuki, and Y. Naito, *Am. Soc.*, **101**, 5070 (1979).
[2] J. Tsuji, Y. Kobayashi, H. Kataoka, and T. Takahashi, *Tetrahedron Letters*, **21**, 3393 (1980).

Organoaluminum compounds.

Diethylaluminum phenylthiolate (**1**).[1] The reagent is prepared from triethyl-aluminum and thiophenol and is used without isolation. The aluminum reagent reacts regio- and stereoselectively with vinyl epoxides (**2**) to afford mainly (Z)-4-phenylthio-2-butene-1-ol derivatives (**3**). In contrast reaction of vinyl epoxides with C_6H_5SH and $N(C_2H_5)_3$ gives mainly the (E)-isomer. Lithium benzenethiolate does not cleave vinyl epoxides at 25°.

Example:

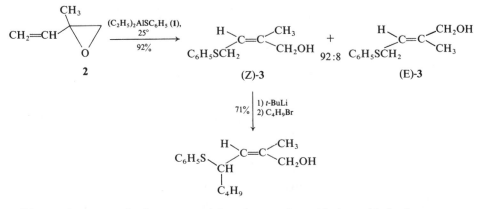

The products can also be converted into furanes by oxidation with lead tetra-acetate (equation I).

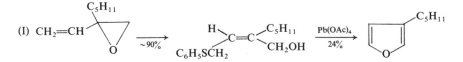

[1] A. Yasuda, M. Takahashi, and H. Takoya, *Tetrahedron Letters*, **22**, 2413 (1981).

Organocopper reagents, 9, 328–334.

Coupling with enol esters (**7**, 93). A new synthesis of an alkyl-substituted alkene involves coupling of a lithium dialkyl cuprate with an enol triflate,[1] available from a ketone by reaction with triflic anhydride and 2,6-di-*t*-butylpyridine.[2] A wide variety of organocuprates can be used and the geometry of the enolate is largely retained. Reported yields are in the range 60–100%.

Examples:

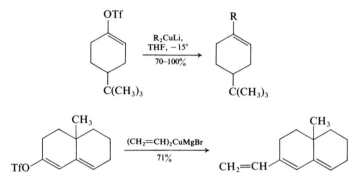

The reaction of the enol diethylphosphate of a *β*-keto ester with a primary dialkyl cuprate provides a useful stereoselective synthesis of *β*-alkyl-*α,β*-unsaturated esters (equation II).[3] This coupling was used in two steps in a recent synthesis of mokupalide, a novel C_{30}-isoprenoid.[4]

(II) $CH_3(CH_2)_2\overset{\overset{\displaystyle O}{\|}}{C}CH_2COOCH_3$ $\xrightarrow{\substack{\text{1) NaH, ether} \\ \text{2) ClPO(OC}_2\text{H}_5)_2}}$

$(C_2H_5O)_2\overset{\overset{\displaystyle O}{\|}}{P}O$, $CH_3(CH_2)_2$ C=C COOCH$_3$, H $\xrightarrow[83\%]{(CH_3)_2CuLi}$ CH_3 , $CH_3(CH_2)_2$ C=C COOCH$_3$, H

(>94% E)

Alkylation of allylic halides and alcohols (**8**, 334–335). Complete details of the reaction of $RCu \cdot BF_3$ with allylic halides and alcohols are now available. The reagent is probably an ate complex, $RBF_3^-Cu^+$, at least at low temperatures. In the case of allylic halides, THF is superior to ether for effecting *γ*-substitution. However use of ether is essential for the direct alkylation. Addition of BF_3 (even 2 equivalents) has no effect on reactions of C_6H_5Cu.[5]

S_N2' *reaction with α,β-epoxy ketones.*[6] The enolate **1** of 2,3-epoxycyclohexanone reacts with methyllithium to give, after acidic work-up, 2-methyl-2-cyclohexenone (**3**), the product of S_N2 addition. Reaction of **1** with lithium dimethyl cuprate on the other hand results in 6-methyl-2-cyclohexenone (**2**), the product of S_N2' addition.

The same S_N2' addition is also observed in reactions of the cuprate with the silyl enol ether or the enol phosphate of the epoxy ketone. This regioselectivity is applicable to five- and six-membered ring systems and, to a lesser extent, to acyclic systems, which are generally less reactive.

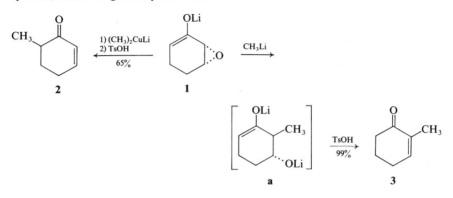

2   **1**

a   **3**

Reaction with propiolactone. The reaction of organocuprates with β-propio-lactone can be used to prepare β-substituted propionic acids.[7,8]
   Examples:

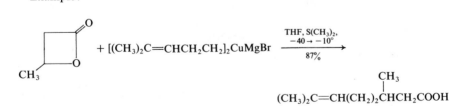

$$+ 2n\text{-BuMgCl} + \text{CuCl} \xrightarrow[\text{95\%}]{\text{S(CH}_3)_2, \text{ THF}} n\text{-Bu(CH}_2)_2\text{COOH}$$

$$+ (n\text{-Bu})_2\text{CuMgBr} \xrightarrow[\text{85\%}]{} \text{CH}_3(\text{CH}_2)_4\text{CHCOOH} \;\; \overset{|}{\text{CH}_3}$$

The reaction has been extended to a synthesis of terpenes from β-methyl-β-propiolactone, available by hydrogenation of ketene dimer.[9]
   Example:

$$+ [(\text{CH}_3)_2\text{C}{=}\text{CHCH}_2\text{CH}_2]_2\text{CuMgBr} \xrightarrow[\text{87\%}]{\substack{\text{THF, S(CH}_3)_2, \\ -40 \to -10°}}$$

$$(\text{CH}_3)_2\text{C}{=}\text{CH(CH}_2)_2\overset{\overset{\displaystyle \text{CH}_3}{|}}{\text{CH}}\text{CH}_2\text{COOH}$$

When β-ethynyl-β-propiolactone (**1**) is the substrate, allenic 3,4-alkadienoic acids (**2**) are obtained in high yield. The reaction has been used for a synthesis of pellitorine (**3**).

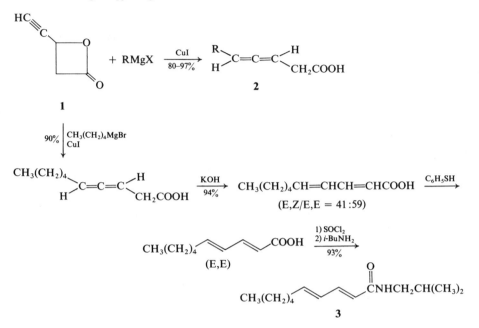

$$CH_3(CH_2)_4CH\!=\!CHCH\!=\!CHCOOH$$
$$(E,Z/E,E = 41:59)$$

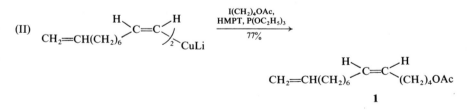

3

(*Z*)-*Alkenes.* Lithium dialkyl cuprates undergo *syn*-addition to acetylene to afford (Z)-dialkenyl cuprates in quantitative yield. The products can be alkylated under conditions by which both alkenyl groups are used. In the case of reactive halides 2 equivalents of HMPT is necessary; with less reactive halides 3 equivalents of triethyl phosphite is also necessary (equation I). The products are essentially only the (Z)-isomers.[11]

(I) $R_2CuLi + 2HC\!\equiv\!CH \longrightarrow$

This synthesis has been applied to several insect pheromones, for example the pheromone (**1**) of *Cossus cossus* (equation II).[12]

(II)

Dialkenyl cuprates react with α,β-unsaturated sulfones to form (Z)-γ,δ-unsaturated sulfones. Desulfonylation proceeds with retention of configuration.[13]

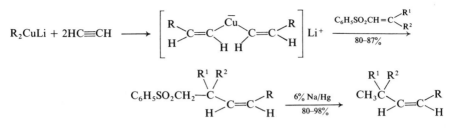

$$R_2CuLi + 2HC\equiv CH \longrightarrow$$

1,3-Dienes.[14]
(Z)-Dialkenyl cuprates couple with alkenyl halides in the presence of $Pd[P(C_6H_5)_3]_4$ to form conjugated dienes in good yield, but with fair stereo-selectivity. If 1 equivalent of zinc bromide (**8**, 86) is present, the configuration of the alkenyl halide is retained ($\sim 99.5\%$).

Substitution of allylic sulfoxides.
Lithium dimethyl cuprate and di-*n*-butyl cuprate react with allylic sulfoxides or sulfones mainly by γ-substitution to give tri-substituted alkenes (equation I).[15]

β,β'-Dialkylation of enones.[16]
The reaction of a lithium dialkyl cuprate (excess) with an α,β-unsaturated ketone substituted by a heteroatom at the β'-position (**1**) results in two successive conjugate additions to introduce an alkyl group at both the β- and β'-position.

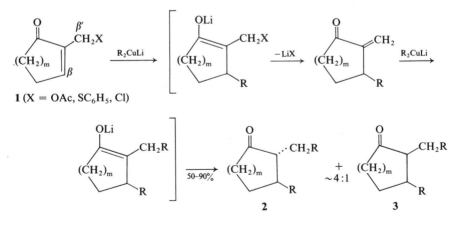

1 (X = OAc, SC$_6$H$_5$, Cl)

Two different alkyl groups can be introduced by a two-step sequence (equation I).

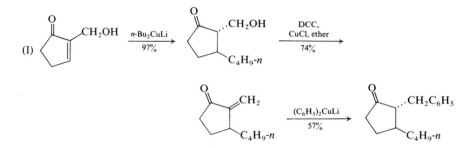

(I)

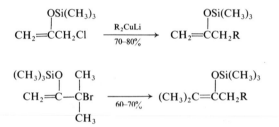

Regiospecific route to silyl enol ethers.[17] Silyl enol ethers are formed regio-specifically by reaction of lithium dialkyl cuprates or diaryl cuprates with 2-trimethylsiloxyallyl halides. An allylic rearrangement occurs in reactions with hindered halides.

Examples:

$$CH_2\!\!=\!\!\underset{\underset{\displaystyle OSi(CH_3)_3}{|}}{C}CH_2Cl \xrightarrow[70-80\%]{R_2CuLi} CH_2\!\!=\!\!\underset{\underset{\displaystyle OSi(CH_3)_3}{|}}{C}CH_2R$$

$$CH_2\!\!=\!\!\underset{\underset{\displaystyle CH_3}{|}}{\overset{\overset{\displaystyle (CH_3)_3SiO}{|}}{C}}\!\!-\!\!\underset{\underset{\displaystyle CH_3}{|}}{\overset{\overset{\displaystyle CH_3}{|}}{C}}Br \xrightarrow[60-70\%]{} (CH_3)_2C\!\!=\!\!\underset{\underset{\displaystyle OSi(CH_3)_3}{|}}{C}CH_2R$$

1,4-Addition to enals.[18] Conjugate addition of lithium dialkylcuprates is favored over 1,2-addition by use of low temperatures and a nonpolar solvent (pentane). However, 1,2-addition becomes more important if the double bond is trisubstituted.

Conjugate additions. Conjugate additions using cuprates of the type R_2CuLi or R_2CuMgX suffer from the drawback that only one R group is utilized. An expedient is addition of a trialkylphosphine[19] or of 1,5-cyclooctadiene[20] as a non-participating complexing agent. These ligands not only permit transfer of both R groups, but also enhance the reactivity and permit transfer of the bulky isopropyl group (but not of the *t*-butyl group).

Complex Cuprates

RCu–AlCl₃. Reaction of $RCu–AlCl_3$ with γ-acetoxy-α,β-unsaturated esters results in substitution of the acetoxy group by the R group (equation I).[21] In contrast the reagent undergoes conjugate addition to γ-acetoxy- or γ-silyloxy-α,β-enones[22] (equation II). Lithium dialkyl cuprates are not useful for these reactions.[23]

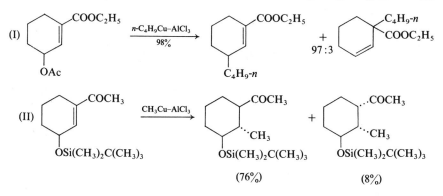

(CH₃)₃Cu₂MgCl. This reagent is prepared according to the stoichiometry shown in equation (I). The inorganic salt functions as a stabilizer.

(I) $3CH_3MgCl + 2CuBr + 6LiBr \rightarrow (CH_3)_3Cu_2MgCl \cdot 6LiBr + 2MgBrCl$

1

This mixed cuprate is useful for preparation of 2-methyl-1-alkenes from 1-alkynes (equation II). In the absence of LiBr the intermediate vinylcuprate **a** decomposes to

(II) $C_6H_5C\equiv CH + 1$ $\xrightarrow[\text{ether, 0°}]{\text{THF,}}$ $\left[\begin{array}{c} C_6H_5 \\ CH_3 \end{array} C=C \begin{array}{c} H \\ Cu_2(CH_3)_2 \end{array} \cdot MgCl \right]$ $\xrightarrow[98\%]{H_2O}$

a

$\begin{array}{c} C_6H_5 \\ CH_3 \end{array} C=CH_2$

the dimer $[C_6H_5(CH_3)C=C-]_2$. One drawback is that only one methyl group is transferred from the cuprate.[24]

Chloromagnesium dimethyl cuprate, $(CH_3)_2CuMgCl$ (**1**). This mixed cuprate is prepared by reaction of CH_3MgCl (2 equivalents) with CuBr (1 equivalent) in THF. Most lithium dialkyl cuprates react with α,β-unsaturated aldehydes to give a mixture of 1,2- and 1,4-adducts, but **1** tends to form mainly the latter adducts.[25]
Example:

$\begin{array}{c} C_2H_5 \\ H \end{array} C=C \begin{array}{c} CH_3 \\ CHO \end{array} + 1$ $\xrightarrow[49\%]{THF}$ $\begin{array}{c} C_2H_5 \\ CH_3 \end{array} CH-\overset{\overset{\displaystyle CH_3}{|}}{C}HCHO$

Mixed cyanocuprates (RCuCN)Li. The earlier synthesis of 2-cyclohexenols (**9**, 329–320) has been extended to provide a general route to highly substituted 2-cyclo-hexenols as shown in equation (I). The crucial step involves a regiospecific 1,4-addition of a cyanocuprate to an α-exo-methylene epoxide.[26]

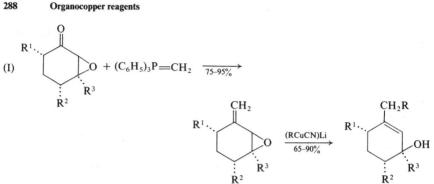

The 1,4-addition of alkyl cyanocuprates to epoxides to form cyclic α,β-unsaturated ketones provides a stereoselective method for conversion of dehydro-epiandrosterone into cholesterol (**4**) and other related steroids. Thus the keto epoxide **2** (obtained from dehydroepiandrosterone in 55% overall yield in three steps) reacts with lithium isohexylcyanocuprate to give the adduct **3**, which can be converted to cholesterol (**4**) in several known steps.[27]

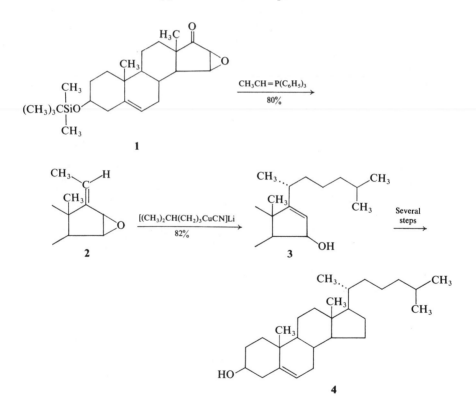

Lithium phenylthio(trimethylstannyl)cuprate, $(CH_3)_3Sn(SC_6H_5)CuLi$ (**1**). Mol. wt. 343.44. The cuprate is obtained by addition of C_6H_5SCu to a solution of $(CH_3)_3SnLi$ in THF. Depending on the experimental conditions, the cuprate (**1**) adds to α,β-acetylenic esters to afford either (E)- or (Z)-β-trimethylstannyl-α,β-unsaturated esters (equation I). At low temperatures and in the presence of methanol, the (E)-product (**2**) is formed under kinetic control. At higher temperatures the more stable (Z)-isomer (**3**) is obtained. The products are attractive precursors to isomerically pure vinyl reagents.[28]

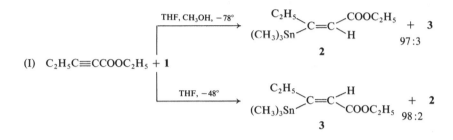

[1] J. E. McMurry and W. J. Scott, *Tetrahedron Letters*, **21**, 4313 (1980).
[2] P. J. Stang and T. E. Fisk, *Synthesis*, 438 (1979); P. J. Stang and W. Treptow, *Synthesis*, **21**, 283 (1980).
[3] F. W. Sum and L. Weiler, *Can. J. Chem.*, **57**, 1431 (1979).
[4] F. W. Sum and L. Weiler, *Am. Soc.*, **101**, 4401 (1979).
[5] Y. Yamamoto, S. Yamamoto, H. Yatagai, and K. Maruyama, *Am. Soc.*, **102**, 2318 (1980).
[6] P. A. Wender, J. M. Erhardt, and L. J. Letendre, *Am. Soc.*, **103**, 2114 (1981).
[7] J. F. Normant, A. Alexakis, and G. Cahiez, *Tetrahedron Letters*, **21**, 935 (1980).
[8] T. Fujisawa, T. Sato, T. Kawara, M. Kawashima, H. Shimizu, and Y. Ito, *Tetrahedron Letters*, **21**, 2181 (1980).
[9] T. Fujisawa, T. Sato, T. Kawara, A. Noda, and T. Obinata, *Tetrahedron Letters*, **21**, 2553 (1980).
[10] T. Sato, M. Kawashima, and T. Fujisawa, *Tetrahedron Letters*, **22**, 2375 (1981).
[11] A. Alexakis, G. Cahiez, and J. F. Normant, *Synthesis*, 826 (1979).
[12] G. Cahiez, A. Alexakis, and J. F. Normant, *Tetrahedron Letters*, **21**, 1433 (1980).
[13] G. De Chirico, V. Fiandanese, G. Marchese, F. Naso, and O. Sciacovelli, *J. C. S. Chem. Comm.*, 523 (1981).
[14] N. Jabri, A. Alexakis, and J. F. Normant, *Tetrahedron Letters*, **22**, 959 (1981).
[15] Y. Masaki, K. Sakuma, and K. Kaji, *J.C.S. Chem. Comm.*, 434 (1980).
[16] A. B. Smith III, B. A. Wexler, and J. S. Slade, *Tetrahedron Letters*, **21**, 3237 (1980).
[17] H. Sakurai, A. Shirahata, Y. Araki, and A. Hosomi, *Tetrahedron Letters*, **21**, 2325 (1980).
[18] C. Chuit, J. P. Foulon, and J. F. Normant, *Tetrahedron*, **36**, 2305 (1980).
[19] M. Suzuki, T. Suzuki, T. Kawagishi, and R. Noyori, *Tetrahedron Letters*, **21**, 1247 (1980).
[20] F. Leyendecker and F. Jesser, *Tetrahedron Letters*, **21**, 1311 (1980).
[21] T. Ibuka and H. Minakata, *Syn. Comm.*, **10**, 119 (1980).
[22] T. Ibuka, H. Minakata, Y. Mitsui, K. Kinoshita, Y. Kawami, and N. Kimura, *Tetrahedron Letters*, **21**, 4073 (1980); T. Ibuka, H. Minakata, Y., Mitsui, K. Kinoshita, and Y. Kawami, *J.C.S. Chem. Comm.*, 1193 (1980).
[23] R. A. Ruden and W. E. Litterer, *Tetrahedron Letters*, 2043 (1975).

[24] H. Westmijze, H. Kleijn, J. Meijer, and P. Vermeer, *Rec. Trav.*, **100**, 98 (1981).
[25] C. Chuit, J. P. Foulon, and J. F. Normant, *Tetrahedron*, **37**, 1385 (1981).
[26] J. P. Marino and H. Abe, *Synthesis*, 872 (1980).
[27] J. P. Marino and H. Abe, *Am. Soc.*, **103**, 2907 (1981).
[28] E. Piers and H. E. Morton, *J. Org.*, **45**, 4263 (1980).

Organolithium reagents.

2-Butenolides.[1] 4-Substituted 2-butenolides can be prepared by reaction of organolithium reagents with 4-hydroxy-2-butenolide (**1**), easily prepared from furfural (equation I). The reaction of **1** with 2 equivalents of RLi or ArLi in THF at

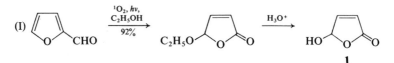

$-70°$ followed by acidification results in an oil (**2**) that cyclizes to the 2-butenolide (**3**) spontaneously at $0°$.

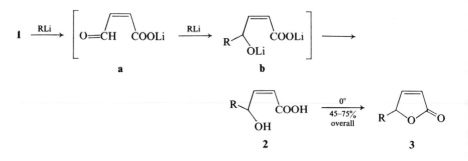

[1] F. W. Machado-Araujo and J. Gore, *Tetrahedron Letters*, **22**, 1969 (1981).

Organomanganese(II) iodides, 7, 222; 8, 312–313; 9, 289.

Ketone synthesis.[1] RMnI reagents react preferentially with acid chloride groups to form ketones even in the presence of formyl, acetyl, or even keto groups when the reaction is conducted in a mixture of ether and methylene chloride at $-30 \rightarrow 20°$.

[1] G. Cahiez, *Tetrahedron Letters*, **22**, 1239 (1981).

Osmium tetroxide, 1, 759–767.

Review.[1] *cis*-Dihydroxylation with osmium tetroxide has been reviewed (246 references).

[1] M. Schröder, *Chem. Rev.*, **80**, 187 (1980).

Osmium tetroxide–Dihydroquinine acetate (1),

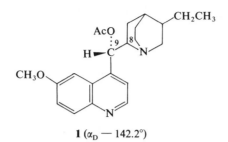

1 (α_D — 142.2°)

Optically active **cis,vic-*diols.*[1]** It is known that pyridine catalyzes the hydroxylation of alkenes with OsO_4 and that the osmate ester intermediates form an isolable complex with pyridine (**1**, 760–761). Hentges and Sharpless reasoned that a similar chiral amine could induce chirality in the diol. And indeed addition of 1 equivalent of **1** or of the C_8-diastereoisomer, dihydroquinidine acetate (**2**), does result in *vic*-diols in fair to high enantiomeric excess, particularly in reactions performed in toluene at —78°. Opposite stereoselectivities are exhibited by **1** and **2**. Optical yields range from 25 to 85%. Use of an amine in which the chiral center is two carbon atoms removed from the coordination site lowers the optical yield to 3–18%.

[1] S. G. Hentges and K. B. Sharpless, *Am. Soc.*, **102**, 4263 (1980).

Osmium tetroxide–Diphenyl selenoxide.

cis-*Dihydroxylation.*[1] Selenoxides oxidize osmium(VI) to osmium(VIII) in an alkaline aqueous solution and thus can be used as a co-oxidant in dihydroxylation reactions with OsO_4 with consequent use of only catalytic amounts of OsO_4. Diphenyl selenoxide and methyl phenyl selenoxide have been used. The latter selenoxide can be generated from the selenide by oxidation with singlet oxygen.

[1] A. G. Abatjoglou and D. R. Bryant, *Tetrahedron Letters*, **22**, 2051 (1981).

Osmium tetroxide–N-Methylmorpholine N-oxide, OsO_4–NMMO, **7**, 256–257; **9**, 334.

α-*Hydroxy ketones.*[1] Ketones can be hydroxylated at the α-position by oxidation of the enol silyl ether with the monohydrate of N-methylmorpholine N-oxide (1 equivalent) and a catalytic amount of OsO_4. An attractive feature is the control of regioselectivity with the method of generation of the enol. The rate of oxidation can be increased by use of *t*-butyl alcohol as solvent and pyridine as catalyst. Hydroxylation at methine and methylene groups proceeds in higher yield than hydroxylation at methyl groups. Olefinic bonds are not affected.

Examples:

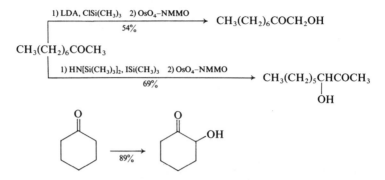

CH₃(CH₂)₆COCH₃

$$1) \text{LDA, ClSi(CH}_3)_3 \quad 2) \text{OsO}_4\text{–NMMO} \atop 54\%$$

CH₃(CH₂)₆COCH₂OH

$$1) \text{HN[Si(CH}_3)_3]_2, \text{ISi(CH}_3)_3 \quad 2) \text{OsO}_4\text{–NMMO} \atop 69\%$$

CH₃(CH₂)₅CHCOCH₃
|
OH

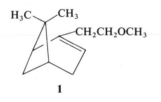

[1] J. P. McCormick, W. Tomasik, and M. W. Johnson, *Tetrahedron Letters*, **22**, 607 (1981).

Osmium tetroxide–Trimethylamine N-oxide–Pyridine.

Dihydroxylation of hindered double bonds.[1] The α-pinene derivative **1** (Nopol) is hydroxylated in low yield by osmium tetroxide and *t*-butyl hydroperoxide. Hydroxylation is effected in 62% yield by use of OsO₄ in combination with trimethylamine N-oxide as oxidant and pyridine as base. This method is generally suitable for hindered alkenes (yields 78–93%).

H₃C   CH₃

CH₂CH₂OCH₃

1

[1] R. Ray and D. S. Matteson, *Tetrahedron Letters*, **21**, 449 (1980).

Oxoperoxobis(N-phenylbenzohydroxamato)molybenum(VI) (1), Mol. wt. 568.38, m.p. 150° (dec.), orange-yellow. The complex is prepared by reaction of MoO₂(acac)₂

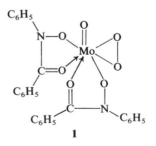

1

with $C_6H_5CON(C_6H_5)OH$. The product is converted into **1** by treatment with excess H_2O_2.

Oxidations.[1] The reagent **1** oxidizes primary and secondary alcohols to carbonyl compounds in fair to good yield. It is not useful for epoxidation of simple alkenes, but it epoxidizes allylic alcohols to form α,β-epoxy alcohols in 60–70% yield. In general, this epoxidation is more stereospecific than that observed with *t*-butyl hydroperoxide in combination with $Mo(CO)_6$ (**9**, 81–82).

[1] H. Tomioka, K. Takai, K. Oshima, and H. Nozaki, *Tetrahedron Letters*, **21**, 4843 (1980).

Oxygen, 4, 362; **5**, 482–486; **6**, 426–430; **7**, 258–260; **8**, 366–367; **9**, 335–337.

Oxidative decyanation (**6**, 430).[1] This reaction can be conducted under phase-transfer conditions with 50% NaOH in DMSO with catalytic amounts of benzyl-triethylammonium chloride (TEBA). The method fails with purely aliphatic nitriles. The highest yields (85–90%) are obtained with aromatic secondary nitriles.

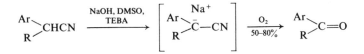

Oxygenation of methyl linolenate.[2] Autoxidation of this triunsaturated acid results in a complex mixture of hydroperoxyepidioxides. A single product of this type has been obtained by lipoxygenase-catalyzed oxidation to give a hydroperoxide **1**, followed by heating in air at 40° for 96 hours. This procedure gives the single hydroperoxyepidioxide **2** in about 30% yield.

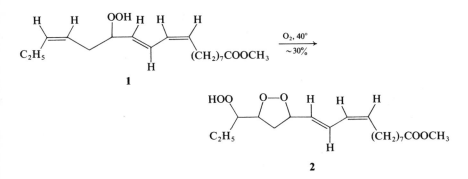

α-Hydroxylation of amides and esters (**6**, 427–428). Full details are now available.[3] The report cautions that the *n*-butyllithium used to generate LDA should be completely consumed before oxygenation of the enolate, since oxygen can convert this reagent into explosive peroxides. Fourteen examples of the reaction are cited (yields 70–90%).

[1] A. Donetti, O. Boniardi, and A. Ezhaya, *Synthesis*, 1009 (1980).
[2] H. W.-S. Chan, J. A. Matthew, and D. T. Coxon, *J.C.S. Chem. Comm.*, 235 (1980).
[3] H. H. Wasserman and B. H. Lipshutz, *Org. Syn.*, submitted (1981).

Oxygen, singlet, 4, 362–363; **5**, 486–491; **6**, 431–436; **7**, 261–269; **8**, 367–374; **9**, 338–341.

Reaction with α,β-unsaturated ketones and lactones.[1] The reactivity of α,β-enones to singlet oxygen depends on the conformation. Systems that exist in *s-trans*-conformations (*e.g.*, Δ⁴-3-ketosteroids) react slowly if at all. However, *s-cis*-enones react readily. For example, (R)-(+)-pulegone (**1**) reacts to give the products **2–4**. The same products are obtained by oxidation with triphenyl phosphite ozonide (**3**, 324–325).

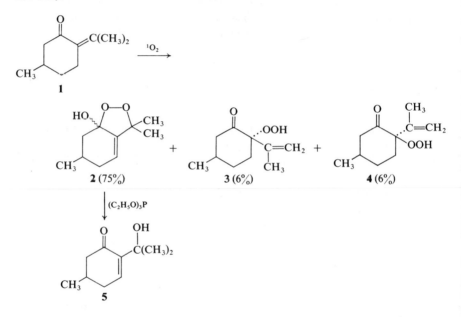

Macrolides. A new route to macrolides[2] is based on earlier studies[3] that revealed that oxazoles can serve as a protecting group for carboxylic acids. Thus photo-oxidation of trisubstituted oxazoles (**1**) produces triamides (**2**) in 55–85%

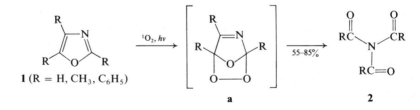

yield. This photo-oxidation has now been used to effect intramolecular acylation in a general synthesis of lactones, illustrated for a synthesis of (E)- and (Z)-recifeiolide (**3**) (equation I).

(I)

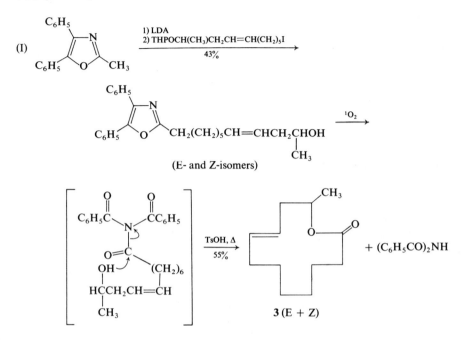

(E- and Z-isomers)

3 (E + Z)

+ $(C_6H_5CO)_2NH$

[1] H. E. Ensley, R. V. C. Carr, R. S. Martin, and T. E. Pierce, *Am. Soc.*, **102**, 2836 (1980).
[2] H. H. Wasserman, R. J. Gambale, and M. J. Pulwer, *Tetrahedron Letters*, **22**, 1737 (1981).
[3] H. H. Wasserman and M. B. Floyd, *Tetrahedron Suppl.*, **7**, 441 (1966); H. H. Wasserman, F. J. Vinick, and Y. C. Chang, *Am. Soc.*, **94**, 7180 (1972).

Ozone, 1, 773–777; **4**, 363–364; **5**, 491–495; **6**, 436–441; **7**, 269–271; **8**, 374–377; **9**, 341–343.

Alicyclic 1,2-diones.[1] An attractive route to cyclopentane-1,2-dione (**2**) is ozonolysis of 2-cyclopentylidenecyclopentanone (**1**).[2] The main by-product is cyclopentanone, which can be recycled to **1**.

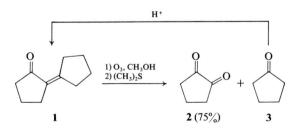

Selenoxide elimination.[3] A short synthesis of royal jelly acid (2) from a commercially available starting material (1) involves phenylselenenylation followed by concurrent ozonation of a double bond and selenoxide elimination (equation I).

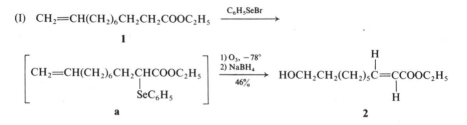

(I) $CH_2{=}CH(CH_2)_6CH_2CH_2COOC_2H_5 \xrightarrow{C_6H_5SeBr}$

1

$$\left[CH_2{=}CH(CH_2)_6CH_2\underset{\underset{SeC_6H_5}{|}}{CH}COOC_2H_5 \right] \xrightarrow[46\%]{\underset{2)\,NaBH_4}{1)\,O_3,\,-78°}} HOCH_2CH_2(CH_2)_5\underset{\underset{H}{|}}{\overset{\overset{H}{|}}{C}}{=}CCOOC_2H_5$$

a   **2**

[1] J. Wrobel and J. M. Cook, *Syn. Comm.*, **10**, 333 (1980).
[2] A. T. Nielsen and W. J. Houlihan, *Org. React.*, **16**, 114 (1968).
[3] T. A. Hase and R. Kivikari, *Acta Chem. Scand.*, B33, 589 (1979).

Ozone–Silica gel, 6, 440; **7**, 271–273; **8**, 375–377; **9**, 343.

Hydroxylation of t-carbon atoms.[1] Detailed directions including necessary precautions for dry ozonation have been published. Seven examples for preparation of tertiary alcohols are included (yields 72–90%).

[1] Z. Cohen, H. Varkony, E. Keinan, and Y. Mazur, *Org. Syn.*, **59**, 176 (1979).

P

Palladium–Graphite, $C_{16}Pd$. The reagent is prepared by treatment of $PdCl_2$ in DME with C_8K at 100°.

Hydrogenation catalyst.[1] $C_{16}Pd$ catalyzes hydrogenation of aromatic nitro compounds to anilines and of nitroalkanes to amines (both quantitative). It is also effective for hydrogenation of alkynes to (Z)-alkenes in 90–99% yield, particularly when ethylenediamine is also present. In fact it is superior to palladium on carbon in the presence of EDA, even though the rate is somewhat slower.

Arylation or vinylation of activated C=C bonds.[2] This reaction in the Heck method is catalyzed by $Pd(OAc)_2$, which is believed to be reduced to Pd(0). In fact $C_{16}Pd$ in the presence of a tertiary amine is equally effective (equations I and II).

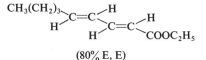

[1] D. Savoia, C. Trombini, A. Umani-Ronchi, and G. Verardo, *J.C.S. Chem. Comm.*, 540 (1981).

[2] D. Savoia, C. Trombini, A. Umani-Ronchi, and G. Verardo, *J.C.S. Chem. Comm.*, 541 (1981).

Palladium(II) acetate, 1, 778; **2,** 303; **4,** 365; **5,** 496–497; **6,** 442–443; **7,** 274–277; **8,** 378–382; **9,** 344–349.

Intramolecular cyclization via π-allylpalladium complexes.[1] The unsaturated β-keto ester **1** cyclizes to two ketones in the presence of $Pd(OAc)_2$ in combination with a phosphine. The ratio of **2** to **3** is independent of the phosphine, but is dependent on the solvent as indicated.

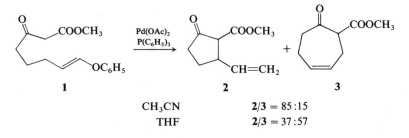

$$CH_3CN \qquad 2/3 = 85:15$$
$$THF \qquad 2/3 = 37:57$$

β-Aryl-α,β-unsaturated carbonyl compounds.[2] In the presence of Pd(OAc)$_2$–LiCl (1:5) aryl iodides react with allyl trimethylsilyl ethers in DMF to form (E)-β-aryl-α,β-unsaturated aldehydes and ketones. A similar reaction has been reported for allyl alcohols (7, 274).

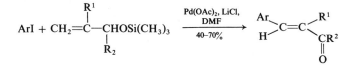

Arenecarboxylic acids.[3] Aryl diazonium tetrafluoroborates react with carbon monoxide in the presence of catalytic amounts of Pd(OAc)$_2$ to give aryl carboxylic acids in 70–85% yield.

2,2′-Bipyrroles.[4] Oxidation of 1-aroylpyrroles (1) with Pd(OAc)$_2$ in HOAc results in 1,1′-diaroyl-2,2′-bipyrroles (2), which are hydrolyzed by acidic aqueous methanol to 2,2′-bipyrroles (3).

Example:

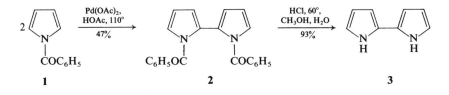

[1] J. Tsuji, Y. Kobayashi, H. Kataoka, and T. Takahashi, *Tetrahedron Letters*, **21**, 1475 (1980).
[2] T. Hirao, J. Enda, Y. Ohshiro, and T. Agawa, *Chem. Letters*, 403 (1981).
[3] K. Nagira, K. Kikukawa, F. Wada, and T. Matsuda, *J. Org.*, **45**, 2365 (1980).
[4] T. Itahara, *J.C.S. Chem. Comm.*, 49 (1980).

Palladium(II) acetate–Triphenylphosphine, 5, 497–498; **9,** 349–350.

Oxidation of secondary alcohols. Yoshida et al.[1] have oxidized secondary alcohols to the corresponding ketone using bromobenzene as oxidant and [(C$_6$H$_5$)$_3$P]$_4$Pd or Pd(OAc)$_2$ in conjunction with P(C$_6$H$_5$)$_3$ or

$(C_6H_5)_2PCH_2CH_2P(C_6H_5)_2$ as the catalyst. A base is also necessary; K_2CO_3 is more satisfactory than NaH, although oxidations with the latter base are faster. Isolated yields vary from 75 to 100%.

With some modifications the method can be extended to Δ^2-, Δ^3-, and Δ^4-unsaturated alcohols.[2] Bromomesitylene is superior to bromobenzene as oxidant, and $Pd(OAc)_2$ and $P(C_6H_5)_3$ in the ratio 1:2 is the most satisfactory catalyst. Either NaH or K_2CO_3 can be used as base, but yields are generally higher with the former reagent. Yields are in the range 65–100% Unfortunately the method is not generally applicable to primary allylic alcohols.

[1] Y. Tamaru, Y. Yamamoto, Y. Yamada, and Z. Yoshida, *Tetrahedron Letters*, 1401 (1979).
[2] Y. Tamaru, Y. Yamamoto, Y. Yamada, and Z. Yoshida, *Tetrahedron Letters*, **22**, 1801 (1981).

Palladium *t*-butyl peroxide trifluoroacetate (PPT), $CF_3CO_2PdOOC(CH_3)_3$ (1). Mol. wt. 296.52, stable orange crystals. This material can be prepared from $Pd(OCOCF_3)_2$ and *t*-BuOOH or from $Pd(OAc)_2$, *t*-BuOOH, and CF_3COOH (85% yield).

Oxidation of —CH=CH$_2$ *to* —COCH$_3$. Mimoun *et al.*[1] have prepared a number of reagents in which the CF_3 group of **1** is replaced by other groups. However **1** is the most effective for conversion of terminal alkenes to methyl ketones. Yields are high and the reaction is usually complete within an hour. The reaction can be catalytic with respect to **1** if *t*-BuOOH is present (equation I).

$$\text{(I)} \quad t\text{-BuOOH} + RCH{=}CH_2 \xrightarrow{\quad \mathbf{1} \quad} t\text{-BuOH} + RCOCH_3$$
$$(\geqslant 98\%)$$

[1] H. Mimoun, R. Charpentier, A. Mitschler, J. Fischer, and R. Weiss, *Am. Soc.*, **102**, 1047 (1980).

Palladium catalysts, 1, 778–782; **2**, 203; **4**, 368–369; **5**, 499; **6**, 445–446; **7**, 275–277; **8**, 382–383; **9**, 351–352.

Isomerization of allyl ethers.[1] Allyl ethers (and methallyl ethers) are isomerized to 1-propenyl ethers when refluxed in benzene with palladium on carbon (4–140 hours). The reaction is compatible with various functional groups (epoxide, aldehyde, hydroxyl). Isolated yields are about 80%.

Cleavage of protective groups of amino acids (**6**, 445). An example of use of catalytic hydrogenation with palladium black in liquid ammonia for cleavage of protective groups of a sulfur-containing peptide has been published (equation I). In this case only the terminal CBO group is cleaved; the *t*-butyl protective groups are not affected. Addition of DMF and triethylamine improves the yield.[2]

(I) $C_6H_5CH_2OCONHCHCONHCHCOOC(CH_3)_3$ $\xrightarrow[\text{82-86\%}]{\substack{H_2, \text{ Pd, } N(C_2H_5)_3, \\ \text{DMF, } NH_3, -33°}}$

$\quad\quad$ | $\quad\quad\quad\quad$ |
$\quad(CH_3)_3COCH_2 \quad\quad CH_2SC(CH_3)_3$

$\quad\quad\quad\quad\quad\quad\quad\quad H_2NCHCONHCHCOOC(CH_3)_3$
$\quad\quad\quad\quad\quad\quad\quad\quad\quad$ | $\quad\quad\quad\quad$ |
$\quad\quad\quad\quad\quad\quad\quad (CH_3)_3COCH_2 \quad\quad CH_2SC(CH_3)_3$

Cleavage of benzyl ethers.[3] Benzyl ethers are cleaved in high yield by catalytic transfer hydrogenation with 20% palladium on carbon (**1**, 782)[4] as catalyst and cyclohexene as hydrogen donor.

[1] H. A. J. Carless and D. J. Haywood, *J.C.S. Chem. Comm.*, 980 (1980).
[2] A. M. Felix, M. H. Jimenez, and J. Meienhofer, *Org. Syn.*, **59**, 159 (1979).
[3] S. Hanessian, T. J. Liak, and B. Vanasse, *Synthesis*, 396 (1981).
[4] W. M. Pearlman, *Tetrahedron Letters*, 1663 (1967).

Palladium(II) chloride, 1, 782; **3**, 303–305; **4**, 367–370; **5**, 500–503; **6**, 447–450; **7**, 277; **8**, 384–385; **9**, 352.

1,5-Dienes from π-allylpalladium chloride dimers.[1] Irradiation of π-allylpalladium chloride dimers (**4**, 369; **5**, 500–501; **6**, 447–448) at 366 nm results in dimerization to 1,5-dienes.

Examples:

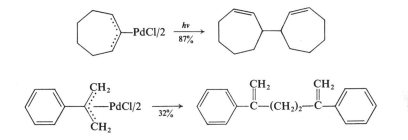

Thallation–carbonylation.[2] Catalytic amounts of $PdCl_2$ permit carbonylation of arylthallium ditrifluoroacetates (**3**, 287) at room temperature and atmospheric pressure. This sequence permits synthesis of esters, lactones, anhydrides, and various heterocycles.

Examples:

$C_6H_6 \xrightarrow{Tl(OCOCF_3)_3} [C_6H_5Tl(OCOCF_3)_2] \xrightarrow[\text{52\%}]{\substack{CO, PdCl_2, LiCl, \\ MgO, CH_3OH,}} C_6H_5COOCH_3$

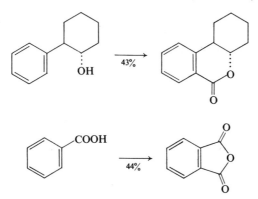

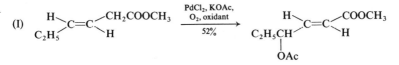

Acetoxylation of β,γ-unsaturated esters.[3] β,γ-Unsaturated esters are oxidized by $PdCl_2$–$CuCl$–O_2 to γ-keto esters in 20–40% yield. Reaction of these esters with $PdCl_2$ and KOAc and pentyl nitrite [oxidizes Pd(0)] under oxygen leads to γ-acetoxy-α,β-unsaturated esters (45–55% yield). The double bond has the (E)-configuration (equation I).

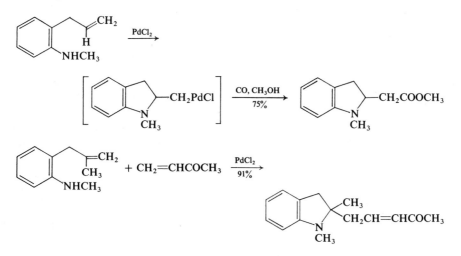

Cyclization–insertion. Hegedus *et al.*[4] have reported synthesis of some nitrogen-containing heterocycles by cyclization–insertion reactions proceeding through σ-alkylpalladium(II) intermediates.

Examples:

Wacker oxidation. Tsuji *et al.*[5] have developed two procedures for oxidation of 1-alkenes to methyl ketones with oxygen that are catalyzed by $PdCl_2$ (**7**, 278; **9**, 327). The solvent in both cases is aqueous DMF. One method uses $PdCl_2$–CuCl (molar ratio 1:10); the other uses $PdCl_2$ and *p*-benzoquinone (molar ratio 1:100). Both procedures are about equivalent for oxidation of simple 1-alkenes to methyl ketones, but the former method is usually more effective for oxidation of more complex 1-alkenes.

[1] J. Muzart and J.-P. Pete, *J.C.S. Chem. Comm.*, 257 (1980).
[2] R. C. Larock and C. A. Fellows, *J. Org.*, **45**, (1980).
[3] J. Tsuji, K. Sakai, H. Nagashima, and I. Shimizu, *Tetrahedron Letters*, **22**, 131 (1981).
[4] L. S. Hegedus, G. F. Allen, and D. J. Olsen, *Am. Soc.*, **102**, 3583 (1980).
[5] J. Tsuji, H. Nagashima, and H. Nemoto, *Org. Syn.*, submitted (1980).

Palladium(II) chloride–Copper(II) chloride, 9, 353.

Oxidative carbonylation. Terminal acetylenes are converted into acetylene-carboxylates by CO (1 atm.) and an alcohol in the presence of a catalytic amount of $PdCl_2$ and 1 equivalent of $CuCl_2$. In addition a base such as sodium acetate is necessary. The reaction is an oxidative carbonylation with $PdCl_2$, which is reduced to Pd(0). $PdCl_2$ can be used in catalytic amounts if $CuCl_2$ is available for oxidation of Pd(0) to $PdCl_2$. The base is needed to neutralize the HCl formed (equation I). The yields of the carboxylate are 60–75%.[1]

$$\text{(I)}\quad RC{\equiv}CH + CO + R^1OH + PdCl_2 \xrightarrow[60-70\%]{} RC{=}CCOOR^1 + Pd + 2HCl$$

$$Pd + 2CuCl_2 \longrightarrow PdCl_2 + 2CuCl$$

$$HCl + AcONa \longrightarrow NaCl + AcOH$$

[1] J. Tsuji, M. Takahashi, and T. Takahashi, *Tetrahedron Letters*, **21**, 849 (1980).

Palladium(II) trifluoroacetate, $Pd(OCOCF_3)_2$. Mol. wt. 332.44, brown powder, m.p. 210°(dec.). The salt can be prepared by reaction of $Pd(OAc)_2$ with excess TFA at 80°.[1]

π-Allylpalladium complexes.[2] This Pd salt reacts rapidly with acyclic alkenes to form π-allylpalladium dimers. The complexes are isolable but somewhat unstable. However they can be converted into the corresponding chlorides by tetra-*n*-butyl-ammonium chloride. In fact this sequence is often the method of choice for preparation of the latter complexes. Valuable features are that an excess of the alkene is not necessary and that even monosubstituted alkenes are reactive.

Exocyclic double bonds generally also form the expected complexes. However methylenecyclohexane and 1-methylcyclohexene are converted mainly into toluene.

This dehydrogenation with Pd(II) can become catalytic in the presence of maleic anhydride or dimethyl fumarate as hydride acceptor.

[1] T. A. Stephenson, S. M. Morehouse, A. R. Powell, J. P. Heffler, and G. Wilkinson, *J. Chem. Soc.*, 3632 (1965).

[2] B. M. Trost and T. J. Metzner, *Am. Soc.*, **102**, 3572 (1980).

(E)-(2,4-Pentadienyl)trimethylsilane (1). Mol. wt. 140.30, b.p. 41°/36 mm, stable.
Preparation[1]:

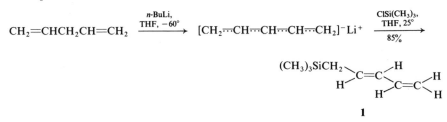

1

Pentadienylation. This allylic silane in the presence of $TiCl_4$ adds to aldehydes and ketones to give (E)-2,4-pentadienyl substituted alcohols of the type $CH_2=CHCH=CHCH_2C(OH)R^1R^2$.

Examples:

$$CH_3CH_2CH=O + 1 \xrightarrow[79\%]{TiCl_4, CH_2Cl_2} \underset{(E)}{CH_2=CHCH=CHCH_2}\overset{OH}{\overset{|}{C}}HCH_2CH_3$$

$$(C_2H_5)_2C=O + 1 \xrightarrow[55\%]{} \underset{(E)}{CH_2=CHCH=CHCH_2}\overset{OH}{\overset{|}{C}}(C_2H_5)_2$$

The reaction cannot be used with α,β-enones or α,β-enals because of a competitive Diels-Alder reaction induced by $TiCl_4$.

Pentadienyllithium is less useful than **1** for pentadienylation because a mixture of isomeric alcohols is formed.

The (Z)-isomer of **1** has also been prepared.[2]

[1] D. Seyferth and J. Pornet, *J. Org.*, **45**, 1721 (1980).

[2] H. Yasuda, M. Yamauchi, and A. Nakamura, *J. Organometal. Chem.*, **202**, C1 (1980).

Pentafluorophenylcopper, 5, 504–505, m.p. 200°(dec.). Detailed directions for preparation of the reagent, which exists as the tetramer, and of the dioxane complex, have been published (equation I). The report includes use for the preparation of (pentafluorophenyl)benzene (79% yield) and 1-(pentafluorophenyl)-adamantane (93% yield).[1]

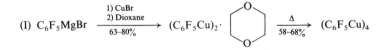

$$\text{(I)} \ C_6F_5MgBr \ \xrightarrow[\substack{63-80\%}]{\substack{1) \ CuBr \\ 2) \ Dioxane}} \ (C_6F_5Cu)_2 \cdot \quad \xrightarrow[58-68\%]{\Delta} \ (C_6F_5Cu)_4$$

[1] A. Cairncross, W. A. Sheppard, and E. Wonchoba, *Org. Syn.*, **59**, 122 (1979).

Perchloryl fluoride, 1, 802–803; **2,** 310–311; **6,** 454; **7,** 280.

Hydroxylation of an enone.[1] The enol ether **2** of the α,β-enone (**1**) reacts with this reagent to form the γ-hydroxy derivative (**3**) of the enone.

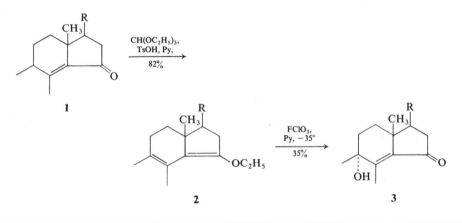

Caution: On one occasion of several repetitions of this reaction a violent explosion occurred.

[1] G. J. Schroepfer, E. J. Parish, M. Tsuda, and A. A. Kandutsch, *Biochem. Biophys. Res. Comm.*, **91**, 606 (1979).

Periodic acid, 1, 815–819; **2,** 313–315; **3,** 220; **4,** 374–375; **5,** 508–510.

Oxidative cleavage of ethylene thioketals.[1] Steroidal ketones can be regenerated from ethylene thioketals by oxidation with HIO_4. The reaction is rapid, and yields are almost quantitative.

[1] J. Cairns and R. T. Logan, *J.C.S. Chem. Comm.*, 886 (1980).

Permonophosphoric acid, $(HO)_2\overset{\overset{\displaystyle O}{\|}}{P}OOH$. This peracid is prepared by oxidation of P_2O_5 in CH_3CN with 90% H_2O_2.

Oxidations. This peracid has been known since 1910 but a report of its use in oxidation of organic compounds has appeared only recently. It hydroxylates phenol, anisole, and toluene in CH_3CN at the *ortho*- and *para*-positions, with the former predominating.[1] It converts alkenes into epoxides, but the H_3PO_4 formed can cleave the epoxide ring.[2] It is superior to the usual peroxycarboxylic acids for Baeyer-Villiger oxidation of acetophenones to arylacetates.[3]

[1] Y. Ogata, I. Urasaki, K. Nagura, and N. Satomi, *Tetrahedron*, **30**, 3021 (1974).
[2] Y. Ogata, K. Tomizawa, and T. Ikeda, *J. Org.*, **44**, 2362 (1979).
[3] Y. Ogata, K. Tomizawa, and T. Ikeda, *J. Org.*, **44**, 2417 (1979).

Phase-transfer catalysts, 8, 387–391; **9**, 356–361.

β-Lactams.[1] β-Amino acids are converted to β-lactams by dehydration with methanesulfonyl chloride under phase-transfer conditions. Tetra-*n*-butylammonium hydrogen sulfate is the most useful ammonium salt (equation I).

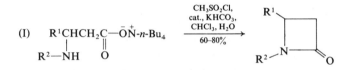

$$
\text{(I)} \quad \underset{\underset{R^2-NH}{|}}{R^1CHCH_2}\underset{\overset{\|}{O}}{C}-\overset{+}{O}\overline{N}\text{-}n\text{-}Bu_4 \quad \xrightarrow[60-80\%]{\substack{CH_3SO_2Cl,\\ cat., KHCO_3,\\ CHCl_3, H_2O}} \quad
$$

Ethyl N-alkylcarbamates via hydroboration. The reaction of trialkylboranes with carboethoxynitrene is best conducted under phase-transfer conditions (equation I). Either benzyltriethylammonium chloride or methyl trioctylammonium chloride is a satisfactory catalyst.[2]

$$
\text{(I)} \quad O_2N-\text{⟨aryl⟩}-SO_3NHCOOC_2H_5 \quad \xrightarrow{\substack{Na_2CO_3,\\ H_2O}}
$$

$$
[:\ddot{N}COOC_2H_5] \quad \xrightarrow[CH_2Cl_2]{R_3B,} \quad R_2B\overset{\overset{R}{|}}{N}COOC_2H_5 \quad \xrightarrow[\substack{70-97\%\\ \text{overall}}]{H_2O} \quad RNHCOOC_2H_5
$$

Amides from nitriles. One classical reagent for this reaction is H_2O_2–NaOH in a suitable solvent.[3] This reaction can be carried out advantageously under phase-transfer catalyzed conditions.[4] Tetra-*n*-butylammonium hydrogen sulfate is satisfactory; the effectiveness varies with the structure of the nitrile. An excess of 30% H_2O_2 is used; the solvent system is CH_2Cl_2–20% NaOH. Yields are 85–95%.

Furanones.[5] The first step in a simplified synthesis of bislactones involves condensation of an α-halo aldehyde (**1**) and potassium ethyl malonate (**2**) in benzene–water in the presence of tetra-*n*-butylammonium bromide. The furanone **3** is converted to the bislactone **4** by bromination followed by treatment with HBr. The product is a precursor to *dl*-avenaciolide (**5**), a fungicide from *Aspergillus avenaceaus.*

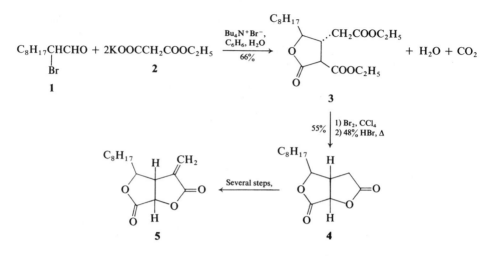

Oxidations with chromium trioxide.[6] Secondary alcohols can be oxidized to ketones in good yields by CrO_3 in the presence of catalytic amounts of tetraalkylammonium halides. Yields from oxidation of primary alcohols are moderate.

Permethylation of carbohydrates.[7] Permethylation of carbohydrates in a solid-liquid system is not possible because of lack of solubility in benzene or methylene chloride. The difficulty can be circumvented by reaction of peracetylated sugars with methyl bromide in C_6H_6–aqueous NaOH with tetrabutylammonium hydrogen sulfate as catalyst.

Azo coupling.[8] Coupling between 4-nitrobenzenediazonium chloride and various arylamines in an H_2O–CH_2Cl_2 system is markedly catalyzed by sodium 4-dodecylbenzenesulfonate. Crown ethers are not effective catalysts.

[1] Y. Watanabe and T. Mukaiyama, *Chem. Letters*, 443 (1981).
[2] I. Akimoto and A. Suzuki, *Syn. Comm.*, **11**, 475 (1981).
[3] C. R. Noller, *Org. Syn. Coll. Vol.*, **2**, 586 (1943); J. S. Buck, *Org. Syn. Coll. Vol.*, **2**, 44 (1943).
[4] C. Cacchi, D. Misiti, and F. La Torre, *Synthesis*, 243 (1980).
[5] T. Sakai, H. Horikawa, and A. Takeda, *J. Org.*, **45**, 2039 (1980).
[6] G. Gelbard, T. Brunelet, and C. Jouitteau, *Tetrahedron Letters*, **21**, 4653 (1980).
[7] P. DiCesare, P. Duchaussoy, and B. Gross, *Synthesis*, 953 (1980).
[8] M. Ellwood, J. Griffiths, and P. Gregory, *J.C.S. Chem. Comm.*, 181 (1980).

$$S$$
$$\|$$
Phenyl chlorothionocarbonate, C_6H_5OCCl (**1**). Mol. wt. 218.11, b.p. 81–83°/6 mm. The reagent is prepared by reaction of phenol with thiophosgene in aqueous NaOH.[1]

Deoxygenation of alcohols.[2] Secondary alcohols react with **1** in the presence of pyridine or 4-dimethylaminopyridine to form thiono esters (**2**), which are cleaved

$$\text{ROH} + 1 \xrightarrow{\text{Py}} \underset{2}{\text{ROCOC}_6\text{H}_5} \xrightarrow[\text{AIBN}]{n\text{-Bu}_3\text{SnH,}} \text{RH}$$

reductively by tri-n-butyltin hydride.[3] For example, this sequence converts cholesterol to Δ^5-cholestene in 85% yield. It is particularly useful for conversion of ribonucleosides to 2'-deoxyribonucleosides. The 3'- and 5'-hydroxyl groups are protected by the cyclic disiloxane 1,3-dichloro-1,1,3,3-tetraisopropyldisiloxane (3).[4] The resulting product (5) is thioacylated, reductively cleaved, and deprotected with tetra-n-butylammonium fluoride. An example is the conversion of adenosine (4) to 2'-deoxyadenosine (6) in 78% overall yield.

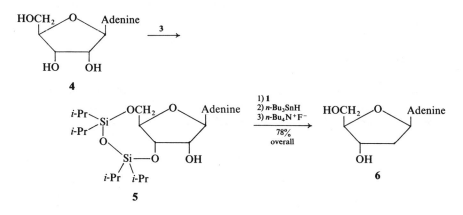

[1] M. Miyazaki, and K. Nakanishi, Japanese Patent 1322 (1957) [C.A., **52**, 4684g (1958)].
[2] M. J. Robins and J. S. Wilson, *Am. Soc.*, **103**, 932 (1981).
[3] D. H. R. Barton *et al.*, *J.C.S. Perkin I*, 1574 (1975); 1718 (1977).
[4] W. T. Markiewicz, *J. Chem. Res. (S)*, 24 (1979).

Phenyl cyanate (1). Mol. wt. 119.12, b.p. 60°/4 mm.
    Preparation:

$$\text{C}_6\text{H}_5\text{OH} + \text{BrCN} \xrightarrow[\underset{80\%}{\text{ether, pentane}}]{\text{N(C}_2\text{H}_5)_3,} \underset{1}{\text{C}_6\text{H}_5\text{OC}{\equiv}\text{N}}$$

α,β-Unsaturated nitriles.[1] The reagent is useful for conversion of lithium acetylides or 1-lithio-1-alkenes to α,β-unsaturated nitriles.
    Examples:

$$\text{RC}{\equiv}\text{CLi} \xrightarrow[85\text{--}95\%]{1,\ \text{ether}} \text{RC}{\equiv}\text{CCN}$$

$$C_6H_{13}-C=C-H \quad \xrightarrow[80\%]{1, \text{ ether}, -70°} \quad C_6H_{13}-C=C-H$$
with Li and CN substituents

[1] R. E. Murray and G. Zweifel, *Synthesis*, 150 (1980).

Phenyldiazomethane, 1, 834. A new method for preparation of this (and other aryl-diazomethanes) involves a vacuum pyrolysis of the sodium salt of benzaldehyde tosylhydrazone, a method introduced for carrying out the Bamford-Stevens reaction. The yield is 80%, the highest yield yet reported. Another advantage is that the reagent is obtained free from solvents. The pyrolysis can also be conducted in ethylene glycol at 80° with extraction of the aryldiazomethane into hexane.[1]

Caution: All diazo compounds are highly toxic and potentially explosive.

[1] X. Creary, *Org. Syn.*, submitted (1981).

α-Phenylglycine methyl ester, 9, 395–396.

β,γ-Unsaturated ketones.[1] The last step in a synthesis of β,γ-enones involves hydrolysis of a 2-allyl-3-oxazoline-5-one, previously conducted with Ba(OH)$_2$ in variable yield. Recent investigations reveal that treatment with Cr(OAc)$_2$–H$_3$PO$_2$ or reduction with NaBH$_4$ in THF–CH$_3$OH followed by a treatment with citric acid is superior.

Examples:

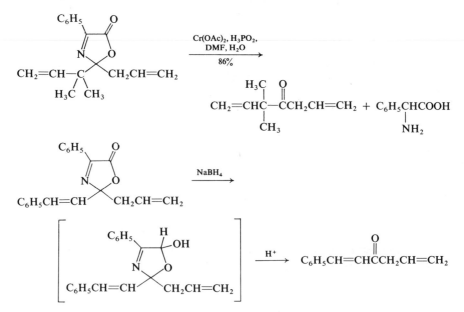

[1] U. Niewohner and W. Steglich, *Angew. Chem. Int. Ed.*, **20**, 395 (1981).

Phenyl isocyanate, 1, 843–844; **2**, 322–323; **4**, 378; **7**, 284–285; **8**, 396.

Nitrile oxide [3 + 2]cycloaddition.[1] A key step in a recent stereospecific synthesis of biotin (**6**) from cycloheptene (**1**) is an intramolecular [3 + 2]cyclo-addition of a nitrile oxide (**a**), obtained by dehydration of a primary nitro compound (**3**), preferably with phenyl isocyanate. This cycloaddition is more efficient than the well-known olefinic nitrone cycloaddition. The carbon atoms in **6** derived from cycloheptene are marked with asterisks.

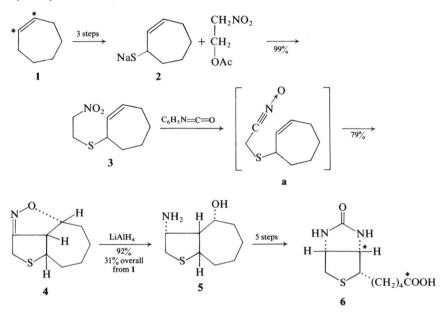

[1] P. N. Confalone, G. Pizzolato, D. L. Confalone, and M. R. Uskokovic, *Am. Soc.*, **102**, 1954 (1980).

Phenyl isocyanide, 1, 843–844.

α-Amino acids from α-methoxyurethanes.[1] Reaction of phenyl isocyanide in the presence of TiCl$_4$ with α-methoxyurethanes (**1**), available by anodic oxidation of urethanes in methanol,[2] results in amides (**2**) of α-amino acids in reasonable yield.

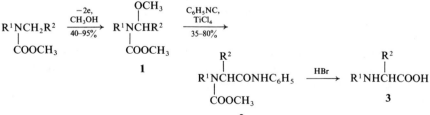

[1] T. Shono, Y. Matsumura, and K. Tsubata, *Tetrahedron Letters*, **22**, 2411 (1981).
[2] T. Shono, Y. Matsumura, and H. Hamaguchi, *Am. Soc.*, **97**, 4264 (1975).

Phenyl N-phenylphosphoramidochloridate, $\begin{array}{c} C_6H_5O \\ C_6H_5HN \end{array} P \begin{array}{c} O \\ Cl \end{array}$ (**1**), 7, 286.

Carboxylic acid anhydrides.[1] Carboxylic acids are converted to the symmetrical anhydride by reaction with **1** and triethylamine in acetone or methylene chloride at room temperature (equation I).

$$(I) \ RCOOH + 1 \xrightarrow{N(C_2H_5)_3} \begin{array}{c} C_6H_5O \\ C_6H_5HN \end{array} P \begin{array}{c} O \quad O \\ \| \\ O-CR \end{array} \xrightarrow{RCOOH}$$

$$\begin{array}{c} O \\ \| \\ (RC)_2O \\ 90-98\% \end{array} + \begin{array}{c} C_6H_5O \\ C_6H_5HN \end{array} P \begin{array}{c} O \\ OH \end{array}$$

$$70\% \downarrow PCl_5$$

$$1$$

Diphenyl phosphorochloridate can be used, but it is more sensitive to moisture than **1**.

[1] R. Mastres and C. Palomo, *Synthesis*, 218 (1981).

Phenylselenoacetaldehyde, $OHCCH_2SeC_6H_5$ (**1**), **9**, 365–366. Two preparations have been reported (equations I[1] and II[2]).

$$(I) \ C_2H_5OCH=CH_2 \xrightarrow[96\%]{C_6H_5SeBr,\ C_2H_5OH} (C_2H_5O)_2CHCH_2SeC_6H_5 \xrightarrow[98\%]{H_3O^+} 1$$

$$(II) \ C_2H_5OCH=CH_2 \xrightarrow[80\%]{\substack{1)\ C_6H_5SeBr,\ 0° \\ 2)\ H_3O^+}} 1$$

α-Vinylation of ketones.[2] This reaction can be accomplished by reaction of **1** with the enolate of a ketone such as **2**, reduction of the product (**3**) to a diol (**4**), and finally elimination of C_6H_5SeOH by the method of Krief *et al.*[3] Overall yields are 65–78%.

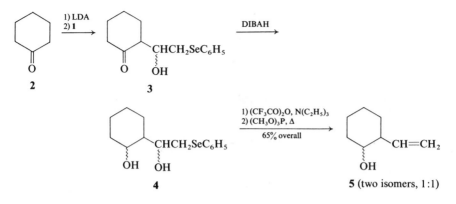

2   3

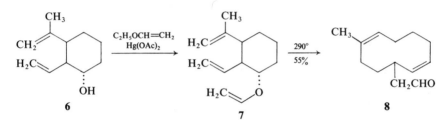

4

5 (two isomers, 1:1)

This vinylation reaction was used in a two-step synthesis of **6** in about 60% yield from cyclohexenone. This product is of interest because the derived triene (**7**) is converted upon heating into the 1,6-cyclododecadiene **8** by a Cope rearrangement followed by a Claisen rearrangement.[4]

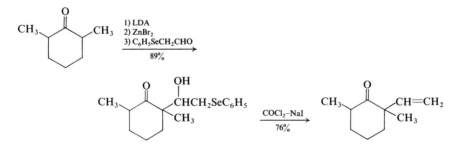

6   7   8

β,γ-Unsaturated ketones. Zinc (or boron) enolates of ketones undergo condensation with phenylselenoacetaldehyde to form β-hydroxy selenides. These products are usually converted to β,γ-unsaturated ketones by treatment with methanesulfonyl chloride and triethylamine.[5] α-Isopropenylation of a ketone is possible with phenylselenoacetone.[6]

Examples:

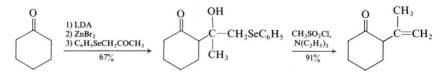

[1] M. Petrzilka and R. Baudat, *Helv.*, **62**, 1406 (1979).

[2] C. J. Kowalski and J.-S. Sung, *Am. Soc.*, **102**, 7950 (1980).

[3] J. Rémion, W. Dumont, and A. Krief, *Tetrahedron Letters*, 1385 (1976).

[4] F. E. Ziegler and J. J. Piwinski, *Am. Soc.*, **101**, 1611 (1979).

[5] D. L. J. Clive and C. G. Russell, *J.C.S. Chem. Comm.*, 434 (1981).

[6] K. B. Sharpless, R. F. Lauer, and A. Y. Teranishi, *Am. Soc.*, **95**, 6137 (1973); I. Ryu, S. Murai, I. Niwa, and N. Sonoda, *Synthesis*, 874 (1977).

N-Phenylselenophthalimide, 9, 366–367.

Cyclization of some β-dicarbonyl compounds.[1] In the presence of this reagent and a catalyst (I_2, *p*-TsOH, ZnI_2), β-dicarbonyl compounds containing an alkenyl substituent undergo cyclization with incorporation of a phenylseleno group.

Examples:

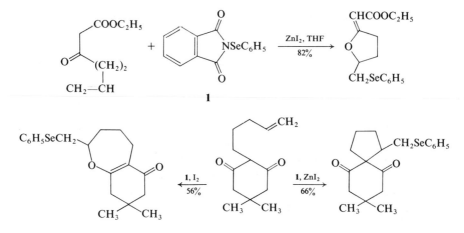

Alkyl phenyl selenides; selenol esters. N-Phenylselenophthalimide is superior to aryl selenocyanates for conversion of alcohols to alkyl phenyl selenides (**6**, 252–253) and of carboxylic acids to selenol esters (**7**, 396–397). When conducted in the presence of an amine the latter reaction provides amides in high yield (equation I).[2]

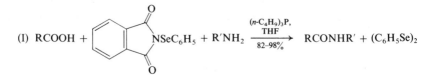

[1] W. P. Jackson, S. V. Ley, and J. A. Morton, *J.C.S. Chem. Comm.*, 1028 (1980).
[2] P. A. Grieco, J. Y. Jaw, D. A. Claremon, and K. C. Nicolaou, *J. Org.*, **46**, 1215 (1981).

3-Phenylsulfonyl-1(3*H*)-isobenzofuranone, 9, 368–370.

Anthraquinones.[1] The 4- or 7-methoxy derivatives of this phthalide (**1**) undergo regioselective conjugate addition to quinone monoacetals **2** to give an enolate that undergoes ring closure followed by elimination of the phenylsulfonyl group to give a triketone, which loses methanol to form an anthraquinone. The sequence constitutes essentially a one-step synthesis. An example is the synthesis of the dimethyl ether of digitopurpone (**3**).

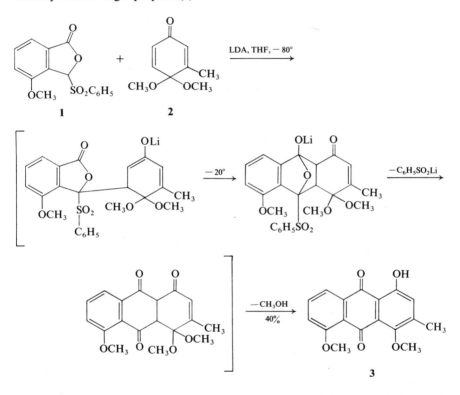

This synthesis is more versatile than a related reaction of lithium phthalide with benzynes (**9**, 285).

[1] R. A. Russell and R. N. Warrener, *J.C.S. Chem. Comm.*, 108 (1981).

Phenylthiomethyl(trimethyl)silane, $C_6H_5SCH_2Si(CH_3)_3$ (**1**). Mol. wt. 196.39, b.p. 158.5°/52 mm. The silane can be prepared by reaction of phenylthiomethyllithium (from thioanisole) with $ClSi(CH_3)_3$; 95% yield.[1]

Formaldehyde anion synthon ($^-$CHO). The anion of **1** (*n*-BuLi, THF, 0°) is readily alkylated, particularly by primary halides. The products can be converted into aldehydes under very mild conditions. Oxidation with *m*-chloroperbenzoic acid gives an unstable sulfoxide, which undergoes an sila-Pummerer rearrangement to an acetal. Addition of water liberates the free aldehyde. Epoxides can also be used as electrophiles.[2,3]

Example:

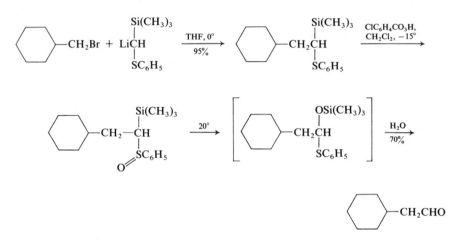

[1] H. Gilman and F. J. Webb, *Am. Soc.*, **62**, 98 (1940); G. D. Cooper, *Am. Soc.* **76**, 3713 (1954).
[2] P. J. Kocienski, *Tetrahedron Letters*, **21**, 1559 (1980).
[3] D. J. Ager and R. C. Cookson, *Tetrahedron Letters*, **21**, 1677 (1980).

Phenylthiophenyl(trimethylsilyl)methane (1). Mol. wt. 252.48.
  Preparation:

$$C_6H_5CH_2Br \xrightarrow[95\%]{C_6H_5SNa} C_6H_5CH_2SC_6H_5 \xrightarrow[89\%]{\substack{1)\ n\text{-BuLi, TMEDA} \\ 2)\ ClSi(CH_3)_3}} C_6H_5CH{\overset{SC_6H_5}{\underset{Si(CH_3)_3}{\big<}}}$$

1

Alkyl phenyl ketones.[1] Alkylation of the anion of **1** followed by oxidation and rearrangement gives a mixed acetal, which is hydrolyzed by acid to a ketone (*cf.* Phenylthiomethyl(trimethyl)silane, this volume).

[1] D. J. Ager, *Tetrahedron Letters*, **21**, 4759 (1980).

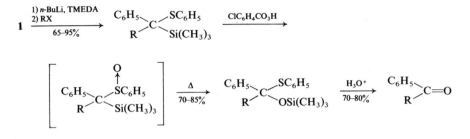

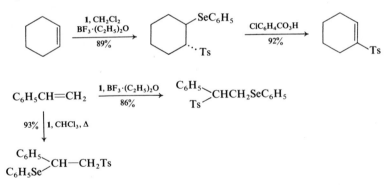

Se-Phenyl *p*-tolueneselenosulfonate, $CH_3C_6H_4SO_2SeC_6H_5$ (1). Mol. wt. 311.26, m.p. 80°.

Preparation[1]:

$$CH_3C_6H_4SO_2NHNH_2 + C_6H_5SeO_2H \xrightarrow[92\%]{} 1 + N_2 + 2H_2O$$

Addition to alkenes.[2] The reagent (1) adds to unhindered alkenes to form anti-Markownikoff adducts; the same reaction catalyzed by BF_3 etherate gives the Markownikoff adduct. The products on oxidation with *m*-chloroperbenzoic acid give vinyl sulfones.

Examples:

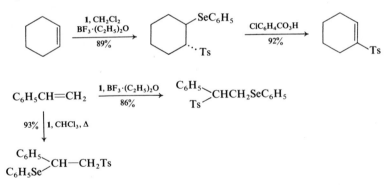

[1] T. G. Back and S. Collins, *Tetrahedron Letters*, **21**, 2213 (1980).
[2] T. G. Back and S. Collins, *Tetrahedron Letters*, **21**, 2215 (1980); *J. Org.*, **46**, 3249 (1981).

Phenyl vinyl sulfone, $CH_2=CHSO_2C_6H_5$ (1). Mol. wt. 168.21, m.p. 66–67°.

Preparation[1]:

$$CH_3CH_2SC_6H_5 \xrightarrow[]{\substack{1)\ SOCl_2,\ CH_2Cl_2 \\ 2)\ Py,\ \Delta}} CH_2=CHSC_6H_5 \xrightarrow[\substack{60\% \\ overall}]{H_2O_2,\ HOAc} 1$$

Diels-Alder reactions.[2] This sulfone can serve as an equivalent in Diels-Alder reactions to ethylene, which is not reactive. An added attraction of this reaction is that the adducts can be alkylated regiospecifically.

Examples:

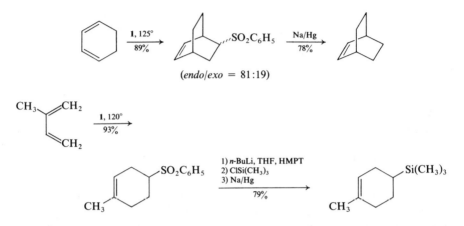

$(endo/exo = 81:19)$

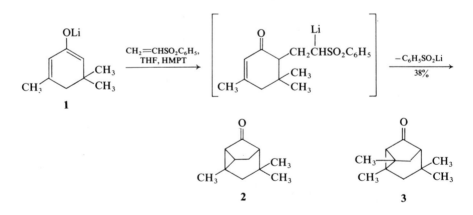

Bicycloannelation.[3] The 2'-enolate (**1**) of isophorone reacts with phenyl vinyl sulfone in THF containing 4 equivalents of HMPT (essential) to give the bicyclo-annelation product **2** in 38% yield. The yield is considerably greater than that obtained from the same reaction using vinyltriphenylphosphonium bromide. Reaction of **1** with isopropenyl phenyl sulfone gives **3** in 21% yield.

[1] F. G. Bordwell and B. M. Pitt, *Am. Soc.*, **77**, 572 (1955); H. Böhme and H. Benther, *Ber.*, **89**, 1464 (1956).
[2] R. V. C. Carr and L. A. Paquette, *Am. Soc.*, **102**, 853 (1980).
[3] R. M. Cory and R. M. Renneboog, *J.C.S. Chem. Comm.*, 1081 (1980).

Phosphoric acid, 1, 860; **4**, 387; **7**, 255.

Isoprenylation of polyhydroxyacetophenones.[1] Direct condensation of isoprene with some polyhydroxyacetophenones is possible in this acid medium. The resulting 2,2-dimethylchromanes can be dehydrogenated to chromenes with DDQ. A typical reaction is outlined in equation (I).

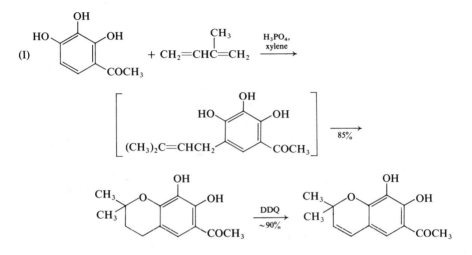

[1] V. K. Ahluwalia and K. K. Arora, *Tetrahedron*, **37**, 1437 (1981).

Phosphoric acid–Formic acid.

2,3-Disubstituted 2-cyclopentenones.[1] Cross-conjugated dienones, or the corresponding ethylene ketals, are cyclized almost entirely to 2,3-disubstituted 2-cyclopentenones by H_3PO_4–HCOOH (1:1) or HBr–CH_3COOH (1:3). 3,4-Disubstituted 2-cyclopentenones are formed, if at all, in low yield.

Examples:

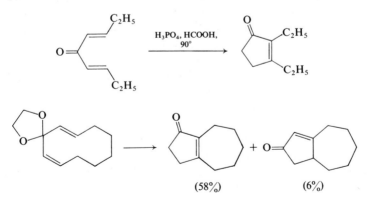

The suggested mechanism includes dehydration of a 2-hydroxycyclopentanone and, indeed, dihydrojasmone (2) was obtained in 74% yield by simple distillation of 1.

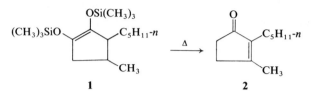

1   2

[1] S. Hirano, S. Takagi, T. Hiyama, and N. Nozaki, *Bull. Chem. Soc. Japan*, **53**, 169 (1980).

Phosphorus(III) chloride, 1, 875–876; 8, 400.

Fisher indole synthesis.[1] 2,3-Disubstituted indoles are formed in 70–90% yield when a ketone phenylhydrazone is treated in benzene with PCl_3 at 25°. The same indoles are formed when the ketone and phenylhydrazine are treated with PCl_3. PCl_5 is less satisfactory than PCl_3. The method is not applicable to aldehydes.

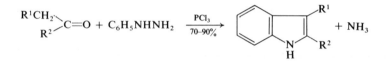

[1] G. Baccolini and P. E. Todesco, *J.C.S. Chem. Comm.*, 563 (1981).

Phosphorus(III) iodide, PI_3. Mol. wt. 411.71. The reagent is obtained by reaction of white phosphorus with 1.3 moles of I_2 in CS_2. Supplier: Alfa.

Deoxygenation of epoxides.[1] Di-, tri-, and tetrasubstituted epoxides are converted into alkenes by treatment with PI_3 (or P_2I_4). The reaction occurs most readily with α,β-disubstituted epoxides and with $> 98\%$ retention of configuration. With more substituted epoxides addition of pyridine or triethylamine improves yields by preventing formation of alkyl iodides.

Deoxygenation of sulfoxides, selenoxides, oximes, and nitroalkanes.[2] The reduction of sulfoxides and selenoxides is possible with PI_3 or P_2I_4 in CH_2Cl_2 even at temperatures of $-78°$. Yields are 70–95%.

Aldehyde oximes and primary nitroalkanes are converted to nitriles by PI_3 or P_2I_4[3] in combination with triethylamine.

Alkylidene cyclopropanes.[4] These compounds (3) can be prepared by reaction of aldehydes and ketones with 1-methylseleno-1-lithiocyclopropane (1) to give β-hydroxycyclopropyl selenides (2). Elimination of CH_3SeOH is then effected with N,N'-carbonyldiimidazole or PI_3 for substrates derived from ketones.

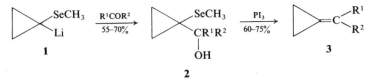

1 → R^{1}COR^2 / 55–70% → **2** → PI$_3$ / 60–75% → **3**

Reduction of α-halo ketones.[5] α-Iodo and α-bromo ketones are reduced by either PI$_3$ or P$_2$I$_4$ at room temperature in yields of about 75–90%.

[1] J. N. Denis, R. Magnane, M. Van Eenoo, and A. Krief, *Nouv. J. Chim.*, **3**, 705 (1979).
[2] J. N. Denis and A. Krief, *J.C.S. Chem. Comm.*, 544 (1980).
[3] J. N. Denis and A. Krief, *Tetrahedron Letters*, 3995 (1979).
[4] S. Halazy and A. Krief, *J.C.S. Chem. Comm.*, 1136 (1979).
[5] J. N. Denis and A. Krief, *Tetrahedron Letters*, **22**, 1431 (1981).

Phosphorus(V) oxide, 1, 871–872; **7,** 291.

Tetralins; indanes. P$_2$O$_5$ can be used for cyclodehydration for substrates such as 1-aryl-1-pentanols or 2-aryl-2-hexanols. In the case of secondary alcohols, yields are sometimes improved by dehydration with TsOH to a styrene followed by cyclization with P$_2$O$_5$.[1] The product of the second example is a known sesquiterpene, calamenene.

Examples:

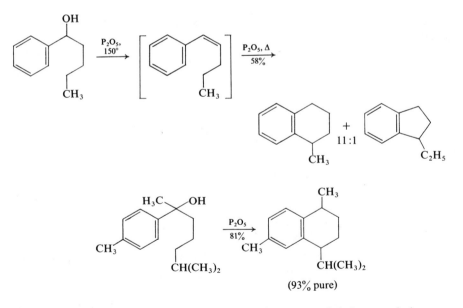

(93% pure)

This cyclodehydration can also be used to obtain some naphthalenes and phenanthrenes (equation I).[2]

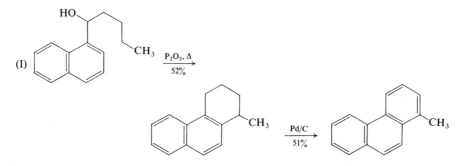

[1] F. E. Condon and D. L. West, *J. Org.*, **45**, 2006 (1980).
[2] F. E. Condon and G. Mitchell, *J. Org.*, **45**, 2009 (1980).

Phosphorus(V) sulfide, 1, 870–871; **3**, 226–228; **4**, 389; **5**, 534–535; **8**, 401; **9**, 374.

Deoxygenation of sulfines.[1] Sulfines are converted to thiones by reaction with P_4S_{10} in CH_2Cl_2 for 1–48 hours at 25°. The homogeneous reagent thiophosphoryl bromide, $PSBr_3$,[2] is somewhat more reactive, but work-up is more difficult.

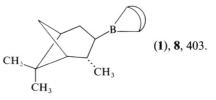

[1] J. A. M. Kuipers, B. H. M. Lammerink, I. W. J. Still, and B. Zwanenburg, *Synthesis*, 295 (1981).
[2] I. W. J. Still and K. Turnbull, *Tetrahedron Letters*, **22**, 1481 (1981).

B-(3)-α-Pinanyl-9-borabicyclo[3.3.1]nonane,   **(1)**, **8**, 403.

The borane is prepared from (+)-α-pinene and 9-BBN.

Asymmetric reduction of α,β-acetylenic ketones. This borane can be used to reduce 1-deuterio aldehydes to chiral (S)-1-deuterio primary alcohols in 90% optical yields. It also reduces α,β-acetylenic ketones to (R)-propargylic alcohols with enantiomeric purity of 73–100%. The ee value is increased by an increase in the size of the group attached to the carbonyl group. The value is also higher in reductions of terminal ynones. Alcohols of the opposite configuration can be obtained with the reagent prepared from (−)-α-pinene.

Johnson *et al.*[3] used this reagent for reduction of **2**, an intermediate in a synthesis of 11α-hydroxy steroids. The desired enantiomer (**3**) was obtained with an optical yield of 97%. Asymmetric reduction of a similar intermediate had been effected previously with LiAlH$_4$–Darvon alcohol with an optical yield of at most 84% (**8**, 185–186).

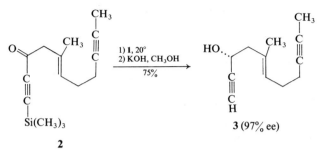

2

3 (97% ee)

This asymmetric reduction has been used for synthesis of optically active 4-alkyl-γ-lactones (equation I).[4]

(I) $C_2H_5\overset{O}{\underset{\|}{C}}C\equiv CCOOC_2H_5 \xrightarrow[58\%]{1} C_2H_5\overset{OH}{\underset{*|}{C}}HC\equiv CCOOC_2H_5 \xrightarrow[\substack{1)\ H_2,\ Pd/C \\ 2)\ H_3O^+}]{}$

(88% ee)

(87% ee)

[1] M. M. Midland, D. C. McDowell, R. L. Hatch, and A. Tramontano, *Am. Soc.*, **102**, 867 (1980).

[2] M. M. Midland, S. Greer, A. Tramontano, and S. A. Zderic, *Am. Soc.*, **101**, 2352 (1979).

[3] W. S. Johnson, B. Frei, and A. S. Gopalan, *J. Org.*, **46**, 1512 (1981).

[4] M. M. Midland and A. Tramontano, *Tetrahedron Letters*, **21**, 3549 (1980).

Platinum catalysts, 1, 890–892; **3**, 332–333; **8**, 405.

Dehydrogenation of reducing sugars.[1] In an alkaline medium and in the presence of Pt/C, Pt black, or Rh/C, reducing sugars produce hydrogen with formation of aldonic acids. The ease of dehydrogenation is influenced by structural features, which can be summarized as follows: 4a-OH > 4e-OH > 2e-OH > 2-H > 2a-OH > 3e-OH ⩾ 3a-OH; at the 6-position CH$_2$OH > CH$_3$ > H > CO$_2^-$. In addition, D-glucose can serve as a hydrogen donor in transfer hydrogenations.

[1] G. de Wit, J. J. de Vlieger, A. C. Kock-van Dalen, R. Heus, R. Laroy, A. J. van Hengstum, A. P. G. Kieboom, and H. van Bekkum, *Carbohydrate Res.*, **91**, 125 (1981).

Polyphosphoric acid, 1, 894–895; **2**, 334–336; **3**, 231–233; **4**, 395–397; **5**, 590–592; **6**, 474–475; **7**, 294–295.

Heterogeneous reactions. Work-up of reactions conducted wth PPA is greatly facilitated by use of a cosolvent for the organic substrate and by vigorous stirring, which also permits better control of the temperature.[1] Of various solvents examined xylene is most satisfactory. Yields are the same or even higher than those obtained with PPA alone. Cyclodehydration, Beckmann rearrangement, and the Fischer indole synthesis can be conducted satisfactorily by this modification.

[1] A. Guy, J.-P. Guetté, and G. Lang, *Synthesis*, 222 (1980) .

Potassium–Alumina. The reagent is prepared by melting potassium over alumina with vigorous stirring. The material is sensitive to oxygen and moisture.

Reductive decyanation.[1] K–Al$_2$O$_3$ reduces alkyl nitriles in hexane at 25° in 70–90% yield. The reaction is rapid for secondary and tertiary nitriles but requires about 1 hour for primary nitriles. Potassium alone is less efficient. The reaction was useful in a synthesis of the sex pheromone **1**.

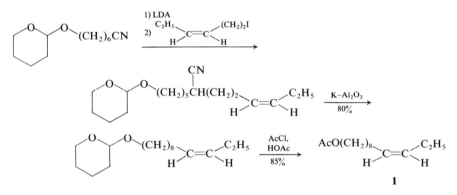

[1] D. Savoia, E. Tagliavini, C. Trombini, and A. Umani-Ronchi, *J. Org.*, **45**, 3227 (1980).

Potassium–18-Crown-6, 9, 387.

Deoxygenation of alcohols.[1] Thiocarbonate derivatives of alcohols are conveniently reduced to alkanes by potassium and 18-crown-6 in *t*-butylamine.

Example:

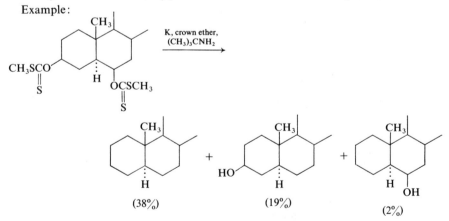

[1] A. G. M. Barrett, P. A. Prokopiou, and D. H. R. Barton, *J.C.S. Perkin I*, 1510 (1981).

Potassium *t*-butoxide, 1, 911–927; **2,** 338–339; **3,** 233–234; **4,** 399–405; **5,** 544–553; **6,** 477–479; **7,** 296–298; **8,** 407–408; **9,** 380–381.

Dehydrocyanation.[1] Enamines of acetone (**2**) can be obtained by dehydrocyanation of aminonitriles (**1**) with potassium *t*-butoxide in *t*-butyl methyl ether or benzene (equation I).

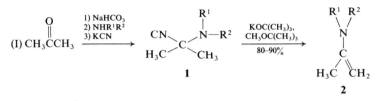

Oxidative hydrolysis of nitriles.[2] Potassium *t*-butoxide catalyzes the oxidation of nitriles to carboxylic acids with loss of the cyano carbon (equation I). The presence of 18-crown-6 improves yields, which vary from 21 to 93%, being highest with long-chain aliphatic nitriles.

$$(I) \quad RCH_2CN \xrightarrow{O_2, \, t\text{-BuOK}} \left[\underset{\overset{\parallel}{RCCN}}{\overset{O}{}} \right] \xrightarrow[21-93\%]{H_2O} RCOOH$$

[1] H. Ahlbrecht and W. Raab, *Synthesis*, 320 (1980).
[2] S. A. DiBiase, R. P. Wolak, Jr., D. M., Dishong, and G. W. Gokel, *J. Org.*, **45,** 3630 (1980).

Potassium carbonate, 5, 552–553; **8,** 408; **9,** 382–383.

γ-Alkylation of allylic halides.[1] The reaction of allylic bromides with α-methyl-thioketones proceeds by an S_N2' mechanism in aqueous K_2CO_3. It probably does not involve the enolate of the ketone, but rather the sequence shown in the specific example given by equation (I). The reaction probably proceeds by formation of a

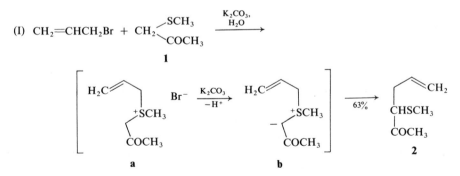

sulfonium salt **a**, which is deprotonated by K_2CO_3 to the ylide **b**. The final step is a [2.3]sigmatropic rearrangement to **2**. The $COCH_3$ group in **1** can be replaced by other electron-withdrawing groups: $CONH_2$, $COOC_2H_5$, $SOCH_3$.

[1] K. Ogura, S. Furakawa, and G. Tsuchihashi, *Am. Soc.*, **102,** 2125 (1980).

Potassium cyanide, 5, 553; **7,** 299; **8,** 409.

   Imidazole synthesis.[1] A new synthesis of 4,5-diarylimidazoles (**2**) involves reaction of catalytic amounts of aqueous ethanolic KCN with N-methyl-C-aryl nitrones (**1**), prepared by condensation of aryl aldehydes with N-methyl-hydroxylamine. The reaction involves an intermediate cyanoimine (**a**).

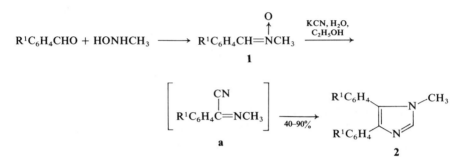

   —CH₂OH → —CH₂CN.[2] This transformation is possible by treatment of a primary alcohol with tri-*n*-butylphosphine, carbon tetrachloride, potassium cyanide, and 18-crown-6 at room temperature. No reaction occurs in the absence of the crown ether. Yields of nitriles are 70–85%.

[1] E. Cawkill and N. G. Clark, *J.C.S. Perkin I,* 244 (1980).
[2] A. Mizuno, Y. Hamada, and T. Shioiri, *Synthesis,* 1007 (1980).

Potassium dichromate, 8, 410; **9,** 383.

   Oxidation.[1] Allylic and benzylic alcohols are oxidized to carbonyl compounds in 70–80% yield by $K_2Cr_2O_7$ in either DMSO or polyethylene glycol, in which the reagent is reasonably soluble.

[1] E. Santaniello, P. Ferraboschi, and P. Sozzani, *Synthesis,* 646 (1980).

Potassium diisopropylamide (KDA), $KN[CH(CH_3)_2]_2$. Mol. wt. 139.28. The base can be prepared by addition of *n*-BuLi to a solution of $KOC(CH_3)_3$ and diisopropyl-amine in THF at $-78°$. It can also be prepared by addition of the amine to a mix-ture of *n*-BuK and $LiOC(CH_3)_3$ in hexane at $0°$.[1]

   Gawley and co-workers[2] report that KDA is superior to LDA and to *n*-BuLi for deprotonation of dimethylhydrazones (**7,** 126–130) or of oxime ethers. The reaction is generally complete in THF at $-78°$ in 15 minutes or less. The potassium coun-terion does not interfere with cuprate formation or conjugate addition.

[1] S. Raucher and G. A. Koolpe, *J. Org.,* **43,** 3794 (1978).
[2] R. E. Gawley, E. J. Termine, and J. Aube, *Tetrahedron Letters,* **21,** 3115 (1980).

Potassium ethanethiolate, KSCOCH₃.

Tetrahydrothiophenes.[1] Reaction of $KSCOCH_3$ (2.4 equivalents) with **1** in DMF in the presence of air results in formation of **2** in 90% yield. In the absence of oxygen the yield of **2** is negligible. The yield is also low if only 1 equivalent of the salt is present.

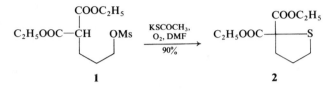

1   **2**

The reaction was used to convert the β-lactam **3** into the isopenam **4**.

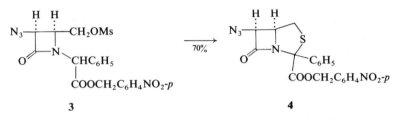

3   **4**

[1] G. H. Hakimelahi and G. Just, *Tetrahedron Letters*, **21**, 2119 (1980).

Potassium fluoride, 1, 933–935; **2**, 346; **5**, 555–556; **6**, 481–482; **8**, 410–412.

Desilylation of acylsilanes. Two groups have reported that KF in combination with 18-crown-6,[1] DMSO, or HMPT[2] converts acylsilanes into acyl anion equivalents. The reaction can be used to obtain aldehydes, ketones, and hydroxy ketones in moderate to good yield.

Examples:

$$C_6H_5\overset{O}{\overset{\|}{C}}-\overset{OH}{\overset{|}{C}}(CH_3)_2 \xleftarrow[35\%]{\begin{array}{l}1)\,Bu_4NF\\2)\,CH_3COCH_3\end{array}} C_6H_5COSi(CH_3)_3 \xrightarrow[75\%]{\begin{array}{l}1)\,KF,\,DMSO\\2)\,H_2O\end{array}} C_6H_5CHO$$

$$90\% \downarrow \begin{array}{l}1)\,KF,\,18\text{-crown-6}\\2)\,C_6H_5CH_2Br\end{array}$$

$$C_6H_5COCH_2C_6H_5$$

Desilylation may involve initial attack of F^- on the carbonyl group followed by C to O migration of the silyl group to give $C_6H_5\bar{C}(F)OSi(CH_3)_3$ or direct displacement of the benzoyl anion.

Desilylation of silyl enol ethers.[3] Hydrolysis of silyl enol ethers of aldehydes with oxalic acid in aqueous THF can result in partial degradation to a ketone (equation

I). Hydrolysis wth KF in methanol furnishes only the desired aldehyde in high yield.

$$\text{(I)} \quad \underset{C_6H_5}{\overset{C_2H_5}{>}}CH-C\underset{CHOSi(CH_3)_3}{\overset{CH_3}{<}} \quad \xrightarrow{H_3O^+}$$

$$\underset{C_6H_5}{\overset{C_2H_5}{>}}CHC\underset{O}{\overset{CH_3}{<}} \quad + \quad \underset{C_6H_5}{\overset{C_2H_5}{>}}CHCHCHO \overset{CH_3}{\underset{|}{}}$$

$$15:85$$

KF–Celite.[4] Celite coated with KF facilitates C-, N-, O-, and S-alkylations in aprotic solvents (DMF, THF, CH_3CN). Yields are generally satisfactory. Selectivity for C-alkylation is somewhat lower than that observed with a tetraalkylammonium fluoride catalyst.

[1] A. Degl'Innocenti, S. Pike, D. R. M. Walton, G. Seconi, A. Ricci, and M. Fiorenza, *J.C.S. Chem. Comm.*, 1021 (1980).
[2] D. Schinzer and C. H. Heathcock, *Tetrahedron Letters*, **22**, 1881 (1981).
[3] C. Chuit, J. P. Foulon and J. F. Normant, *Tetrahedron*, **37**, 1385 (1981).
[4] T. Ando and J. Yamawaki, *Chem. Letters*, 45 (1979).

Potassium–Graphite, 4, 397; **7**, 296; **8**, 405–406; **9**, 378.

Reduction of α,β-unsaturated sulfones.[1] Sodium amalgam (**7**, 326–327) is not satisfactory for reductive cleavage of alkenyl sulfones, $R^1CH=CR^2SO_2C_6H_5$. These substrates are best reduced to alkenes wth potassium–graphite (65–85% yields).

[1] P. O. Ellingssen and K. Undheim, *Acta Chem. Scand.*, **B33**, 528 (1979).

Potassium hexamethyldisilazide, $KN[Si(CH_3)_3]_2$. Mol. wt. 199.49. The base is prepared by reaction of KH with $NH[Si(CH_3)_3]_2$, **6**, 482.

(Z)-Trisubstituted allylic alcohols.[1] The conditions used by Bestmann *et al.* (**7**, 329) for preparation of (Z)-disubstituted alkenes via the Wittig reaction also can provide a stereoselective route to (Z)-trisubstituted allylic alcohols. An example is the reaction of ethylidenetriphenylphosphorane with the THP ether of hydroxyacetone (equation I). The stereoselectivity is decreased with other protecting

$$\text{(I)} \quad \underset{THPOCH_2}{\overset{CH_3}{>}}C=O + (C_6H_5)_3\overset{+}{P}CH_2CH_3\overset{-}{Br} \quad \xrightarrow[83\%]{\underset{THF, HMPT, -78°}{KN[Si(CH_3)_3]_2,}}$$

1

$$\underset{H}{\overset{CH_3}{>}}C=C\underset{CH_3}{\overset{CH_2OTHP}{<}}$$

$$(Z/E = 41:1)$$

groups, other solvents, and other bases. It can be marginally increased by use of a phosphonium tetrafluoroborate rather than a halide. This Wittig reaction is fairly general, but stereoselectivity is decreased by substitution at the carbon β to phosphorus or by substitution at the α′- position of the α-alkoxy ketone. Chemical yields are decreased when the two carbons being joined are hindered. The reaction was used for a synthesis of α-santalol (**2**) of 99% stereoisomeric purity.

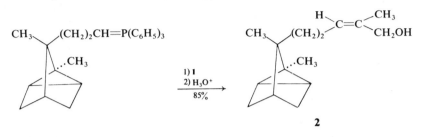

¹ C. Sreckumar, K. P. Darst, and W. C. Still, *J. Org.*, **45**, 4262 (1980).

Potassium hydride, 1, 935; **2**, 346; **4**, 409; **5**, 557; **6**, 482–483; **7**, 302–303; **8**, 412–415; **9**, 386–387.

   Oxy-Cope rearrangement (**7**, 302–303; **8**, 412–414). Further studies indicate that the rearrangement of 1,5-diene alkoxides can be used for stereoselective generation of asymmetry. An application is the stereoselective synthesis of *erythro*-juvabione (**1**), a sesquiterpene from *Abies balsamea*, shown in equation (I).¹

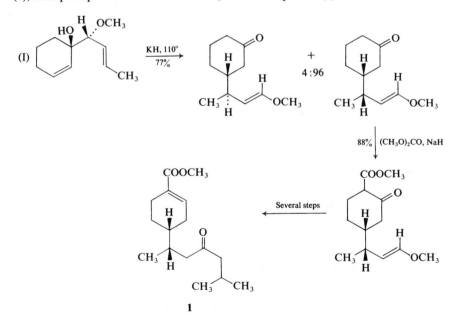

The oxy-Cope rearrangement can be used to obtain bicyclic bridgehead olefins from spirocyclic precursors (equation II).[2,3]

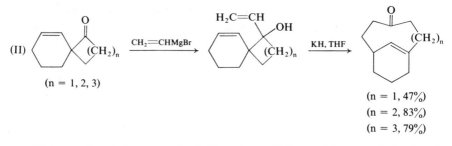

(II)

(n = 1, 2, 3)

(n = 1, 47%)
(n = 2, 83%)
(n = 3, 79%)

Eight-membered ring expansion.[4] Treatment of the cyclohexane derivative **1** (*trans/cis* = 9:1) with KH at room temperature results in the 14-membered ring trienone **2** in 90% yield. The reaction may involve a [5.5]vinylogous oxy-Cope rearrangement or two consecutive [3.3]sigmatropic rearrangements as shown.

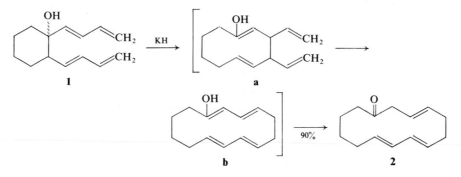

[1] D. A. Evans and J. V. Nelson, *Am. Soc.*, **102**, 774 (1980).
[2] M. Kahn, *Tetrahedron Letters*, **21**, 4547 (1980).
[3] S. G. Levine and R. L. McDaniel, Jr., *J. Org.*, **46**, 2199 (1981).
[4] P. A. Wender and S. M. Sieburth, *Tetrahedron Letters*, **22**, 2471 (1981).

Potassium hydrogen persulfate, $KHSO_5$. Supplier: Alfa (Oxone).

Oxidation of sulfides to sulfones.[1] Sulfides are oxidized chemoselectively to sulfones by $KHSO_5$ (3 equivalents) in high yield. Peracids are usually used for this oxidation, but can also oxidize olefinic groups. Oxidation of sulfides to sulfoxides is also possible with 1 equivalent of reagent.

[1] B. M. Trost and D. P. Curran, *Tetrahedron Letters*, **21**, 1287 (1981).

Potassium hydroxide–Alumina.

Hydrolysis of esters.[1] KOH crushed with neutral alumina effects hydrolysis of even hindered esters at room temperature (4–48 hours).

[1] S. L. Regen and A. K. Mehrotra, *Syn. Comm.*, **11**, 413 (1981).

Potassium iodide–Boron(III) iodide.

Reduction of sulfonic acids to disulfides.[1] This combination of reagents reduces aryl sulfonyl halides and sulfonic acids to disulfides in yields generally $> 90\%$.

[1] G. A. Olah, S. C. Narang, L. D. Field, and R. Karpeles, *J. Org.*, **46**, 2408 (1981).

Potassium iodide–Zinc–Phosphorus(V) oxide.

Alkenes from epoxides.[1] Epoxides in the hexapyranose series are converted into alkenes by reaction with KI, Zn, and P_2O_5 (all in large excess) in DMF at 90°. Yields in the two reported examples were 86 and 83%.

[1] P. J. Garegg, D. Papadimas, and B. Samuelsson, *Carbohydrate Res.*, **80**, 354 (1980).

Potassium methylsulfinylmethylide (Dimsylpotassium), 4, 409.

Permethylation of carbohydrates. One widely used method for this reaction employs dimsylsodium in DMSO and methyl iodide.[1] The purity of the permethylated products is significantly improved if dimsylpotassium is used as the base.[2]

[1] S.-I. Hakomori, *J. Biochem.* (*Tokyo*), **55**, 205 (1964).
[2] L. R. Phillips and B. A. Fraser, *Carbohydrate Res.*, **90**, 149 (1981).

Potassium nitrite, KNO_2.

Inversion of secondary alcohols. Lattrell and Lohaus[1] reported a few years ago that 3-sulfonyloxy-2-azetidinones react with various nucleophiles with inversion. Thus reaction of *trans*-**1** with KNO_2 in DMSO at 55° results in *cis*-**2** in 77% yield. Presumably an intermediate nitrous ester undergoes hydrolysis under the reaction conditions. Use of potassium formate leads to the same result, but requires higher temperatures.

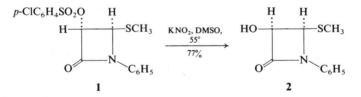

The method has been extended to inversion of steroid and prostaglandin alcohols. In some cases it is superior to inversion with potassium superoxide (**6**, 488–489).[2]

[1] R. Lattrell and G. Lohaus, *Ann.*, 901 (1974).
[2] B. Radüchel, *Synthesis*, 292 (1980).

Potassium nitrosodisulfonate (Fremy's salt), 1, 940–942; **2,** 347–348; **4,** 411; **5,** 562.

Morpholinoquinones.[1] The morpholinoquinone **2** has been obtained by oxidation of **1** with Fremy's salt. The oxidation is markedly dependent on the group attached to the nitrogen. Thus if the acetyl group in **1** is replaced by a benzyl group, no quinone is formed on oxidation. This oxidation was explored during studies aimed at synthesizing rubradirin antibiotics, which contain a morpholino-naphthoquinone chromophore.

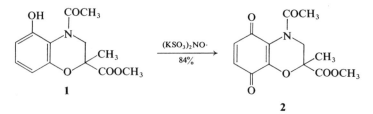

Dehydrogenation of 1,2-dehydropyrrolizidines. Oxidation of retronecine (**1**) with Fremy's salt gives dehydroretronecine (**2**) in satisfactory yield.[2] This oxidation has been conducted with chloranil, manganese dioxide, and potassium permanganate, but yields are lower.[3]

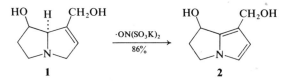

[1] A. P. Kozikowski, K. Sugiyama, and J. P. Springer, *J. Org.*, **46**, 2426 (1981).
[2] A. R. Mattocks, *Chem. Ind.*, 251 (1981).
[3] C.C. J. Culvenor, J. A. Edgar, L. W. Smith, and H. J. Tweeddale, *Aust. J. Chem.*, **23**, 1853 (1970).

Potassium permanganate, 1, 942–952; **2**, 348; **4**, 412–413; **5**, 562–563; **8**, 416–417; **9**, 388–391.

Quaternary aldehydes. Tertiary nitro comounds (**1**) can be converted to primary nitro compounds with a quaternary center (**2**) by reaction with the sodium salt of nitromethane in the presence of sodium hydride in DMSO (equation I).[1]

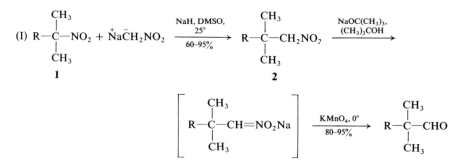

Several methods have been used to convert primary nitro compounds into aldehydes, but the most effective one in the present case is oxidation of the corresponding sodium nitronate with potassium permanganate (**1**, 950–951).

[1] N. Kornblum and A. S. Erickson, *J. Org.*, **46**, 1037 (1981).

Potassium peroxodisulfate, 1, 952–954; 8, 417.

Oxidation of isatin. Two products are possible from Baeyer-Villiger oxidation of isatin (**1**). Oxidation with $K_2S_2O_8$ in sulfuric acid gives only the 1,4-benzoxazine **2**, whereas oxidation with H_2O_2–HOAc gives the anhydride (**3**) of isatoic acid.[1] An earlier preparation of compounds of type **2** involves reaction of *o*-aminophenols and oxalyl chloride.[2]

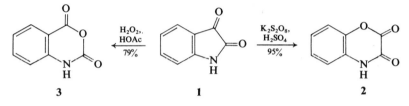

[1] G. Reissenweber and D. Mangold, *Angew. Chem. Int. Ed.*, **19**, 222 (1980) .
[2] K. Dickore, K. Sasse, and K.-D. Bode, *Ann.*, **733**, 70 (1970).

(S)-(−)-Proline, 6, 492–493; 7, 307: 8, 421–424.

Asymmetric bromolactonization (**8**, 421–423). Full details for synthesis of α,α-disubstituted α-hydroxy acids from α,β-unsaturated acids by bromolactonization have been published.[1]

A modification of this asymmetric bromolactonization affords optically active α,β-epoxy aldehydes (equation I).[2] Thus treatment of the bromolactone **1** with

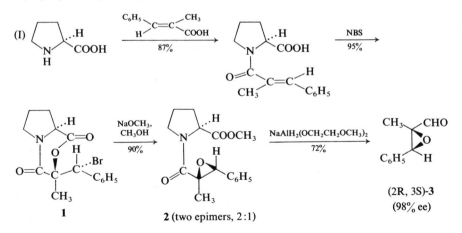

sodium methoxide results in an epoxy ester (2), which gives the chiral α,β-epoxy aldehyde 3 on reductive cleavage of the proline group.

[1] S.-S. Jew, S. Terashima, and K. Koga, *Tetrahedron*, **35**, 2337, 2345 (1980).
[2] S. Terashima, M. Hayashi, and K. Koga, *Tetrahedron Letters*, **21**, 2733 (1980).

(S)-Prolinol, 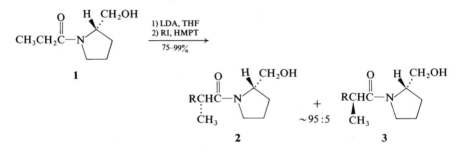 . Mol. wt. 101.15, b.p. 86°, α_D + 3.4°. Supplier:

Aldrich.

Enantioselective alkylation of amides. Two laboratories[1,2] have used (S)-prolinol as the chiral auxiliary for a synthesis of chiral amides. Alkylation of the enolate of the amide 1 (prepared with LDA or *t*-butyllithium) proceeds with pronounced

asymmetric induction. The amides are useful for preparation of optically active carboxylic acids with the (R)-configuration. The optical yields compare favorably with those obtained using chiral oxazolines (**7**, 229–231) or ephedrine (**9**, 209).

The hydroxyl group plays a crucial role in this asymmetric alkylation, since O-alkyl ethers shows the opposite enantiotropic bias, but somewhat less pronounced.

[1] P. E. Sonnet and R. R. Heath, *J. Org.*, **45**, 3137 (1980).
[2] D. A. Evans and J. M. Takacs, *Tetrahedron Letters*, **21**, 4233 (1980).

Propargyl bromide, 6, 493.

α,β-Enones. A new route to these enones involves Reformatsky condensation of propargyl bromide with an aldehyde or ketone followed by acid-catalyzed rearrangement of the β-hydroxyacetylene to an enone (equation I).[1]

(I) $(C_6H_5)_2C{=}O + BrCH_2C{\equiv}CH \xrightarrow{\text{Zn, THF}}$

$(C_6H_5)_2\underset{\underset{OH}{|}}{C}CH_2C{\equiv}CH \xrightarrow[\text{70\% overall}]{\text{H}_2\text{SO}_4,\ \text{HOAc, 70°}} (C_6H_5)_2C{=}CHCOCH_3$

[1] L. E. Friedrich, N. de Vera, and M. Hamilton, *Syn. Comm.*, **10**, 637 (1980).

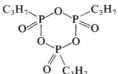

n-Propylphosphonic anhydride,   (1). The reagent exists at

least as a trimer, b.p. 200°/0.3 mm, stable at room temperature, soluble in CH_2Cl_2, dioxane, THF, and DMF. The anhydride is prepared in high yield by hydrolysis of *n*-propylphosphonic dichloride to the phosphonic acid, which is then heated under vacuum. Aldrich supplies a solution of **1** in CH_2Cl_2.

Peptide synthesis.[1] The anhydride effects peptide synthesis as shown in equation (I). Free acid groups are buffered with a tertiary amine such as N-ethylmorpholine. The reaction is complete after 24 hours at room temperature. No racemization is detected in the Anderson-Young test. The resulting alkylphosphonic acid derivative is soluble in water and easily removed.

$$\text{(I)} \quad \underset{O}{\overset{O}{\parallel}} \underset{R^1}{\overset{R^1}{|}} \text{RCNHCHCOOH} + \underset{R^2}{\overset{R^2}{|}} \text{H}_2\text{NCHCOOR}^3 + \mathbf{1} \quad \xrightarrow[\substack{70-95\%}]{\substack{\text{DMF} \\ \text{CH}_2\text{Cl}_2}}$$

$$\underset{O}{\overset{O}{\parallel}} \underset{R^1}{\overset{R^1}{|}} \text{RCNHCHCO} - \underset{R^2}{\overset{R^2}{|}} \text{NHCHCOOR}^3 +$$

[1] H. Wissman and H.-J. Kleiner, *Angew. Chem. Int. Ed.*, **19**, 133 (1980).

Pyridinium bromide perbromide, 1, 967–970; **5**, 568; **6**, 499–500; **9**, 399.

Dethioketalization.[1] Various thioacetals and thioketals are readily hydrolyzed by pyridinium bromide perbromide (1 equivalent) under phase-transfer conditions. Tetrabutylammonium bromide is used as catalyst and aqueous methylene chloride as solvent. The reaction is more efficient in the presence of pyridine as buffer. Yields are generally 75–90%.[1]

[1] G. S. Bates and J. O'Doherty, *J. Org.*, **46**, 1745 (1981).

Pyridinium chloride, 1, 964–966; **2**, 352–353; **3**, 239–240; **4**, 415–418; **5**, 566–567; **6**, 497–498; **7**, 308.

Chlorohydrins. Epoxides can be cleaved to *trans*-chlorohydrins with pyridinium hydrochloride in chloroform or pyridine.[1] The reagent also cleaves α-cyclopropyl ketones to mixtures of γ-chloro ketones obtained by the two possible modes of cyclopropyl ring fission.[2]

[1] M. A. Loreto, L. Pellacani, and P. A. Tardella, *Syn. Comm.*, **11**, 287 (1981).
[2] E. Giacomini, M. A. Loreto, L. Pellacani, and P. A. Tardella, *J. Org.*, **45**, 519 (1980).

Pyridinium chlorochromate (PCC), **6**, 498–499; **7**, 308–309; **8**, 425–427; **9**, 397–399.

Oxidative cleavage of furanes.[1] Alkylfuranes of type **1** are oxidatively cleaved by PCC to *cis*-1,4-enediones (**a**), which isomerize in the acidic medium to the more stable *trans*-isomers (**2**). These products are converted into 4-methoxycyclo-pentenones (**3**) in methanolic sodium hydroxide. The yields are lower (50–60%) in this sequence when $R^1 = H$.

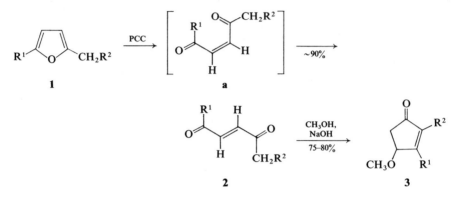

1   a

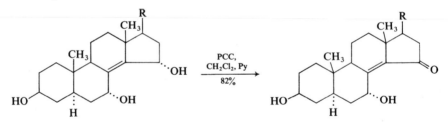

2   3

Selective oxidation of allylic alcohols.[2] PCC in CH_2Cl_2 containing pyridine (2%) selectively oxidizes steroidal equatorial allylic hydroxyl groups. PCC buffered with sodium acetate is nonselective.

Example:

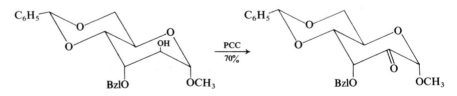

Carbohydrate ketones.[3] This reagent is recommended for oxidation of carbo-hydrate alcohols to ketones without epimerization of axial α-substituents to the more stable epimers. RuO_4 has been used successfully, but is expensive when used in stoichiometric amounts as required for consistent results.

Examples:

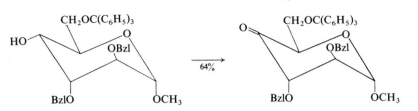

PCC–Alumina.[4] PCC adsorbed on alumina is particularly effective for oxidation of alcohols, giving yields generally higher than those obtained by usual conditions.

[1] G. Piancatelli, A. Screttri, and M. D'Auria, *Tetrahedron,* **36,** 661 (1980).
[2] E. J. Parish and G. J. Schroepfer, Jr., *Chem. Phys. Lipids,* **27,** 281 (1980).
[3] B. B. Bissember and R. H. Wightman, *Carbohydrate Res.,* **81,** 187 (1980).
[4] Y.-S. Cheng, W.-L. Liu, and S. Chen, *Synthesis,* 223 (1980).

Pyridinium chlorochromate–Hydrogen peroxide.

Deoximation (**8,** 427). Oximes can be oxidatively cleaved by PCC alone, but a long reaction time is necessary. Cleavage occurs within minutes at 0–10° when 30% H_2O_2 is added. Yields are generally 65–85%. The actual oxidant may be pyridinium oxodiperoxychlorochromate (**1**).[1]

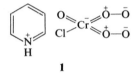

1

[1] J. Drabowicz, *Synthesis,* 125 (1980).

Pyridinium dichromate (PDC), **9,** 399.

Δ^2-Butenolides.[1] A simple route to Δ^2-butenolides from α,β-enals involves conversion to the O-trimethylsilylcyanohydrin (cyanotrimethylsilane, **4,** 542–543; **5,** 720–722; **6,** 632–633), which is then oxidized by PDC (3 equivalents) in DMF. The sequence from geraniol is typical (equation I).

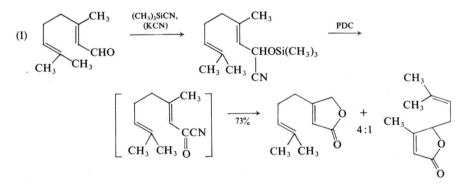

Cyanohydrins of nonconjugated aldehydes are oxidized in high yield to carboxylic acids.

$>CHCH_2OH \rightarrow >C=O$. The diol **1** can be oxidized by PDC to the dione **3**. A short-term reaction gives the keto alcohol (lactol) **2**. This is apparently the first example of PDC oxidation of a primary alcohol beyond the aldehyde stage. Other oxidants based on Cr also oxidize **1** to **3**, but in lower yields. The loss of the CH_2OH group may be a result of facile enolization of the intermediate aldehyde.[2]

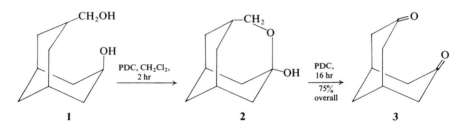

1   **2**   **3**

α-Iodo-α,β-enals. PDC converts complexes of α-ynols with iodine into α-iodo-α,β-enals.[3]

Example:

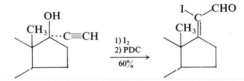

[1] E. J. Corey and G. Schmidt, *Tetrahedron Letters*, **21**, 731 (1980).
[2] J. A. Zalikowski, K. E. Gilbert, and W. T. Borden, *J. Org.*, **45**, 346 (1980).
[3] R. Antonioletti, M. D'Auria, G. Piancatelli, and A. Scretti, *Tetrahedron Letters*, **22**, 1041 (1981).

4-Pyrrolidinopyridine (1), 9, 400.

Ansapeptides. A group of cyclopeptide alkaloids is characterized by a 13-, 14-, or 15-membered *para*-ansa bridge containing a dipeptide unit. The most difficult step in the synthesis is the final ring closure via an amide linkage, since formation of the cyclopeptide dimer is usually favored over formation of the monomer.[1] The highest yields are obtained by use of a pentafluorophenyl ester and a Cbz-protected amino group (*eg.*, **2**). Hydrogenolysis (palladium on charcoal) in dioxane containing a trace of alcohol and with **1** as catalyst results in ring closure to **3** as the major product (equation I). Presumably ring closure takes place on the catalyst surface. In the absence of **1** or in the presence of other bases, the yield of **3** drops dramatically.[2]

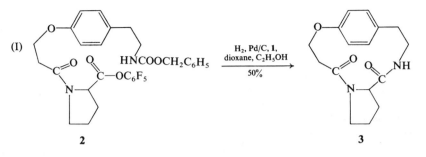

(I)

2   3

This methodology was used for the synthesis of the natural peptide alkaloid dihydrozizyhin G (**4**).[3] The ring closure in this case proceeded in 67% yield.

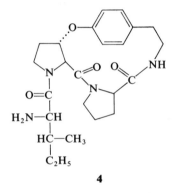

4

[1] J. C. Lagarias, R. A. Houghten, and H. Rapoport, *Am. Soc.*, **100**, 8202 (1978).

[2] U. Schmidt, H. Griesser, A. Lieberknecht, and J. Talbiersky, *Angew. Chem. Int. Ed.*, **20**, 280 (1981).

[3] U. Schmidt, A. Lieberknecht, H. Griesser, and J. Hansler, *Angew. Chem. Int. Ed.*, **20**, 281 (1981).

Q

Quinine, 6, 501; **7**, 311; **8**, 430–431; **9**, 403.

Asymmetric epoxidation (7, 311; **8**, 430) Optically active epoxides of cyclohexenones can be obtained by epoxidation with *t*-butyl hydroperoxide in toluene with solid NaOH and (−)-benzylquininium chloride as the chiral catalyst. In the case of a cyclohexenone the chemical yield is 60% and the optical yield is 20 ± 3%.[1]

[1] H. Wynberg and B. Marsman, *J. Org.*, **45**, 158 (1980).

R

Raney nickel, **1**, 723–731; **2**, 293–294; **5**, 570–571; **6**, 502; **7**, 312; **8**, 433; **9**, 405–406.

Hydrogenation–dehydrogenation of allylic and homoallylic alcohols.[1] Treatment of cholesterol and Δ^4-cholestenol with Raney nickel (previously washed with alcohol) and cyclohexanone in refluxing toluene results mainly in formation of 5β-3-cholestanone (**1**). This hydrogenation–dehydrogenation proceeds much more rapidly with the allylic alcohol **4**.

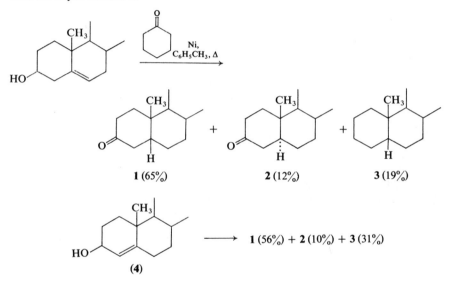

1 (65%)   **2** (12%)   **3** (19%)

1 (56%) + **2** (10%) + **3** (31%)

(**4**)

Reduction of aryl carbonyl compounds.[2] This reduction has been conducted traditionally by the Clemmenson or Wolff-Kishner method or by reduction of dithioketals. Actually it can be conducted in high yield with W-7 Raney nickel in 50% aqueous ethanol (2–5 hours). Nitro, halo, and cyano groups are also reduced. Examples:

$$C_6H_5COC_6H_5 \xrightarrow[97\%]{\text{Raney Ni}} C_6H_5CH_2C_6H_5$$

$$p\text{-}HOOCC_6H_4CHO \xrightarrow[89\%]{} p\text{-}HOOCC_6H_4CH_3$$

$$\text{fluorenone} \xrightarrow[94\%]{} \text{fluorene}$$

Hydrogenolysis of epoxides.[3] Hydrogenation of the Δ^7-14α,15α epoxide **1** over Raney nickel results in the desired alcohol **2** and the hydroxyl-free olefin **3** in about equal amounts. When triethylamine (4%) is added the alcohol **2** is obtained in high yield. 15-Oxygenated sterols are of interest because a number are potent inhibitors of sterol biosynthesis in animal cells in culture.

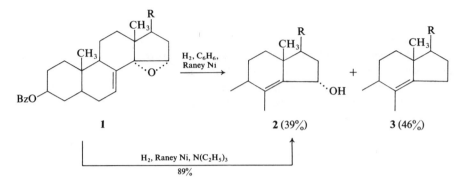

[1] J. Foršek, *Tetrahedron Letters*, **21**, 1071 (1980).
[2] R. H. Mitchell and Y.-H. Lai, *Tetrahedron Letters*, **21**, 2637 (1980).
[3] E. J. Parish and G. J. Schroepfer, Jr., *Chem. Phys. Lipids*, **26**, 141 (1980).

Rhodium(II) carboxylates, 5, 571–572; **7**, 313; **8**, 434–435; **9**, 406–408.

Buchner reaction (**1**, 368–369).[1] The reaction of benzene with an alkyl diazoacetate is catalyzed efficiently by rhodium(II) carboxylates, particularly $Rh(OCOCF_3)_2$. Cycloheptatrienes are formed at room temperture. Of more importance, the 1-carboalkoxy-2,4,6-cycloheptatriene is formed in almost quantita-

tive yield. Similar high selectivity for the kinetic nonconjugated isomer is observed with substituted derivatives of benzene.[2]

Cyclopropanation of $\diagup C{=}C\diagdown$. Hubert, Noels, and their co-workers[3] have compared the effectiveness of $Rh(OCOR)_2$ (**7**, 313), $Pd(OCOR)_2$ (**5**, 469; **6**, 442–443), CuOTf (**5**, 151–152), and $Cu(OTf)_2$ (**7**, 76) as catalysts for cyclopropanation with diazo esters. Pd(II) is particularly efficient for strained (cyclopentene) or conjugated styrene-type olefins. Rh(II) is efficient for both mono- and polyolefins. Cu(I) is intermediate in efficiency. Pd(II) catalysis depends on coordination to the double bond; Rh(II) promotes a carbenoid reaction with the olefin. Copper derivatives are probably effective by one or another of these effects, depending on the structure of the ligand and the olefin.

2-Carbapenems. Merck chemists[4] have reported a highly efficient synthesis of this ring system by a carbene insertion reaction to form the $N{-}C_3$ bond. Thus decomposition of the diazo ketone **1** in the presence of rhodium(II) acetate leads to **2**. Photolysis of **1** results mainly in the imide isomer of **2**.

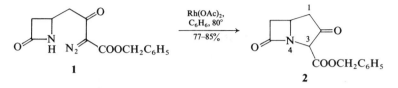

This insertion reaction has provided a stereocontrolled synthesis of (+)-thienamycin (**3**), an important carbapenem antibiotic.[5]

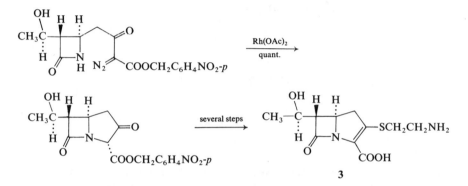

A similar carbenoid route to the fused β-lactam **5** involves reaction of ethyl diazoacetoacetate with 4-acetoxyazetidin-2-one (**4**) catalyzed by rhodium(II) acetate. Rhodium(II) acetate may also promote the cyclization step.[6] For another route to compounds related to **5** *see* **8**, 36–37.

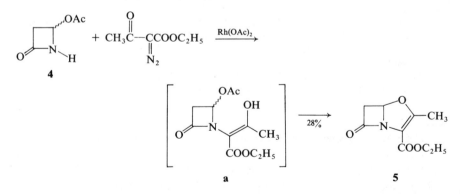

Alkylation of β-lactams.[7] Alkylation of the base- and acid-sensitive β-lactam **1** can be effected efficiently by carbenoid insertion into the N—H bond using

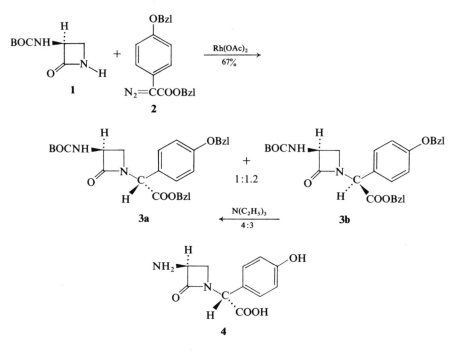

rhodium(II) acetate as catalyst. Thus the catalyzed reaction of **2** with **1** gives the diastereomers **3a** and **3b** in 67% yield. The desired isomer (**3a**) is a precursor to 3-aminonocardicinic acid (**4**). This method is superior to alkylation in the presence of KOH and benzyltriethylammonium chloride, which results in yields of 35–40%.

[1] E. W. Warnhoff, *Org. React.*, **17**, 239 (1970).
[2] A. J. Anciaux, A. Demonceau, A. F. Noels, A. J. Hubert, R. Warin, and P. Teyssié, *J. Org.*, **46**, 873 (1981).
[3] A. J. Anciaux, A. J. Hubert, A. F. Noels, H. Petiniot, and P. Teyssié, *J. Org.*, **45**, 695 (1980).
[4] R. W. Ratcliffe, T. N. Salzmann, and B. G. Christensen, *Tetrahedron Letters*, **21**, 31 (1980).
[5] T. N. Salzmann, R. W. Ratcliffe, B. G. Christensen, and F. A. Bouffard, *Am. Soc.*, **102**, 6161 (1980).
[6] J. Cuffee and A. E. A. Porter, *J.C.S. Chem. Comm.*, 1257 (1980).
[7] P. G. Mattingly and M. J. Miller, *J. Org.*, **46**, 1557 (1981).

Ruthenium–Silica.

Selective hydrogenation of a diketone.[1] 4-Hydroxycyclohexanone has not been obtained in satisfactory yield by selective hydrogenation of 1,4-cyclohexanedione under homogeneous conditions. However it can be obtained in 70% yield by hydrogenation of the dione in 2-propanol with 5% ruthenium on silica. Some other metals (platinum, iridium) are more reactive, but less selective.

[1] M. Bonnet, P. Geneste, and M. Rodriguez, *J. Org.*, **45**, 40 (1980).

Ruthenium(III) chloride, 4, 421; **8,** 437–438.

Rearrangement of allylic alcohols.[1] $RuCl_3$ in combination with NaOH is a useful catalyst for isomerization of allylic alcohols of the type $RCHOHCH=CH_2$ to saturated ketones, $RCOCH_2CH_3$. In the isomerization of a chiral allylic alcohol such as **1,** the product is optically active. The 1,3-hydrogen transfer is 37% stereoselective.

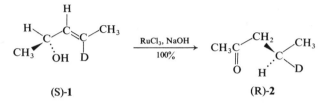

(S)-1   (R)-2

[1] W. Smadja, G. Ville, and C. Georgoulis, *J.C.S. Chem. Comm.,* 594 (1980).

Ruthenium tetroxide, 1, 986–989; **2,** 357–359; **3,** 243–244; **4,** 420–421; **6,** 504–506; **7,** 315; **8,** 438.

Oxidation of cyclic ethers to lactones (**1,** 988). A systematic study of this reaction has been reported. In general, yields are good to excellent by either stoichiometric or catalytic procedures. No anhydrides from further oxidation are detected. The oxidation is chemoselective. Oxidation of a secondary position takes precedence over oxidation of a tertiary site. Primary positions are attacked in preference to secondary positions in the oxidation of acyclic ethers.[1]

[1] A. B. Smith III and R. M. Scarborough, Jr., *Syn. Comm.,* **10,** 205 (1980).

S

Samarium(II) iodide, SmI_2, **8**, 439.

Reductions with SmI_2. SmI_2 (2 equivalents) can be used to reduce sulfoxides to sulfides (80–90% yield) and to convert epoxides to alkenes (65–90%). YbI_2 is somewhat less effective in these reductions.

Both SmI_2 and YbI_2 selectively reduce double bonds conjugated with carboxyl or ester groups, groups that are not reduced by either reagent. Both reagents in the presence of CH_3OH as a proton donor reduce aldehydes and ketones. Aldehydes are reduced particularly readily; selective reduction of an aldehyde in the presence of a ketone is possible. SmI_2 in THF reduces primary bromides, iodides, or tosylates to alkanes in 80–90% yield. Benzylic or allylic halides undergo coupling; head to head coupling is favored.

SmI_2 (stoichiometric amount) promotes alkylation of ketones in THF by alkyl bromides, iodides, or tosylates. The reaction is slow, but can be accelerated by a catalytic amount of $FeCl_3$. This reaction is analogous to reaction of ketones with Grignard reagents. Vinylic and aryl halides do not undergo this alkylation.[1]

Examples:

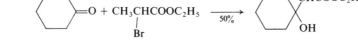

[1] P. Girard, J. L. Namy, and H. B. Kagan, *Am. Soc.*, **102**, 2693 (1980).

Selenium, 1, 990–992; **4**, 222; **5**, 575; **6**, 507–509; **8**, 439.

α-Methylseleno ketones and esters.[1] Selenium metal reacts with lithium enolates of ketones or esters in THF in the presence of 3 equivalents of HMPT at −20 to −10° to form a selenolate ion (30–60 minutes), which can be directly methylated with methyl iodide. Yields are comparable to those obtained by reaction of C_6H_5SeX with enolates, but use of selenium itself reduces the cost.

Examples:

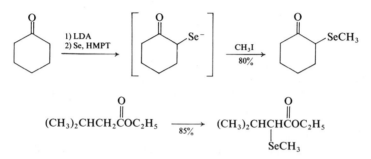

$$(CH_3)_2CHCH_2\overset{\overset{\displaystyle O}{\|}}{C}OC_2H_5 \xrightarrow[85\%]{} (CH_3)_2CHCH\overset{\overset{\displaystyle O}{\|}}{C}OC_2H_5$$
$$\underset{SeCH_3}{|}$$

[1] D. Liotta, G. Zima, C. Barnum, and M. Saindane, *Tetrahedron Letters*, **21**, 3643 (1980).

Selenium dioxide, 1, 992–1000; **2**, 360–362; **3**, 245–247; **4**; 422–424; **5**, 575–576; **6**, 509–510; **8**, 439–440; **9**, 409–410.

Selective oxidation of polymethylpyrimidines.[1] 2,4-Di- and 2,4,6-trimethyl-pyrimidines are selectively oxidized to 4-carboxylic acids by a slight excess of SeO_2 in pyridine. Yields are about 40–65%. This increased reactivity of a 4-methyl group of polymethylpyrimidines is also observed in reaction with ethyl nitrite in liquid ammonia to form 4-aldoximes and with ethyl benzoate in the presence of KOC_2H_5 to form phenacyl derivatives.

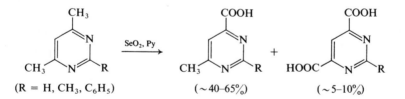

Cinnamyl alcohols.[2] These alcohols can be prepared by oxidation of 3-arylpropenes with selenium dioxide in dioxane. The yield compares favourably with that obtained by $LiAlH_4$ reduction of esters of cinnamic acids.

$$ArCH_2-CH=CH_2 \xrightarrow[24-44\%]{SeO_2} ArCH=CHCH_2OH$$

Oxidation of tetralones.[3] The tetralone **1** is oxidized directly to the *o*-naphthoquinone **2** by selenium dioxide in acetic acid.

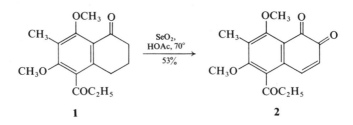

1   **2**

[1] T. Sakasi, T. Sakamoto, and H. Yamanaka, *Heterocycles*, **13**, 235 (1979).
[2] H.-L. Pan, C.-A. Cole, and T. L. Fletcher, *Synthesis*, 813 (1980).
[3] H. Nagaoka, G. Schmid, H. Iio, and Y. Kishi, *Tetrahedron Letters*, **22**, 899 (1981).

Semicarbazide–Silica gel.

Isolation of aldehydes and ketones.[1] Aldehydes and ketones are readily separated from other organic compounds by heating a solution of the mixture in hexane or toluene with semicarbazide supported on silica gel (1:4 or 1:9) at about 70° for 12–18 hours. After filtration the carbonyl compound is regenerated from the solid phase by treatment with oxalic acid in a two-phase system, water and toluene or heptane. Girard T reagent (**1**, 410–411) supported on silica gel can also be used, but offers no advantages over supported semicarbazide.

[1] R. P. Singh, H. N. Subbarao, and S. Dev, *Tetrahedron*, **37**, 843 (1981).

Silica gel, 6, 510; **9**, 410.

Diels-Alder catalyst.[1] Bicyclo[2.2.1]hept-2-ene-2,3-dicarboxylic anhydride (**1**) does not undergo cycloaddition with cyclopentadiene (**2**) in the absence of a catalyst. AlCl₃ in this case does not function as a catalyst. However the two compounds undergo cycloaddition at room temperature when slurried on silica gel. The *anti-* (**3**) and *syn-* (**4**) anhydrides are formed in a ratio of about 3:2. The mixture can be separated readily because **3** is hydrolyzed by KOH in ethanol–water (3:1), whereas **4** is not hydrolyzed under any known conditions.

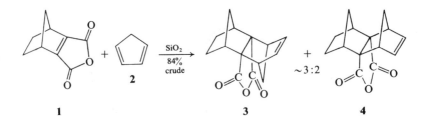

1   **2**   **3**   **4**

2,3-Diphenyloxiranes.[2] Reaction of 2-bromodeoxybenzoin (1) with KCN results in both *cis-* and *trans*-diphenyl-2-cyanooxirane (2 and 3) in about the same amount. When the KCN is absorbed on silica gel, alumina, or a zeolite, the *cis*-isomer is formed almost exclusively. No reaction is observed when carbon or Celite is used as absorbent.

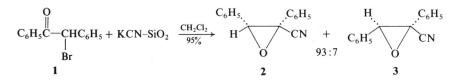

1   **2**   **3**

[1] P. D. Bartlett, A. J. Blakeney, M. Kimura, and W. H. Watson, *Am. Soc.*, **102**, 1383 (1980).
[2] K. Takahashi, T. Nishizauka, and H. Iida, *Tetrahedron Letters*, **22**, 2389 (1981).

Silicon(IV) chloride, SiCl₄. Mol. wt. 169.90, b.p. 57.6°. Supplier: Alfa. Silicon(IV) chloride is the most effective Lewis acid catalyst for dehydrative cyclization of the enamino ketone **1** to julandine (**2**). TiCl₄ is almost as effective. Other Lewis acids are markedly less effective.[1]

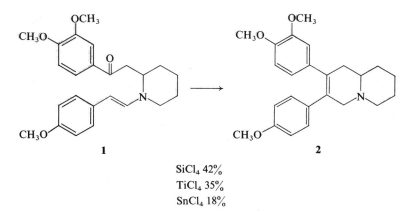

1   **2**

SiCl₄ 42%
TiCl₄ 35%
SnCl₄ 18%

[1] J. E. Cragg, S. H. Hedges, and R. B. Herbert, *Tetrahedron Letters*, **22**, 2127 (1981).

Silver chloride, AgCl. Mol. wt. 143.32.
Amidines. Use of AgCl permits 1:1-addition of a secondary amine to an isocyanide at temperatures below 0°. A single isomer (Z) is formed, which rearranges at temperatures above 25° to the more stable (E)-isomer (equation I). The same reaction with ethyleneimine can be used for synthesis of imidazolines (equation II).[1]

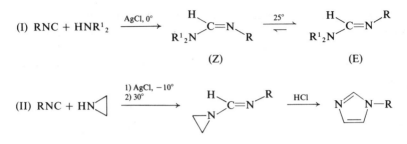

(I) RNC + HNR$^1_2$ $\xrightarrow{\text{AgCl, 0°}}$ (Z) $\underset{}{\overset{25°}{\rightleftharpoons}}$ (E)

(II) RNC + HN⊲ $\xrightarrow[\text{2) 30°}]{\text{1) AgCl, −10°}}$ → $\xrightarrow{\text{HCl}}$

[1] A. F. Hagerty and A. Chandler, *Tetrahedron Letters*, **21**, 885 (1980).

Silver(II) dipicolinate, 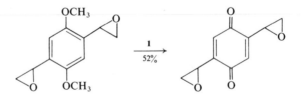 (**1**). Mol. wt. 440.09. The salt is obtained by reaction of dipicolinic acid with silver nitrate and potassium persulfate.[1]

Oxidative demethylation. This reaction is usually conducted with AgO (**4**, 431–432) or CAN (**9**, 99). Syper and co-workers[2] recommended silver(II) dipicolinate (**1**) for oxidation of substrates containing unstable or acid-labile substituents.

Example:

$$\xrightarrow[52\%]{\mathbf{1}}$$

[1] C. W. A. Fowles, R. W. Matthews, and R. A. Walton, *J. Chem. Soc. A*, 1108 (1968); M. G. B. Drew, R. W. Matthews, and R. A. Walton, *J. Chem. Soc. A.*, 1405 (1970).
[2] K. Kloc, J. Młochowski, and L. Syper, *Chem. Letters*, 725 (1980).

Silver fluoride–Pyridine.

Dehydrobromination. Dehydrobromination with AgF–pyridine was first reported some time ago.[1] It has recently proved to be the method of choice in a total synthesis of thienamycin, a carbapenem broad-spectrum antibiotic. For example, attempted dehydrobromination of **1** with DBU in DMSO resulted in elimination of HBr and also carbonate to give a mixture (**2**) of two ene lactams. The desired reaction was effected in 70% yield with AgF in pyridine.[2]

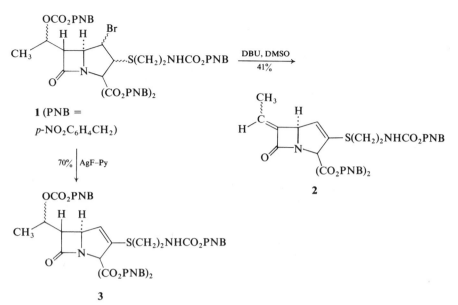

1 (PNB = p-NO$_2$C$_6$H$_4$CH$_2$)

70% | AgF–Py

2

3

This method was superior to use of DBU for dehydrobromination of **4** to **5**; yields were 55% with DBU and 84% for AgF–Py.

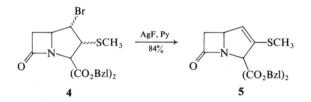

4   **5**

[1] L. Hough and B. Otter, *J.C.S. Chem. Comm.*, 173 (1966).
[2] S. M. Schmitt, D. B. R. Johnston, and B. G. Christensen, *J. Org.*, **45**, 1135, 1142 (1980).

Silver imidazolate, $\overset{N\diagdown}{\underset{\diagdown}{\bigsqcup}}$ NAg   (**1**). Mol. wt. 174.94. The salt precipitates on treat-

ment of imidazole with silver nitrate in aqueous sodium hydroxide.[1]

α-Glucosides. Protected glucals such as **2** react with NIS and a monosaccharide with one free hydroxyl group to form a 2-deoxy-2-iodo-α-D-dissacharide (**4**),[2] presumably by a cyclic iodonium intermediate, in yields of 50–80%. This reaction is improved if the glucal is treated with silver imidazolate and either ZnCl$_2$ or HgCl$_2$ and then with the alcoholic component.[3] Iodine monochloride is formed *in situ* and reacts with **2** to form the 2-iodo-β-pyranosyl chloride (**3**), which can be isolated if

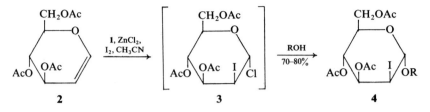

desired. However glycosidation of **3** in the absence of **1** but in the presence of acid acceptors or silver carbonate proceeds slowly and in low yields.

[1] G. Wyss, *Ber.*, **10**, 365 (1877).
[2] J. Thiem, H. Karl, and J. Schwenter, *Synthesis*, 696 (1978).
[3] P. J. Garegg and B. Samuelsson, *Carbohydrate Res.*, **84**, C1 (1980).

Silver(I) nitrate, 1, 1008–1011; **2**, 366–368; **3**, 252; **4**, 429–430; **5**, 582; **7**, 321; **9**, 411.

Solvolysis of methanesulfonates.[1] Epimerization of mestranol (**1**) is best accomplished by conversion to the methanesulfonate (**2**), followed by hydrolysis

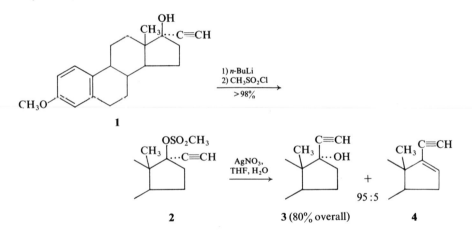

assisted by AgNO₃. Pure epimestranol (**3**) is obtained in 80% overall yield. Epimerization of **1** on alumina gives **3** in 5% yield.

[1] H. Westmijze, H. Kleijn, P. Vermeer, and L. A. van Dijck, *Tetrahedron Letters*, **21**, 2665 (1980).

Silver(I) oxide, 1, 1011; **2**, 368; **3**, 252–254; **4**, 430–431; **5**, 583–585; **6**, 515–518; **7**, 321–322; **8**, 442–443.

Diels-Alder reaction with unstable quinones (**8**, 443). The generation of unstable quinones *in situ* for Diels-Alder reactions has been used to obtain an adduct (**3**), which was transformed to a tricyclic compound (**5**) of interest as a possible intermediate to a quassinoid such as **6**.[1]

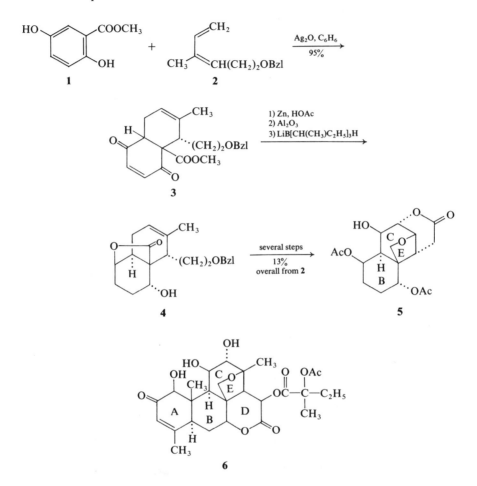

Benzodioxanes (**6**, 516). The earlier synthesis of benzodioxanes by oxidative coupling of catechol derivatives with methoxypropenylphenols has been extended to the first synthesis of the complex benzodioxane silybin (**3**) shown in equation (I).[2] The starting materials are (2R,3R)-dihydroquercetin (**1**) and coniferyl alcohol (**2**). In this case, the reaction is not regioselective, **3** and the isomeric **4** being obtained in nearly equal amounts.

[1] G. A. Kraus and M. J. Taschner, *J. Org.*, **45**, 1174, 1175 (1980).
[2] L. Merlini, A. Zanarotti, A. Pelter, M. P. Rochefort, and R. Hansel, *J.C.S. Perkin I*, 775 (1980).

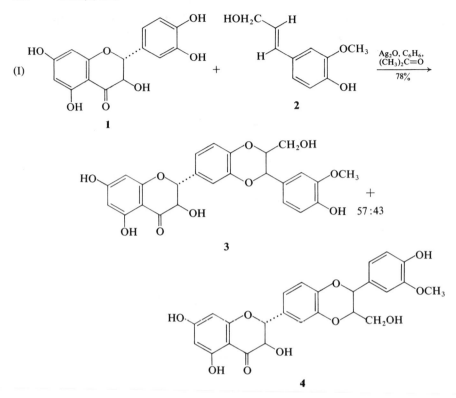

Silver(II) oxide, 1, 1011–1015; **2**, 368; **3**, 252–254; **4**, 430–431; **5**, 583–585; **6**, 515–518; **7**, 321–322; **9**, 412–413.

Oxidative demethylation (**5**, 431–432; **7**, 55, 322). Preparation of quinones by oxidative demethylation of 1,4-dimethoxyarenes with AgO is not satisfactory for ethers containing alkenyl groups because of various side reactions. CAN (**7**, 55), however, can be used, with yields of quinones being in the range 50–85%. Oxidations with AgO can be conducted satisfactorily in the presence of 1 equivalent of 2,4,6-pyridinetricarboxylic acid, which can complex Ag(II) and facilitate electron transfer.[1] Although addition of this acid notably improves yields from oxidations with AgO, it affects only a slight improvement in yield of oxidations with CAN. In general, oxidations with CAN or with AgO in combination with this acid are about equally efficient.

p-Hydroquinone monoacetates are also oxidized to 1,4-quinones by AgO in the presence of nitric acid; this system is more reactive than a 1,4-dimethoxy system (equation I). However, a 1,4-dimethoxy group is more reactive than a 1,4-diacetoxy group (equation II). Hydrolysis of an acetyl group is not involved, since isolated acetyl groups are not hydrolyzed under these conditions.[2]

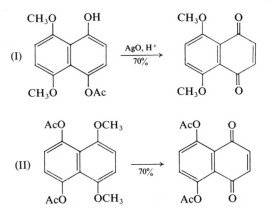

A key step in a synthesis of the 6-deoxyanthracyclinone **4** involves oxidation of **1** with silver(II) oxide to give the crude quinone **a**, which rearranges in the presence of a trace of HCl to the quinone **3** in high yield.[3]

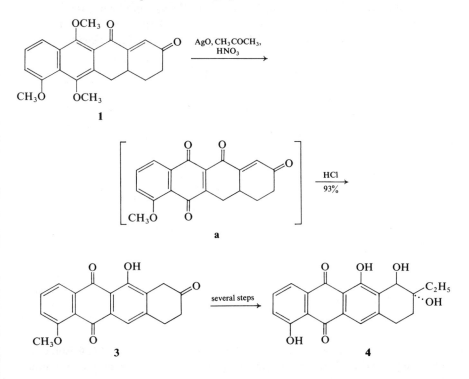

2-Hydroxy-1,4-quinones.[4] Silver oxide oxidizes derivatives of methyl sesamol **1** to quinones of this type (equation I).

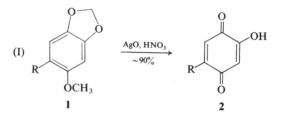

(I)

1   2

[1] L. Syper, K. Kloc, and J. Młochowski, *Tetrahedron*, **36**, 123 (1980).

[2] C. Escobar, F. Farina, R. Martinez-Utrilla, and M. C. Paredes, *J. Chem. Res. (M)*, 3154 (1977); *J. Chem. Res. (S)*, 156 (1980).

[3] A. S. Kende, M.-P. Gesson, and T. P. Demuth, *Tetrahedron Letters*, **21**, 1667 (1981).

[4] G. A. Kraus and K. Neuenschwander, *Syn. Comm.*, **10**, 9 (1980).

Silver perchlorate, 2, 369–370; **4,** 432–435; **5,** 585–587; **6,** 518–519; **7,** 322–323; **9,** 413–414.

Isomerization of cyclopropenes.[1] Irradiation of the cyclopropene **1** results in the alkenes **2** and **3**, formed by cleavage of the *b* bond. Exposure of **1** to silver ion results in two different alkenes, **4** and **5**, formed by cleavage of the *a* bond.

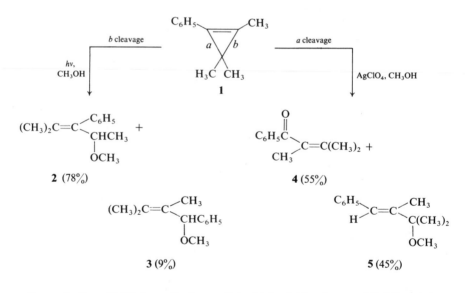

Isomerization of (E)-**6** results in *exo*-**7** in high yield, whereas (Z)-**6** isomerizes exclusively to *endo*-**7**. The retention of configuration is explained on the assumption of the intermediate carbonium ions **a** and **b**.

[1] A. Padwa, T. J. Blacklock, and R. Loza, *Am. Soc.*, **103**, 2404 (1981).

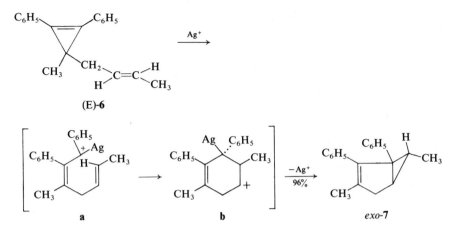

(E)-6

a   b   *exo*-7

Silver(I) trifluoroacetate, 1, 1018–1019; **7**, 323–324; **8**, 444–445.

1,3-Butadienyl acetates. These useful Diels-Alder dienes (**2**) can be obtained by isomerization of 2-propynylic acetates (**1**) with this silver salt or with $PdCl_2$. The substrates (**1**) rearrange to allenyl acetates (**3**) with CuCl.[1]

Example:

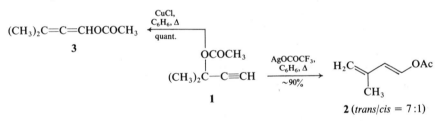

[1] R. C. Cookson, M. C. Cramp, and P. J. Parsons, *J.C.S. Chem. Comm.*, 197 (1980).

Sodium–Ammonia, 1, 1041; **2**, 374–376; **3**, 259; **4**, 438; 589–591; **6**, 523; **7**, 324–325; **9**, 415.

Reduction of aryl esters.[1] Alkyl benzoates are reduced to 1,4-dihydro compounds by sodium in $THF–NH_3$ if 1–2 equivalents of water is present before addition of the metal (equation I). The presence of water can be useful for reduction of other aromatic systems (equation II).

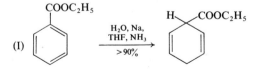

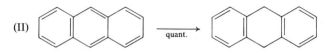

(II)

quant.

¹ P. W. Rabideau, D. L. Huser, and S. J. Nyikos, *Tetrahedron Letters*, **21**, 1401 (1980).

Sodium–Chlorotrimethylsilane, 8, 446.

1-Trimethylsilylcycloalkenes. Some years ago Russian chemists[1] prepared 1-trimethylsilylcyclohexene by coupling 1-chlorocyclohexene with chlorotrimethylsilane mediated by sodium. More recent work indicates that this Wurtz-type coupling is a general and useful reaction (equation I).[2]

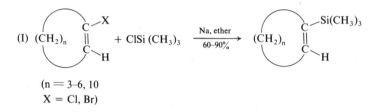

$$(n = 3\text{–}6, 10$$
$$X = Cl, Br)$$

¹ A. D. Petrov, V. F. Mironov, and V. G. Glukhovtsev, *Zh. Obshch. Khim.*, **27**, 1535 (1957) [*C.A.*, **52**, 3668 (1958)].
² G. Nagendrappa, *Synthesis*, 704 (1980).

Sodium benzeneselenoate, 5, 273; 6, 548–549; 8, 447–448.

Cleavage of lactones (**8**, 447–448) *and cyclopropanes.* Details are available from a study of cleavage of lactones with the selenium reagent in DMF at 120–125°. The efficiency of this reaction varies with the substrate. The larger the lactone ring, the longer the reaction time. In the course of this work, cleavage of cyclopropanes with this reagent was examined. If the ring is activated by a keto or nitrile group, cleavage is possible.[1]

Examples:

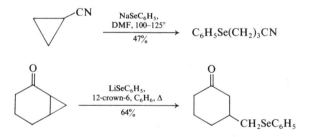

¹ R. M. Scarborough, Jr., B. H. Toder, and A. B. Smith III, *Am. Soc.*, **102**, 3904 (1980).

Sodium bis(2-methoxyethoxy)aluminium hydride (SMEAH), **3**, 260–261; **4**, 441–442; **5**, 596; **6**, 528–529; **7**, 327–329; **8**, 448–449; **9**, 418–420.

Regioselective reduction of an epoxide. A key step in a short synthesis of (+)-muscarine (**6**) from the dibenzoate of 2,5-anhydro-D-glucitol (**1**) involves reduction of the epoxide **2**. Use of SMEAH gives a 12:1 mixture of **3** and **4**; use of LiAlH$_4$ gives the same diols in the ratio 3:1.[1]

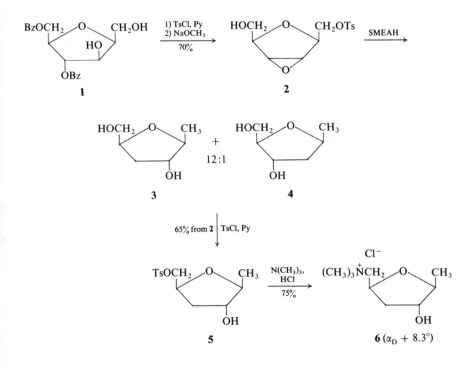

[1] A. M. Mubarak and D. M. Brown, *Tetrahedron Letters*, **21**, 2455 (1980).

Sodium borohydride, 1, 1049–1055; **2**, 377–378; **3**, 262–264; **4**, 443–444; **5**, 597–601; **6**, 530–534; **7**, 329–331; **8**, 449–451; **9**, 420–421.

Reduction of α-alkoxy-β-keto esters.[1] *t*-Butoxy esters of α-alkoxy-β-keto acids are reduced by NaBH$_4$ in 2-propanol selectively to *erythro*-α-alkoxy-β-hydroxy carboxylates. The report suggests that the selectivity results from reduction of an intermediate five-membered chelate involving Na$^+$ and, in fact, no stereoselectivity is observed if a crown ether is present.

Example:

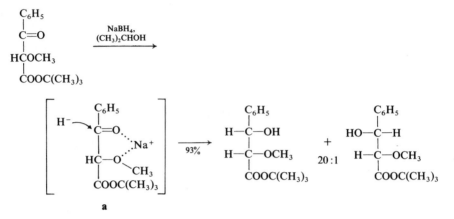

a

Reduction of acid chlorides (8, 461).[2] Acid chlorides can be selectively reduced to aldehydes by slightly less than 1 molar equivalent of NaBH₄ in a mixture of DMF–THF at −70°. It is essential to minimize further reduction by quenching the reaction with a mixture of propionic acid–dilute HCl and ethyl vinyl ether (*caution:* H₂ is evolved). Both aliphatic and aromatic aldehydes can be obtained in 80–95% yield.

Stereoselective reduction of a 2-chloro-1,3-diketone.[3] The reduction of **1** (nonenolizable) proceeds predominately *anti* with respect to the C—Cl bond. The chlorine atom in the product can be removed by zinc dust reduction of the methoxymethyl ether.

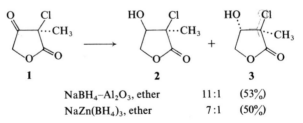

NaBH₄–Al₂O₃, ether	11:1	(53%)
NaZn(BH₄)₃, ether	7:1	(50%)

Polymer-bound reagent. Reaction of a quaternary ammonium anion-exchange resin (Amberlites) wth NaBH₄ results in an immobilized borohydride reducing agent (**1**), which is somewhat less reactive than NaBH₄.[4]

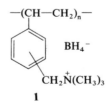

1

A difference in the stereoselectivity of $NaBH_4$ and the polymeric-bound reagent has been observed. Thus the ketone **2** is reduced mainly to the *erythro*-alcohol **3** by $NaBH_4$, but mainly to the *threo*-alcohol **4** by **1**.[5]

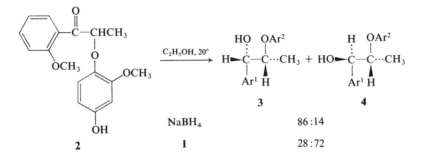

[1] R. S. Glass, D. R. Deardorff, and K. Henegar, *Tetrahedron Letters*, **21**, 2467 (1980).
[2] J. H. Babler and B. J. Invergo, *Tetrahedron Letters*, **22**, 11 (1981).
[3] H. Wyss, U. Vögeli, and R. Scheffold, *Helv.*, **64**, 775 (1981).
[4] H. Gibson and F. Bailey, *J.C.S. Chem. Comm.*, 815 (1977).
[5] G. Brunow, L. Koskinen, and P. Urpilainen, *Acta Chem. Scand.*, **35**, 53 (1981).

Sodium borohydride–Cadmium chloride · 1.5 DMF. The salt is obtained by crystallization of $CdCl_2 \cdot 2.5\ H_2O$ from DMF.

Reduction of RCOCl to RCHO.[1] Sodium borohydride and this cadmium salt reduce acyl chlorides to aldehydes, but only in solvents such as DMF, DMA, and HMPT. DMF is essential, but the amount of DMF in the salt is sufficient. For the most part, yields are 50–90%. Aryl, alkyl, and benzylic halides do not react. Other functional groups (nitrile, nitro, ester, C=C) are also inert.

[1] I. D. Entwistle, P. Boehm, R. A. W. Johnstone, and R. P. Telford, *J.C.S. Perkin I*, 27 (1980).

Sodium borohydride–Cerium(III) chloride.

Stereoselective reduction. 2,3-Epoxycyclohexanones can be reduced predominately to the *trans*-epoxy alcohol by sodium borohydride if catalytic amounts of $CeCl_3$ are present.[1]

[1] G. Rücker, H. Hörster, and W. Gajewski, *Syn. Comm.*, **10**, 623 (1980).

Sodium borohydride–Cobalt(II) chloride.

Reduction of imides.[1] The N-imidotryptamine (**1**) is reduced selectively by $NaBH_4$–$CoCl_2$ (5:1 equivalents) to the hydroxy lactam **2** in 95% yield. The product is converted in quantitative yield to the β-carboline **3** by treatment with conc. HCl at 30°.

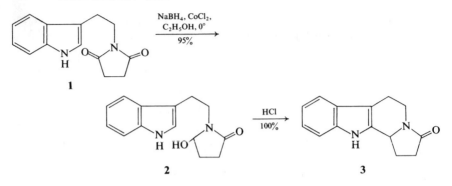

1

2   **3**

¹ G. M. Atta-ur-Rahman, M. Ghazala, N. Sultana, and M. Bashir, *Tetrahedron Letters*, **21**, 1773 (1980).

Sodium borohydride–Pyridine.

Reduction of α,β-enones. NaBH$_4$ forms a nearly homogeneous solution in pyridine at 25°. This solution reduces *l*-carvone (**1**) at 25° (12 hours) mainly to the dihydrocarveols **2**. A minor product is **3**. Jones oxidation of the mixture gives dihydrocarvone (**4**).[1] This method was first reported by Kupfer.[2]

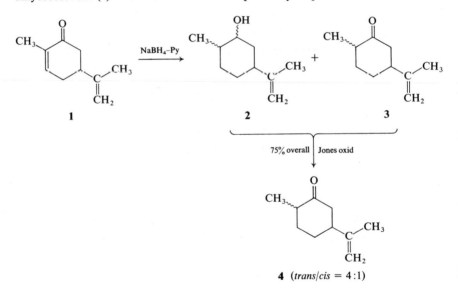

4 (*trans/cis* = 4:1)

¹ S. Raucher and K.-J. Hwang, *Syn. Comm.*, **10**, 133 (1980).
² D. Kupfer, *Tetrahedron*, **15**, 193 (1961).

Sodium cyanoborohydride, 4, 448–451; **5,** 607–609; **6,** 537–538; **7,** 334–335; **8,** 454–455; **9,** 424–426.

2,5-Dialkylpyrrolidines. These compounds can be prepared by reductive amination of 1,4-diketones with sodium cyanoborohydride and ammonium acetate (**4**, 448–449). 1-Pyrrolines are usually formed also, but they can be reduced to pyrrolidines by $NaBH_4$ in a second step.[1]

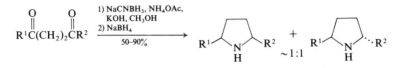

Macrocyclic lactams. Two key steps in a synthesis of the polyamine alkaloid dihydroperiphylline (**4**) involve ring expansion of the cyclic imino ether **1** by reaction with the β-lactam **2** to form **3**. Sodium cyanoborohydride in acetic acid reduces **3** in high yield to **4**, probably by way of intermediates **a** and **b**.[2]

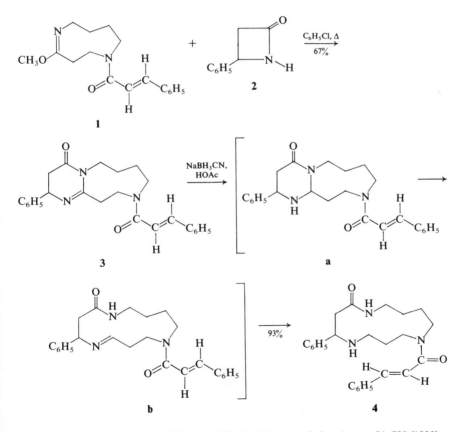

[1] T. H. Jones, J. B. Franko, M. S. Blum, and H. M. Fales. *Tetrahedron Letters*, **21**, 789 (1980).
[2] H. H. Wasserman and H. Matsuyama, *Am. Soc.*, **103**, 461 (1981).

Sodium dicarbonylcyclopentadienylferrate, 5, 610; **6,** 538–539; **8,** 454–455; **9,** 426.

β-Lactams.[1] The synthesis of β-lactams employing organoiron complexes as intermediates (**8,** 454–455) has been extended to the preparation of bicyclic lactams, in particular to the new carbopenams. A synthesis of 2-methylcarbopenam (**1**) is shown in equation (I).

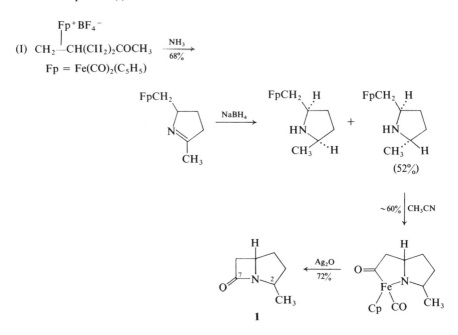

$$\text{(I)} \quad CH_2\text{—}CH(CH_2)_2COCH_3 \xrightarrow[68\%]{NH_3}$$

$$Fp = Fe(CO)_2(C_5H_5)$$

[1] S. R. Berryhill and M. Rosenblum, *J. Org.,* **45,** 1984 (1980).

Sodium O,O-diethyl phosphorotelluroate, 8, 455.

Deoxygenation of epoxides. Complete details for preparation and use of the reagent for deoxygenation of epoxides, particularly of terminal epoxides, are available.[1]

[1] D. L. J. Clive and S. M. Menchen, *J. Org.,* **45,** 2347 (1980).

Sodium α-(N,N-dimethylamino)naphthalenide (1). The radical anion is prepared and used in the same way as sodium naphthalenide. However the derived α-dimethylaminonaphthalene can be separated from products simply by an aqueous wash.[1]

Reductive deacetoxylation.[2] This naphthalenide in HMPT is recommended for reductive cleavage of α-acetoxy esters.[2] Lithium in liquid ammonia can be used, but yields are generally lower. The reaction was developed particularly for a synthesis of diethyl allylmalonates from alkenes via an ene reaction (*see* Diethyl oxomalonate, this volume). An example is formulated in equation (I). Since an ester enolate is an intermediate, a one-pot reductive alkylation is possible.[3]

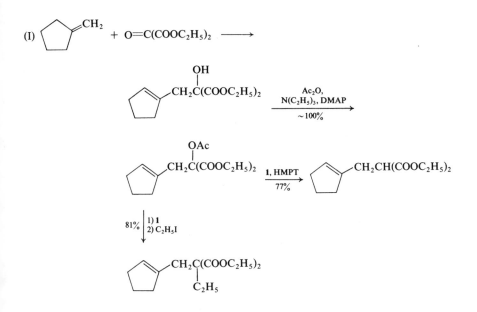

(I)

[1] B. Angelo, *Bull Soc.*, 3855 (1968); S. Bank and M. Platz, *Tetrahedron Letters*, 2097 (1973).
[2] S. N. Pardo, S. Ghosh, and R. G. Salomon, *Tetrahedron Letters*, **22**, 1885 (1981).
[3] M. F. Salomon, S. N. Pardo, and R. G. Salomon, *Am. Soc.*, **102**, 2473 (1980).

Sodium dithionite, 1, 1081–1083; **5,** 615–617; **7,** 336; **8,** 456–458; **9,** 426–427.

Marschalk reaction (**8,** 456–457). Reaction of leucoquinizarin (**1**) with acetaldehyde results in **2**. Products of this type have been considered to be intermediates in the Marschalk reaction.

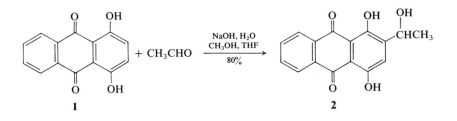

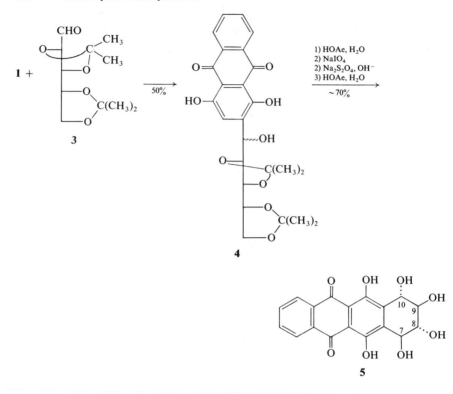

Reaction of **1** with the sugar aldehyde **3** under the same conditions results in **4**. Cleavage of the terminal isopropylidene group followed by periodate gives an aldehyde. This cyclizes when treated with sodium dithionite in an alkaline solution at 0°. Finally the tetracyclic **5** is obtained by acid-catalyzed cleavage of the isopropylidene group. Spectrographic evidence suggests that the C_7 and C_{10} hydroxyl groups are *trans* to those at C_8 and C_9.[1]

    Reduction of imines; cleavage of oximes.[2] $Na_2S_2O_4$ in aqueous DMF reduces imines to N-alkylamines at 110°. Yields are 40–73% (five examples). Carbonyl compounds are regenerated from oximes by cleavage with aqueous $Na_2S_2O_4$ at 25°. Yields are 54–96% (four examples).

[1] D. J. Mincher and G. Shaw, *J.C.S. Chem. Comm.*, 508 (1981).
[2] P. M. Pojer, *Aust. J. Chem.*, **32**, 201 (1979).

Sodium hydride–Dimethyl sulfoxide.

    Cleavage of acetals.[1] Ethylene acetals of β,γ-unsaturated carbonyl compounds are cleaved to dienol ethers by NaH in DMSO.

Example:

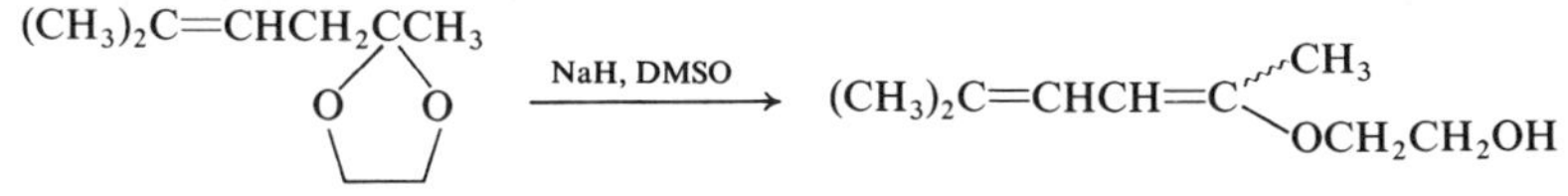

$(CH_3)_2C=CHCH_2CCH_3$ $\xrightarrow{\text{NaH, DMSO}}$ $(CH_3)_2C=CHCH=C\overset{\text{\textit{''}}CH_3}{\underset{OCH_2CH_2OH}{}}$

(E) and (Z)

[1] K. Steinbeck and B. Osterwinter, *Tetrahedron Letters*, **21**, 1515 (1980).

Sodium hydride–Nickel acetate–Sodium *t*-amyl oxide (Nic), NaH–Ni(OAc)$_2$–C$_2$H$_5$C(CH$_3$)$_2$ONa.

Hydrogenation catalyst.[1] The combination of sodium hydride (large excess) and nickel acetate (1 equivalent) when activated by sodium *t*-amyl oxide (2 equivalents) is a highly active, inexpensive, heterogeneous hydrogenation catalyst. It allows selective hydrogenation of internal triple bonds to *cis*-alkenes in less than 15 minutes in yields generally >90%. Moreover Nic also catalyzes hydrogenation of alkenes, although this reaction is sensitive to substituents on the double bond. Cyclo-alkenes are reduced in the following order of reactivity: C$_5$ ~ C$_7$ > C$_6$ > C$_8$. Carbonyl groups are also hydrogenated with this catalyst under ambient conditions.

Preliminary experiments indicate that similar active catalysts are obtained when Ni(OAc)$_2$ is replaced by some other metal salts such as Co(OAc)$_2$ and Pd(OAc)$_2$.

[1] J.-J. Brunet, P. Gallois, and P. Caubere, *J. Org.*, **45**, 1937, 1946 (1980).

Sodium hydrogen telluride, 8, 459–460.

Selective reduction.[1] NaHTe reduces α,β-unsaturated ketones, aldehydes, esters, and lactones at room temperature to the corresponding saturated carbonyl compounds in high yield. It is not satisfactory for reduction of enediones.

[1] M. Yamashita, Y. Kato, and R. Suemitsu, *Chem. Letters*, 847 (1980).

Sodium hypochlorite, 1, 1084–1087; **2**, 67; **3**, 45, 243; **4**, 456; **5**, 617; **6**, 543; **7**, 337–338; **8**, 461–463; **9**, 430.

Oxidation of $>$CHOH (**7**, 337). Sodium hypochlorite solutions[1] oxidize secon-dary alcohols dissolved in acetic acid to ketones in yields of 90–95%. Selective oxi-dation in the presence of a primary alcohol group is possible. The oxidation has been conducted, with suitable precautions, on a large scale.[2]

[1] Commercial solutions (1.8–2.0 *M*) suitable for swimming pool use were used.
[2] R. V. Stevens, K. T. Chapman, and H. N. Weller, *J. Org.*, **45**, 2030 (1980).

Sodium iodide, 1, 1087–1090; **2**, 384; **3**, 267; **4**, 456–457; **6**, 543; **7**, 338.

Dehalogenation of α-halo ketones.[1] This reaction can be effected with NaI in THF–H$_2$O (1:1) containing 5% H$_2$SO$_4$ at 25°. In the case of aliphatic α-halo ketones

NaI in refluxing dioxane–water is used. Hindered bromo ketones are not reduced. In the absence of acid, conversion to the α-iodo ketone is the only observed reaction (**1**, 1087).

Examples:

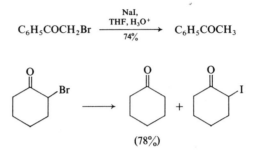

$$C_6H_5COCH_2Br \xrightarrow[74\%]{\substack{NaI, \\ THF, H_3O^+}} C_6H_5COCH_3$$

(78%)

[1] A. L. Gemal and J. L. Luche, *Tetrahedron Letters*, **21**, 3195 (1980).

Sodium iodide–Hydrochloric acid.

Reduction of enediones.[1] 1,4-Enediones and 4-oxo-2-alkenals are reduced to saturated 1,4-dicarbonyl compounds in nearly quantitative yield by a large excess of NaI–HCl (1:1) in acetone. The method is ineffective with butenedioic acids and α,β-enones and α,β-enals.

[1] M. D'Auria, G. Piancatelli, and A. Scettri, *Synthesis*, 245 (1980).

Sodium methoxide–Dimethyl sulfoxide.

Diosphenols.[1] In the first total synthesis of *dl*-quassin (**5**), the two diosphenol methyl ether groups were introduced in a two-step sequence. Vedejs oxygenation (**5**, 269–270) of the diketone **1** results in the bis(α-hydroxy ketone) **2** in 35% yield. Treatment of **2** with sodium methoxide in DMSO under argon effects oxidation to the bis(diosphenol) **3** with concomitant epimerization at C_9. Methylation of **3** is effected with $NaOCH_3$–DMSO and CH_3I to give **4**. Actually **2** can be converted directly into **4** by treatment with $NaOCH_3$, DMSO, and CH_3OH for 1 hour and then CH_3I for 15 minutes in 57% overall yield. Quassin **5** is obtained by mild acid hydrolysis followed by oxidation with Fetizon's reagent. The oxidation of **2** to **3** does not involve base-catalyzed oxygenation. In fact the presence of oxygen is deleterious.

[1] P. A. Grieco, S. Ferriño, and G. Vidari, *Am. Soc.*, **102**, 7586 (1980).

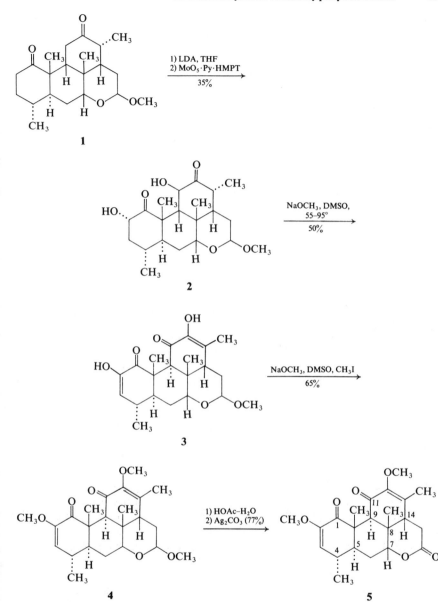

Sodium N-methylanilide–Hexamethylphosphoric triamide.

Selective cleavage of aryl methyl ethers.[1] The sodium salt (NaH) of N-methylaniline in the presence of HMPT cleaves methyl (and benzyl) aryl ethers in high yield. It is particularly useful for selective cleavage of one methoxy group of polymethoxyarenes. The directing effect of a methoxy group is $o > m > p$. For this purpose the present method is superior to use of sodium thioethoxide (**4**, 465).

[1] B. Loubinoux, G. Coudert, and G. Guillaumet, *Synthesis*, 638 (1980).

Sodium methylselenolate, NaSeCH$_3$.

Alkenes from **vic-dihalides.**[1] The reaction of *vic*-dihalides with NaSeCH$_3$ or NaSeC$_6$H$_5$ in ethanol or THF–HMPT (3:1) results in alkenes. The reaction involves a formal *anti*-elimination in the case of *vic*-dibromides or *vic*-chloroiodides, but a formal *syn*-elimination in the case of *vic*-dichlorides. Elimination to form alkenes also occurs with β-haloalkyl phenyl selenides.

[1] M. Sevrin, J. N. Denis, and A. Krief, *Tetrahedron Letters*, **21**, 1877 (1980).

Sodium permanganate monohydrate, NaMnO$_4$·H$_2$O.

Heterogeneous oxidations.[1] A variety of substrates can be oxidized by this solid reagent in refluxing hexane (1.5–24 hours). The reaction proceeds in good yield with alcohols (85–100%), aldehydes (65–80%), and sulfides; but alkenes, alkynes, epoxides, and amides are not oxidized. Surprisingly allylic alcohols are oxidized more slowly than saturated alcohols.

[1] F. M. Menger and C. Lee, *Tetrahedron Letters*, **22**, 1655 (1981).

Sodium pyridylselenate (1). Mol. wt. 180.04.

   Preparation:

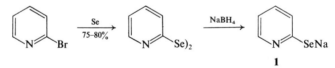

Synthesis of alkenes.[1] Terminal alkenes are formed in good yield by oxidation of primary alkyl 2-pyridyl selenides with a slight excess of H$_2$O$_2$ (equation I). The same reaction with primary alkyl phenyl selenides proceeds in much lower yield.

(I) RCH$_2$CH$_2$X + **1** $\xrightarrow[\text{Quant.}]{}$ ⟨pyridyl⟩—SeCH$_2$CH$_2$R $\xrightarrow[65-85\%]{H_2O_2}$ RCH=CH$_2$

[1] A. Toshimitsu, H. Owada, S. Uemura, and M. Okano, *Tetrahedron Letters*, **21**, 5037 (1980).

Sodium selenophenolate, **5**, 272–276; **6**, 548–549; **7**, 341; **9**, 432–434.

Aromatic α-methylene-δ-lactones.[1] A new approach to this system is outlined in equation (I).

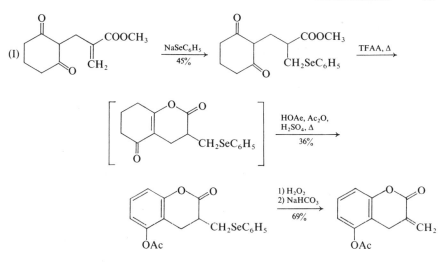

[1] W. C. Groutas, D. Felker, and D. Magnin, *Syn. Comm.*, **10**, 355 (1980).

Sodium triethylborohydride–Iron(II) chloride.

Desulfurization of thiols.[1] The combination of $FeCl_2$ (2 equivalents) and $NaB(C_2H_5)_3H$ (4 equivalents) desulfurizes aryl, benzylic, and alkyl thiols to hydrocarbons at $-78 \rightarrow 25°$. $CoCl_2$ and VCl_3 are less useful metal salts. Yields are in the range 60–85% (four examples).

[1] H. Alper and T. L. Prince, *Angew. Chem. Int. Ed.*, **19**, 315 (1980).

Sodium tris(3,5-dimethylphenoxy)borohydride; Sodium tris(3,5-di-*t*-butylphenoxy)-borohydride, $NaBH(OAr)_3$. These complexes are prepared from $NaBH_4$ and 3 equivalents of the phenol.

Selective reduction of aldehydes.[1] These reagents selectively reduce aldehydes in the presence of ketones; the selectivity is equal to or greater than that of previously known reagents.[1]

[1] S. Yamaguchi, K. Kabuto, and F. Yashuhara, *Chem. Letters*, 461 (1981).

Sodium trithiocarbonate, $Na_2CS_3 \cdot 2H_2O$. Mol. wt. 190.22. Preparation.[1]

Demercuration.[2] A recent synthesis of the Prelog-Djerassi lactone (**5**), a degradation product of some macrolide antibiotics, involves mercury(II) cyclization of **1** to **2**. Usual methods of demercuration ($NaBH_4$, Na/Hg, H_2S–py) favor formation of the 2-epimer (**6**) of **5**. The most useful reagent is sodium trithiocarbonate, which gives **5** and **6** in a ratio of 3.5 : 1. The ratio is 2 : 1 when Na_2S is used.

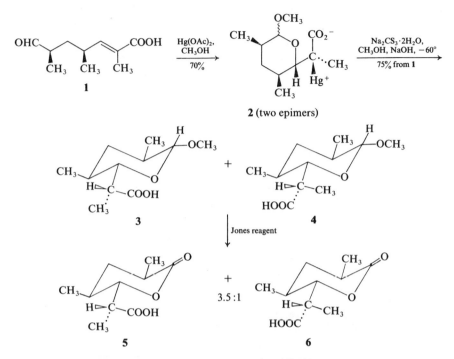

2 (two epimers)

[1] A. Lamotte, M. Posthault, and J.-C. Merlin, *Bull. Soc.*, 915 (1965).
[2] P. A. Bartlett and J. L. Adams, *Am. Soc.*, **102**, 337 (1980).

Stannic chloride, 1, 1111–1113; **3,** 269; **5,** 627–631; **6,** 553–554; **7,** 342–345; **9,** 436–438.

1-Aryltetralones.[1] Cyclopropyl ketones such as **1** cyclize to 1-aryltetralones **2** and **3** in the presence of Lewis acid catalysts such as $SnCl_4$, CF_3COOH, and BF_3 etherate.

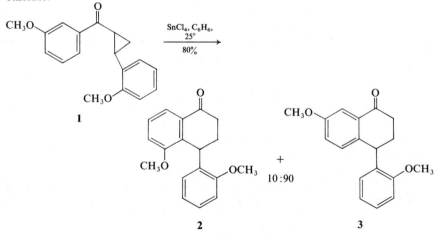

Diels-Alder catalyst. The key step in a recent total synthesis of androstanes is a $SnCl_4$-catalyzed Diels-Alder reaction of **1** with the (Z)-dienophile **2**.[2] The geometry of the diene favors addition *anti* to the C_{10}-methyl group, and the catalyst promotes the desired *endo*-orientation. $AlCl_3$ and BF_3 etherate are less suitable for additions involving aliphatic bifunctional dienophiles. The initial adduct **a** can be isolated, but in only 15–20% yield. The synthesis of the androstane **4** is completed by ketalization of **3** followed by a novel cyclization affected with dimsylsodium.

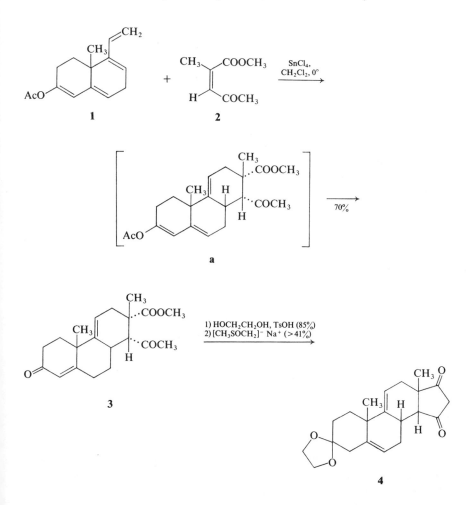

If the ketone **5**, corresponding to the enol acetate **1**, is used, addition *syn* to the C_{10}-methyl group is favored. Thus cycloaddition of **5** to **6** catalyzed by $SnCl_4$ results in preferential formation of **7**. This product can be converted in several steps to the 8α-androstane derivative **8**.[3]

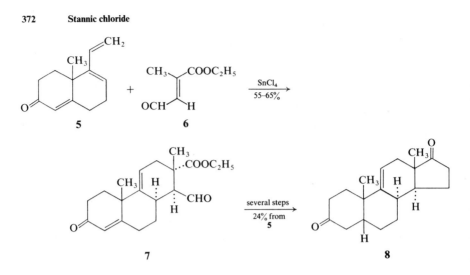

5   6   7   8

Liu and Browne[4] have reported an example of reversal of the regioselectivity of a Diels-Alder reaction solely with a Lewis acid catalyst. Thus by appropriate choice of BF$_3$ etherate or SnCl$_4$ addition of isoprene to **9** can be controlled to give either **10** or **11** as the main product.

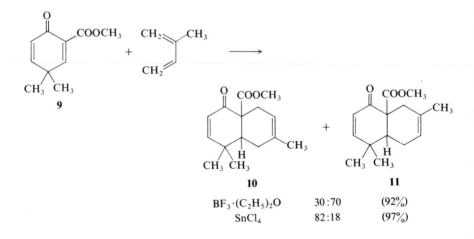

9   10   11

BF$_3$·(C$_2$H$_5$)$_2$O	30:70	(92%)
SnCl$_4$	82:18	(97%)

3,5-Disubstituted cyclohexadienones undergo Diels-Alder reactions more slowly than the unsubstituted counterparts. Thus **12** does not react with piperylene (**13**) at 180°, but in the presence of SnCl$_4$ the reaction proceeds in 85% yield at 25°. Moreover a complete reversal of face selectivity can be achieved by use of a Lewis acid catalyst. Thus **15** reacts thermally with **13** to give **16**, whereas the catalyzed reaction results in **17**. Thus the stereochemistry of four asymmetric centers can be controlled.[5]

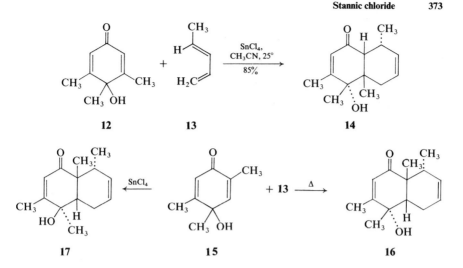

12   13   14

17   15   16

α-Thioalkylation of ketones.[6] This reaction can be effected by reaction of thioketals with silyl enol ethers catalyzed by $SnCl_4$, which is more effective than $FeCl_3$ or $Hg(OCOCF_3)_2$.

Examples:

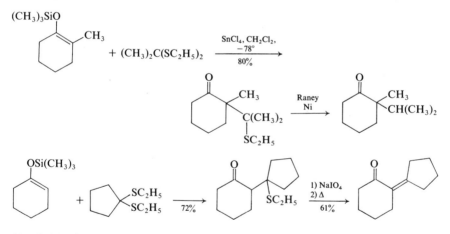

[1] W. S. Murphy and S. Wattanasin, *Tetrahedron Letters*, **21**, 1887 (1980); *J. C. S. Perkin I*, 2920 (1981).

[2] M. Kakushima, J. Das, G. R. Reid, P. S. White, and Z. Valenta, *Can. J. Chem.*, **57**, 3357 (1979).

[3] M. Kakushima, L. Allain, R. A. Dickinson, P. S. White, and Z. Valenta, *Can. J. Chem.*, **57**, 3354 (1979).

[4] H.-J. Liu and E. N. C. Browne, *Can. J. Chem.*, **59**, 601 (1981).

[5] D. Liotta, M. Saindane, and C. Barnum, *Am. Soc.*, **103**, 3224 (1981).

[6] M. T. Reetz and A. Giannis, *Syn. Comm.*, **11**, 315 (1981).

Stannous chloride–Silver perchlorate.

α-Glucosylation. α-Glucosides are formed predominantly by reaction of an alcohol with 2,3,4,6-tetra-O-benzyl-β-glucosyl fluoride (1) in the presence of $SnCl_2$ and $AgClO_4$.[1]

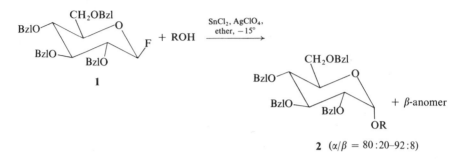

2 $(\alpha/\beta = 80:20\text{–}92:8)$

[1] T. Mukaiyama, Y. Murai, and S. Shoda, *Chem. Letters*, 431 (1981).

Stannous fluoride, SnF_2. Mol. wt. 194.68, m.p. 219°. Supplier: Alfa.

Homoallylic alcohols.[1] Allyl iodide reacts wth SnF_2 to form *in situ* allyltin difluoroiodide, which reacts with aldehydes to form homoallylic alcohols in yields of 80–90%. The reaction with ketones proceeds in lower yield. 1,3-Dimethyl-2-imidazolidinone is the most satisfactory solvent. $SnCl_2$ and $SnBr_2$ can be used, but yields are somewhat lower.

Examples:

$$CH_2{=}CHCH_2I + RCHO \xrightarrow[88\text{–}96\%]{SnF_2} CH_2{=}CHCH_2\overset{\overset{\displaystyle H}{|}}{\underset{\underset{\displaystyle R}{|}}{C}}{-}OH$$

$$CH_2{=}CHCH_2I + RCOCH_3 \xrightarrow[50\text{–}60\%]{} CH_2{=}CHCH_2\overset{\overset{\displaystyle CH_3}{|}}{\underset{\underset{\displaystyle R}{|}}{C}}{-}OH$$

[1] T. Mukaiyama, T. Harada, and S. Shoda, *Chem. Letters*, 1507 (1980).

Sulfoacetic acid, HO_3SCH_2COOH (1). The acid is prepared *in situ* from acetic anhydride and conc. sulfuric acid.

Pyrylium salts.[1] Unlike the usually used pyrylium perchlorate salts, pyrylium sulfoacetates are not explosive. They are readily prepared as shown for the preparation of 2,4,6-trimethylpyrylium sulfoacetate (2).

$$CH_2=C(CH_3)_2 + 3(CH_3CO)_2O + HO_3SCH_2COOH \longrightarrow$$

1

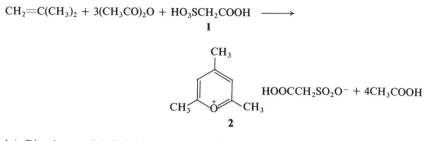

$$HOOCCH_2SO_2O^- + 4CH_3COOH$$

2

[1] A. Dinculescu and A. T. Balaban, *Org. Syn.*, submitted (1980).

Sulfuric acid–Trifluoroacetic anhydride.

Diaryl sulfones.[1] Sulfuric acid and trifluoroacetic anhydride (1:2 equivalents) form bis(trifluoroacetyl) sulfate (1), which converts arenes and phenol ethers into diaryl sulfones in 70–99% yield.

Example:

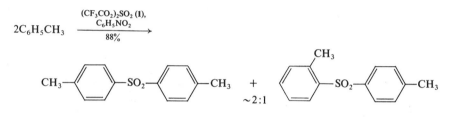

[1] T. E. Tyobeka, R. A. Hancock, and H. Weigel, *J.C.S. Chem. Comm.*, 114 (1980).

Sulfuryl chloride, 1, 1128–1131; **2,** 394–395; **3,** 276; **4,** 474–475; **5,** 641; **6,** 561; **7,** 349–350.

Dehydrogenation of 2,5- and 4,5-dihydrothiophenes. Lederle chemists[1] have converted the 2,5-dihydrothiophene derivative **1** into the thiophene **2** with SO_2Cl_2.

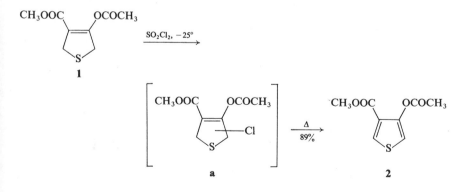

BASF chemists[2] report that this reaction is highly effective for dehydrogenation of both 2,5- and 4,5-dihydrothiophenes. It was examined in the course of a synthesis of an artificial sweetener thiophenesaccharin (**4**), which is 1000 times sweeter than sucrose and which lacks a bitter aftertaste. The commercial synthesis of **4** from **3** includes this dehydrogenation as one step.

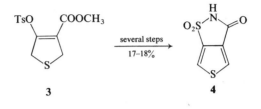

[1] J. B. Press, C. M. Hofmann, and S. R. Safir, *J. Org.*, **44**, 3292 (1979).
[2] P. A. Rossy, W. Hoffman, and N. Müller, *J. Org.*, **45**, 617 (1980).

T

Tellurium chloride, 9, 443.

Olefin inversion.[1] The adduct of TeCl$_4$ with alkenes is reduced by sodium sulfide to an olefin and tellurium. The adduct is formed predominately by *cis*-addition, whereas reduction involves *trans*-elimination via an epitelluride. As a consequence the overall reaction proceeds with inversion of the olefin. For example, (Z)-2-butene can be converted into an 81:19 mixture of (E)- and (Z)-2-butene.

[1] J.-E. Bäckvall and L. Engman, *Tetrahedron Letters*, **22**, 1919 (1981).

2,4,4,6-Tetrabromo-2,5-cyclohexadienone (1), 4, 476–477; **5**, 643–644; **6**, 563; **7**, 351–352.

Brominative cyclization of polyenes.[1] Treatment of nerolidol (2) with **1** results in formation (in low yield) of α- and β-synderol (**3** and **4**) and the bicyclic ethers 3β-bromo-8-epicaparrapi oxide (**5**) and the C$_8$-epimer.

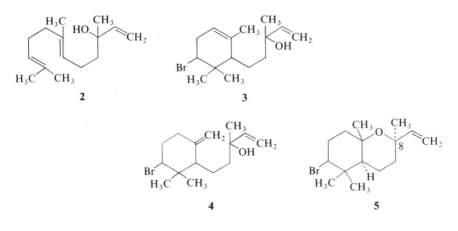

[1] T. Kato, K. Ishii, I. Ichinose, Y. Nakai, and T. Kumagai, *J.C.S. Chem. Comm.*, 1106 (1980).

Tetra-*n*-butylammonium azide, $(n\text{-}C_4H_9)_4\overset{+}{N}N_3{}^-$. This hygroscopic salt (**1**) is prepared by reaction of $(n\text{-}C_4H_9)_4NOH$ with NaN$_3$ in CH$_2$Cl$_2$.[1]

Phenoxysulfonyl azide. This novel azidosulfate is prepared in two steps from phenol (equation I), preferably under homogeneous conditions. Sulfonyl azides are reduced to sulfamates, ArOSO$_2$NH$_2$, which undergo cycloaddition with alkenes to form aziridines.[2]

$$(I)\ C_6H_5OH \xrightarrow{SO_2Cl_2} C_6H_5OSO_2Cl \xrightarrow[\substack{75-80\% \\ \text{overall}}]{1,\ C_6H_6} C_6H_5OSO_2N_3$$

[1] A. Brändström, B. Lamm, and I. Palmertz, *Acta Chem. Scand.*, **28B**, 699 (1974).
[2] M. Hedayatullah and J. C. Hugueny, *Org. Syn.*, submitted (1980); *see also* M. Hedayatullah and A. Guy, *Tetrahedron Letters*, 2455 (1975).

Tetra-*n*-butylammonium borohydride, 6, 564–565; **7,** 352–353.

Reduction of aldehydes.[1] Ketones are reduced so slowly by this reagent at 25° that selective reduction of aldehydes is possible.

Reduction of nitriles and amides.[2] These compounds are selectively reduced to the corresponding amines by this metal hydride in refluxing methylene chloride. Under the same conditions indoles can be reduced to indolenines in moderate yield.

1,4-Reduction of 2-cyclohexenones. 2-Cyclohexenones undergo almost exclusive 1,4-reduction with this hydride in THF or toluene.[3] The regioselectivity is comparable to that observed with $LiBH_4$ in the presence of cryptate [2.1.1].[4]

[1] T. N. Sorrell and P. S. Pearlman, *Tetrahedron Letters*, **21**, 3963 (1980).
[2] T. Wakamatsu, H. Inaki, A. Ogawa, M. Watanabe, and Y. Ban, *Heterocycles*, **14**, 1437, 1441 (1980).
[3] E. D'Incan and A. Loupy, *Tetrahedron*, **37**, 1171 (1981).
[4] A. Loupy and J. Seyden-Penne, *Tetrahedron*, **36**, 1937 (1980).

Tetra-*n*-butylammonium fluoride, 4, 477–478; **5,** 645; **7,** 353–354; **8,** 467–468; **9,** 444–446.

o-Xylylenes. A new route to these intermediates (**2**) involves a 1,4-elimination of salts such as **1**, prepared as shown in equation (I), with $Bu_4N^+F^-$ (equation II).

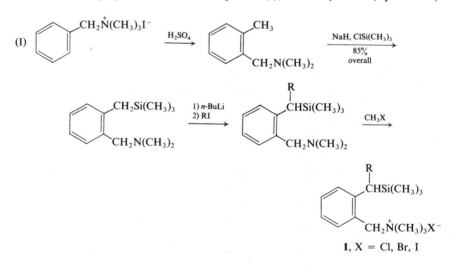

1, X = Cl, Br, I

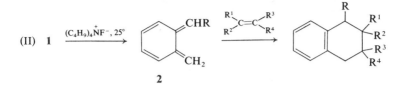

(II)   **1** $\xrightarrow{(C_4H_9)_4\overset{+}{N}F^-,\ 25°}$ **2** $\xrightarrow{\ \ \ }$

These *o*-xylylenes are not isolated as such, but are trapped as cycloadducts with alkenes or alkynes.[1]

Examples:

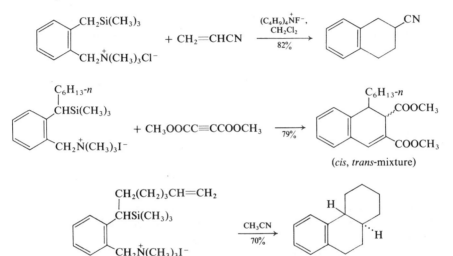

A similar 1,6-elimination of **3**, prepared as shown, induced by fluoride ion results in *p*-quinodimethane (**a**), which dimerizes to [2.2] paracyclophane (**4**).[2]

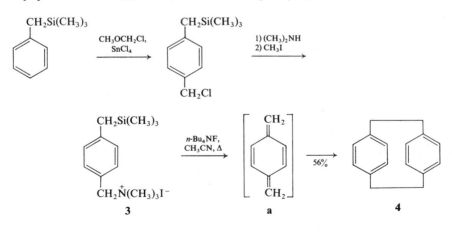

Another example:

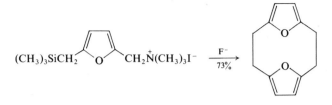

α-Allenic alcohols.[3] The reaction of allylsilanes with carbonyl compounds catalyzed by this salt to give homoallylic alcohols (**9**, 455–446) has been extended to a synthesis of α-allenic alcohols. Use of TiCl₄ as catalyst results in a 2-chloro-1,3-diene.

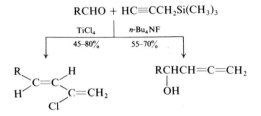

Spirocyclization with 2-silyl-1,3-dithianes.[4] Treatment of the trimethylsilyl-alkanal **1** with fluoride ion liberates an anion that undergoes cyclization to the aldehyde group to yield **2**. Various Lewis acids do not promote this cyclization. Similar desilylation of **3** is followed by an intramolecular Michael addition to give **4** in 64% yield.

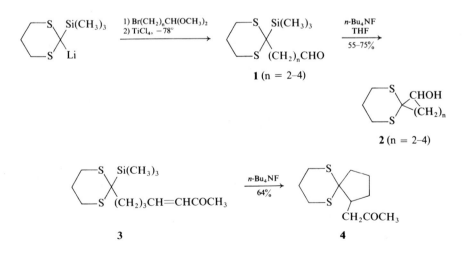

Nitro-aldol reaction (**9**, 444).[5] The β-amino alcohol obtained by fluoride ion catalyzed aldol condensation of aldehydes with silyl nitronates (**1**) is the practically pure (RS,SR)-diastereoisomer (**3**).

$$R^1CHO + R^2CH{=}NO_2Si(CH_3)_2C(CH_3)_3 \quad \xrightarrow{\quad F^- \quad}$$

1

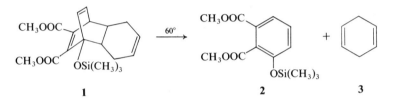

(*erythro*-**2**, >95%)

3

[4 + 2]Cycloreversion.[6] The reverse Diels-Alder reaction of **1** proceeds at 60° with a half-life of 236 minutes. Cycloreversion of the parent diester is even slower.

If **1** is treated with tetra-*n*-butylammonium fluoride to generate the alkoxide ion, fragmentation is accelerated by a factor of 10^6.

[1] Y. Ito, M. Nakatsuka, and T. Saegusa, *Am. Soc.*, **102**, 863 (1980).
[2] Y. Ito, S. Miyata, M. Nakatsuka, and T. Saegusa, *J. Org.*, **46**, 1043 (1981).
[2] J. Pornet, *Tetrahedron Letters*, **22**, 453, 455 (1981).
[4] D. B. Grotjahn and N. H. Andersen, *J.C.S. Chem. Comm.*, 306 (1981).
[5] D. Seebach, A. K. Beck, F. Lehr, T. Weller and E. Colvin, *Angew. Chem. Int. Ed.*, **20**, 397 (1981).
[6] O. Papies and W. Grimme, *Tetrahedron Letters*, **21**, 2799 (1980).

Tetra-*n*-butylammonium periodate, $(C_4H_9)_4\overset{+}{N}IO_4{}^-$. Mol. wt. 433.4, m.p. 158–159°, water insoluble, unstable to light. The reagent is prepared in quantitative yield by reaction of sodium periodate with tetrabutylammonium hydrogen sulfate in aqueous solution.

Oxidations. This reagent oxidizes sulfides to sulfoxides in refluxing chloroform in 70–90% yield. It effects oxidative decarboxylation of α-hydroxy carboxylic acids to the noraldehyde in 85–90% yield (equation I). It also oxidatively cleaves aromatic compounds of the type $ArCOCH_2Br$ to $ArCOOH$ in about 75% yield.[1]

Arylacetic acids, $ArCH_2COOH$, are converted to the corresponding noral-dehydes, ArCHO, by reaction with 1 equivalent of the oxidant in dioxane (reflux,

$$\text{(I)} \quad \underset{\overset{|}{\text{OH}}}{\text{RCHCOOH}} \xrightarrow[85-90\%]{(C_4H_9)_4\overset{+}{N}IO_4{}^-, \, CHCl_3} RCHO + CO_2 + H_2O$$

8–16 hours). Yields are 50–70%. Oxidation of diphenylacetic acid, $(C_6H_5)_2CHCOOH$, to benzophenone, $(C_6H_5)_2C{=}O$, requires 2 equivalents of oxidant (yield 85%). This reaction has been conducted with alkaline sodium hypochlorite (8, 463).[2]

[1] E. Santaniello, A. Manzocchi, and C. Farachi, *Synthesis*, 563 (1980).
[2] E. Santaniello, F. Ponti, and A. Manzocchi, *Tetrahedron Letters*, **21**, 2655 (1980).

Tetracarbonyldi-μ-chlorodirhodium, 8, 469–470.

Isomerization of allylic alcohols.[1] Allylic alcohols can be isomerized to carbonyl compounds by several organometallic reagents at elevated temperatures. The reaction can be conducted at 25–30° overnight with $[Rh(CO)_2Cl]_2$ under phase-transfer conditions. Cleaner reactions obtain if benzyltriethylammonium chloride is used as catalyst.

Example:

$$\underset{\overset{|}{\text{OH}}}{CH_2{=}CHCHC_3H_7\text{-}n} \xrightarrow[100\%]{\substack{[Rh(CO)_2Cl]_2, \, NaOH, \\ CH_2Cl_2, \, C_6H_5CH_2N^+(C_2H_5)_3Cl^-}} C_2H_5COC_3H_7\text{-}n$$

[1] H. Alper and K. Hachem, *J. Org.*, **45**, 2269 (1980).

Tetra-μ^3-carbonyldodecacarbonylhexarhodium, $Rh_6(CO)_{16}$ **(1).** Mol. wt. 1065.59, black, air-stable solid. Preparation.[1]

Cyclopropanation.[2] This metal carbonyl cluster is an effective catalyst for cyclopropanation of alkenes with ethyl diazoacetate. Minor by-products are diethyl maleate and fumarate, but products of allylic C—H insertion are not formed. The yield of the cyclopropane can be increased if the ethyl diazoacetate is added slowly over a period of 6 hours to the olefin and catalyst. Under these conditions yields of cyclopropanes are 85–90%.

[1] R. B. King, *Prog. Inorg. Chem.*, **15**, 439 (1971).
[2] M. P. Doyle, W. H. Tamblyn, W. E. Buhro, and R. L. Dorow, *Tetrahedron Letters*, **22**, 1783 (1981).

Tetrafluoroboric acid, HBF$_4$, 1, 1139.

α-Vinylcyclobutanones.[1] The lithium anion[2] of the mixed ketal **1** of cyclopropanone reacts with an α,β-unsaturated carbonyl compound to form a 1-cyclopropylallyl alcohol **2**. This adduct rearranges to a 2-vinylcyclobutanone (**3**) on treatment with HBF$_4$ in wet THF (*cf.* **8**, 139–141).

Examples:

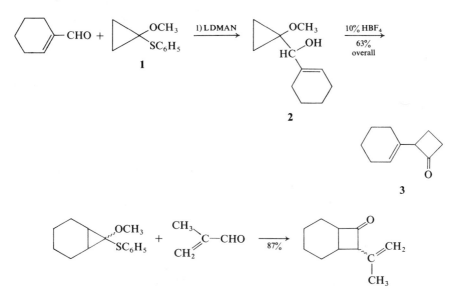

2

3

See methanesulfonic acid, this volume, for acid-catalyzed rearrangement of these 2-vinylcyclobutanones to cyclopentenones and cyclohexenones.

Trost and Jungheim[3] have developed an alternative route to 2-vinylcyclo-butanones from the reagent **4**, which involves a dehydrative rearrangement (**8**, 138–139).

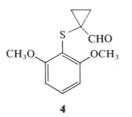

4

[1] T. Cohen and J. R. Matz, *Tetrahedron Letters*, **22**, 2455 (1981).
[2] Prepared with lithium 1-dimethylaminonaphthalenide (this volume).
[3] B. M. Trost and L. N. Jungheim, *Am. Soc.*, **102**, 7910 (1980).

Tetra-*n*-hexylammonium bromide, $(n\text{-}C_6H_{13})_4N^+Br^-$ (**1**).

Wittig-Horner reaction.[1] Phthalides do not undergo the Wittig-Horner reaction in satisfactory or reproducible yield. Addition of 1 mol % or less of tetrahexyl-ammonium bromide notably improved the reaction of **2** with **3**.

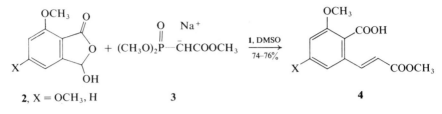

2, X = OCH₃, H   **3**   **4**

¹ B. M. Trost, G. T. Rivers, and J. M. Gold, *J. Org.*, **45**, 1835 (1980).

1,1,2,3-Tetrakis(trimethylsilyloxy)-1,3-butadiene (1). Mol. wt. 406.82, b.p. 76–78°/
0.07 mm.
   Preparation:

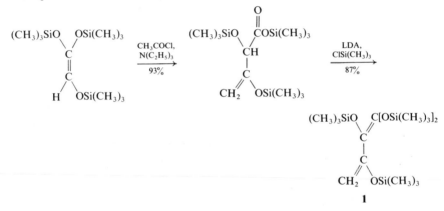

   *Anthragallols.*¹ The reagent is used to convert chloronaphthoquinones to
anthragallols. An example is the conversion of 3-chloro-7-methyljuglone (**2**) to 7-
hydroxyemodin (**3**).

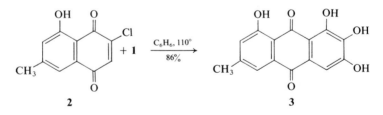

2   **3**

¹ G. Roberge and P. Brassard, *Synthesis*, 381 (1981).

Tetrakis(triphenylphosphine)palladium(0), 6, 571–573; **7,** 357–358; **8,** 472–476; **9,**
451–458.
   Rearrangement of allylic acetates (*cf.* **9,** 44). The rearrangement of an α-cyano

allylic acetate to a dienenitrile catalyzed by a salt has been used to effect a simple synthesis of pellitorine (**1**) (equation I).[1]

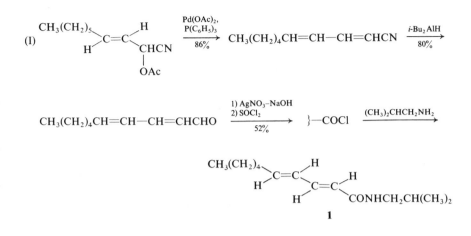

In the presence of tetrakis(triphenylphosphine)palladium α-cyano allylic acetates rearrange to γ-acetoxy-α,β-unsaturated nitriles. The products can be converted into furane derivatives (equation II).[2]

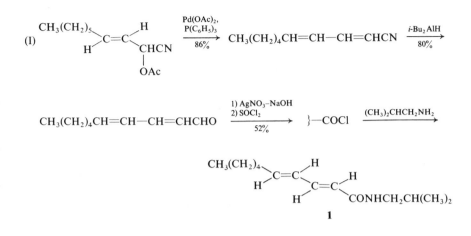

Reduction of allylic acetates.[3] Allylic acetates can be reduced to alkenes after activation by conversion to a π-allylpalladium complex with Pd(0) complexes. The complex can then be reduced by either $NaBH_3CN$ or $NaBH_4$. The former reagent is more chemoselective.

Examples:

$C_6H_5CH=CHCH_2OAc$ $\xrightarrow[\text{90\%}]{\substack{\text{1) Pd(0), P(C}_6\text{H}_5)_3\text{, THF} \\ \text{2) NaBH}_3\text{CN}}}$

$$C_6H_5CH=CHCH_3 + C_6H_5CH_2CH=CH_2$$
$$99:1$$

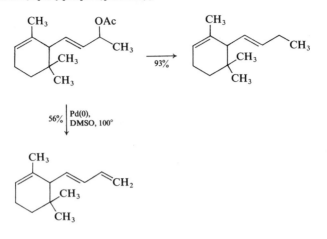

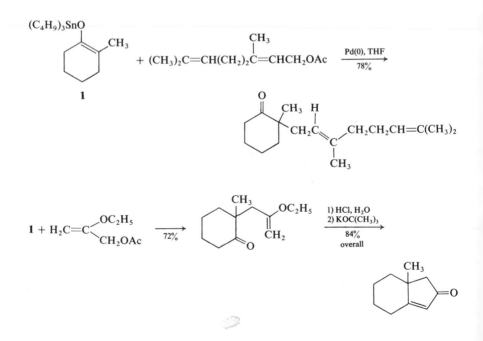

The second example illustrates a method for conversion of allylic acetates to dienes.

Alkylation of allylic acetates. Regioselective monoalkylation of allylic acetates is possible by use of enol stannanes (prepared by reaction of lithium enolates with chlorotri-*n*-butyltin) in the presence of this Pd complex. The less substituted end of the allyl group is alkylated with formation of the (E)-isomer.[4]

Examples:

Coupling of allylstannanes and allyl acetates.[5] This reaction is also catalyzed by tetrakis(triphenylphosphine)palladium.

Examples:

$$C_6H_5CH=CHCH_2OAc + CH_2=\overset{\overset{\displaystyle CH_3}{|}}{C}CH_2Sn(C_4H_9)_3 \xrightarrow[69\%]{Pd(0), THF} C_6H_5CH=CHCH_2CH_2\overset{\overset{\displaystyle CH_3}{|}}{C}=CH_2$$

$$\downarrow {\scriptstyle (C_4H_9)_3SnSn(C_4H_9)_3,\; Pd(0)}$$

$$[C_6H_5CH=CHCH_2Sn(C_4H_9)_3] \xrightarrow[51\%]{\substack{C_6H_5CH=CHCH_2OAc,\\ Pd(0)}} C_6H_5CH=CHCH_2\overset{\overset{\displaystyle C_6H_5}{|}}{C}HCH=CH_2$$

Diene synthesis.[6] β-Acetoxy carboxylic acids undergo loss of CH_3COOH and CO_2 when refluxed in THF or DMSO in the presence of triethylamine (1 equivalent) and catalytic amounts of Pd(0). This fragmentation is highly stereo-selective; the (E)-alkene is formed predominately, irrespective of the stereo-chemistry of the substrate. The method is particularly useful for stereocontrolled synthesis of 1,3-dienes from stereoisomeric mixtures.

Examples:

$$THPOCH_2(CH_2)_7\overset{\overset{\displaystyle OAc}{|}}{C}H\overset{\overset{\displaystyle H}{|}}{C}HC=\overset{\overset{\displaystyle H}{|}}{C}CH_3 \xrightarrow[77-82\%]{\substack{Pd(0),\; N(C_2H_5)_3,\\ THF,\; \Delta}}$$

with the $COOH$ and H below the first two carbons.

$$THPOCH_2(CH_2)_7\overset{\overset{\displaystyle H}{|}}{\underset{\underset{\displaystyle H}{|}}{C}}=\overset{\overset{\displaystyle H}{|}}{\underset{\underset{\displaystyle H}{|}}{C}}C=CCH_3 + \quad (Z,E)\text{-isomer}$$
$$78:22$$

Cyclopentenone annelation.[7] A new method for this reaction involves Pd(0)-directed C-alkylation of 2-methyl-1,3-cyclopentanedione or 2-methyl-1,3-cyclohexa-nedione with 2-ethoxy-3-acetoxy-1-propene followed by an intramolecular Wittig reaction (equation I). Of special interest, the cyclopentenone 1 can be obtained in optically active form by use of an optically active phosphine. Thus use of (R)-DIOP (**4**, 273; **5**, 360–361; **6**, 309) leads to **1** as a 70:30 mixture of (+)- and (−)-enantiomers. Similar results are obtained with (R)-(−)-methylphenylpropyl-phosphine.

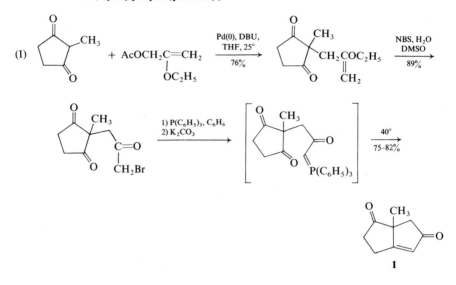

Stereoselective synthesis of 1,4-dienes.[8] This Pd complex markedly catalyzes the cross-coupling of alkenyl and aryl metals containing Al, Zr, or Zn with allylic halides, but not of alkyl metals. The coupling is markedly stereo- and regioselective. Examples:

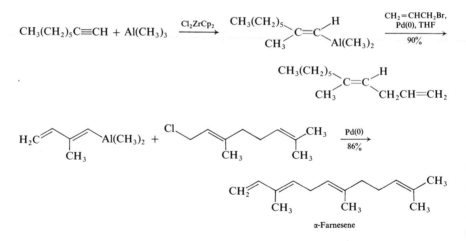

1,5-Dienes and 1,5-enynes. Negishi *et al.*[9] have extended the synthesis of 1,3-enynes by palladium-catalyzed cross-coupling of alkynylzinc chlorides with an alkenyl halide (**8,** 472) to a similar stereospecific synthesis of 1,5-dienes or 1,5-enynes by coupling an alkenyl halide with a homoallyl- or homopropargylzinc chloride. An example is the synthesis of the 1,5-enyne **1** (equation I). This reaction

(I) $(CH_3)_2C$=$CH(CH_2)_2$ $\overset{CH_3}{\underset{H}{\diagup}}C$=$C\overset{I}{\diagdown}$ $+ (CH_3)_3SiC$≡$C(CH_2)_2ZnCl$ $\xrightarrow[80\%]{\substack{1)\ Pd(0),\ THF,\ 28° \\ 2)\ KF,\ DMF}}$

$(CH_3)_2C$=$CH(CH_2)_2$ $\overset{CH_3}{\diagup}C$=$C\overset{(CH_2)_2C≡CH}{\diagdown}_H$

1

provides an attractive route to natural terpenoids and indeed **1** was converted into (E,E)-farnesol (**2**) by a known modification of the terminal triple bond (equation II).[10]

(II) RC≡$CH + Cl_2ZrCp_2 + (CH_3)_3Al$ $\longrightarrow$

1

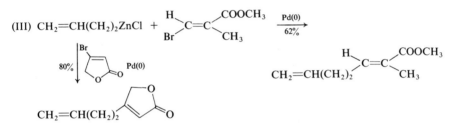

$\left[\begin{array}{c} CH_3 \\ \diagup C=C \diagdown \\ R \quad\quad H \end{array} Al(CH_3)_2 \right]$ $\xrightarrow[85\%]{\substack{1)\ n\text{-BuLi} \\ 2)\ (CH_2O)_n,\ THF}}$ $\overset{CH_3}{\underset{R}{\diagup}}C$=$C\overset{CH_2OH}{\diagdown}_H$

2

The palladium-catalyzed coupling of homoallylzinc halides can also be used with β-bromo-α,β-unsaturated carbonyl derivatives (equation III).[11]

(III) CH_2=$CH(CH_2)_2ZnCl$ $+$ $\overset{H}{\underset{Br}{\diagup}}C$=$C\overset{COOCH_3}{\diagdown}_{CH_3}$ $\xrightarrow[62\%]{Pd(0)}$

$\overset{H}{\underset{CH_2=CH(CH_2)_2}{\diagup}}C$=$C\overset{COOCH_3}{\diagdown}_{CH_3}$

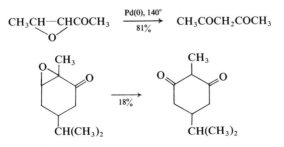

β-Diketones from α,β-epoxy ketones.[12]

In the presence of this Pd(0) complex α,β-epoxy ketones isomerize to β-diketones. An added ligand is usually necessary to avoid precipitation of palladium. The most satisfactory adjunct is 1,2-bis(diphenylphosphino)ethane (dpe). The reaction is conducted in toluene at 80–140° for 10–100 hours. The isomerization is facile with strained epoxides; it is sluggish with epoxides bearing an α-alkyl group.

Examples:

CH_3CH—$CHCOCH_3$ $\xrightarrow[81\%]{Pd(0),\ 140°}$ $CH_3COCH_2COCH_3$

Dialkyl vinylphosphonates.[13] Pd(0) catalyzes the reaction of vinyl bromides and dialkyl phosphites in the presence of triethylamine (1 equivalent) to form dialkyl vinylphosphonates. The configuration of the starting material is retained. An example is the synthesis of phosphonomycin (**1**, equation I).

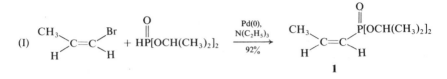

(I)

Coupling of enol phosphates and alkylaluminum compounds.[14] Enol phosphates and R_3Al couple in the presence of 3 mole % of this Pd(0) complex to give alkenes with retention of configuration.

Example:

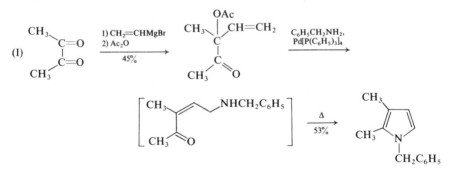

Pyrrole synthesis from α-dicarbonyl compounds.[15] A new approach to N-benzylpyrroles is formulated for biacetyl as starting material (equation I). The method is also suitable for annelation of a pyrrole group to an α-methylene carbonyl compound.

(I)

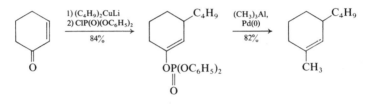

[1] T. Mandai, J. Goto, J. Otera, and M. Kawada, *Chem. Letters*, 313 (1980).
[2] T. Mandai, S. Hashio, J. Goto, and M. Kawada, *Tetrahedron Letters*, **22**, 2187 (1981).
[3] R. O. Hutchins, K. Learn, and R. P. Fulton, *Tetrahedron Letters*, **21**, 27 (1980).
[4] B. M. Trost and E. Keinan, *Tetrahedron Letters*, **21**, 2591 (1980).
[5] B. M. Trost and E. Keinan, *Tetrahedron Letters*, **21**, 2595 (1980).
[6] B. M. Trost and J. M. Fortunak, *Am. Soc.*, **102**, 2841 (1980).

[7] B. M. Trost and D. P. Curran, *Am. Soc.*, **102**, 5699 (1980).

[8] H. Matsushita and E. Negishi, *Am. Soc.*, **103**, 2882 (1981).

[9] E. Negishi, L. F. Valente, and M. Kobayashi, *Am. Soc.*, **102**, 3298 (1980).

[10] N. Okukado and E. Negishi, *Tetrahedron Letters*, 2357 (1978).

[11] M Kobayashi and E. Negishi, *J. Org.*, **45**, 5223 (1980).

[12] M. Suzuki, A. Watanabe, and R. Noyori, *Am. Soc.*, **102**, 2095 (1980).

[13] T. Hirao, T. Masunaga, Y. Ohshiro, and T. Agawa, *Tetrahedron Letters*, **21**, 3595 (1980).

[14] K. Takai, K. Oshima, and H. Nozaki, *Tetrahedron Letters*, **21**, 2531 (1980).

[15] B. M. Trost and E. Keinan, *J. Org.*, **45**, 2741 (1980).

2,4,4,6-Tetramethyl-5,6-dihydro-1,3-(4*H*)-oxazine (1), 3, 280–284; 4, 481–484.

α-Amino ketones.[1] α-Lactams (**2**) react with the lithium salt (**3**) of **1** in THF to afford **4**, the product of selective cleavage of the acyl–nitrogen bond. The products are cleaved under acidic conditions with loss of CO_2 to give α-amino ketones (**5**).

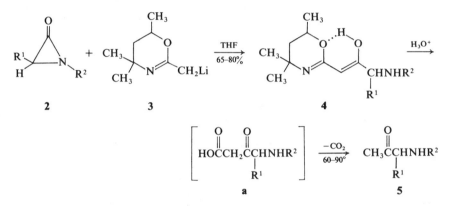

[1] E. R. Talaty, K. C. Bengtsson, and L. M. Pankow, *Syn. Comm.*, **10**, 99 (1980).

2,3,5,6-Tetramethylidene-7-oxanorbornane (1). Mol. wt. 146.18, m.p. 23–25°.

Preparation[1]:

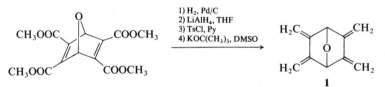

Diels-Alder reactions. This reagent is a useful partner in Diels-Alder reactions because it is possible to obtain monoadducts in good yield and then add a different dienophile to the monoadduct. Such a sequence has been used for a synthesis of **4**, a known precursor to (±)-4-demethoxydaunomycinone (**5**) as outlined in scheme (I) The two dienophiles are methyl vinyl ketone and dehydrobenzene, used in this

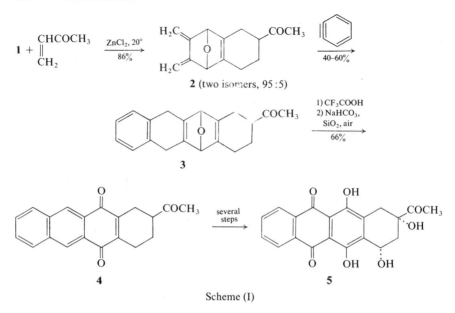

Scheme (I)

order. Initial addition of dehydrobenzene proceeds with somewhat low yield of the monoadduct (~ 20%) and with formation of about 5% of the bis-adduct.[2]

(±)-7,9-Bis(desoxy)-1-methoxydaunomycinone (**6**) has been prepared from **1** in a reaction sequence starting with two Diels-Alder reactions, first with benzoquinone and then methyl vinyl ketone.[3]

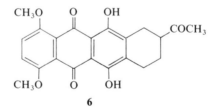

6

[1] P. Vogel and A. Florey, *Helv.*, **57**, 200 (1974).
[2] Y. Bessière and P. Vogel, *Helv.*, **63**, 232 (1980).
[3] P. A. Carrupt and P. Vogel, *Tetrahedron Letters*, 4533 (1979).

Tetranitromethane, 1, 1147; **2**, 404.

1-Nitroalkenes (*cf.* **9**, 292–293). 1-Nitrocycloalkenes can be prepared from cycloalkanones by conversion to the 2,4,6-triisopropylbenzenesulfonyl hydrazone, which is metallated (*sec*-BuLi, 2 equivalents) and then treated with chloro-trimethyltin to give a 1-trimethylstannylcycloalkene. Reaction of the vinylstannane with tetranitromethane in DMSO effects nitrodestannylation to form a nitroalkene. Use of DMSO (or HMPT) is critical for success.[1]

Example:

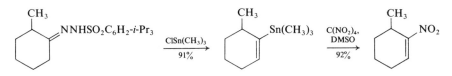

[1] E. J. Corey and H. Estreicher, *Tetrahedron Letters*, **21**, 1113 (1980).

Tetraphenylbismuth trifluoroacetate, $(C_6H_5)_4BiOCOCF_3$ **(1)**. Mol. wt. 630.40. The reagent is prepared by addition of CF_3COOH (1 equivalent) to $Bi(C_6H_5)_5$ in benzene.

Aryl ethers.[1] Unlike other Bi(V) arylating reagents, this reagent converts phenols into unsymmetrical diaryl ethers (equation I). Yields are in the range 50–75%. Yields are improved if the phenol is treated first with $Bi(C_6H_5)_5$ and then with TFA. The reagent also converts aryl thiols into mixed diaryl sulfides.

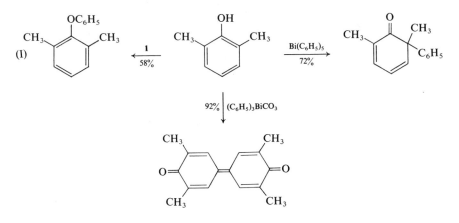

[1] D. H. R. Barton, J.-C. Blazejewski, B. Charpiot, and W. B. Motherwell, *J.C.S. Chem. Comm.*, 503 (1981).

Thallium(III) acetate, 1, 1150–1151; **2**, 406; **3**, 286; **4**, 492; **5**, 655; **7**, 360–361; **9**, 459.

2-Acetoxy-3(2H)-benzofuranones.[1] Enolizable *o*-hydroxyphenyl or *o*-methoxy-phenyl ketones are oxidatively cyclized to these benzofuranones by $Tl(OAc)_3$ (3 equivalents) in HOAc.

Examples:

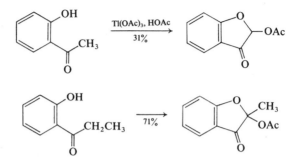

Oxidation of **p-***toluenesulfonylhydrazones.*[2] Carbonyl compounds can be regenerated in good yield from the tosylhydrazones by treatment with 1 equivalent of Tl(OAc)$_3$ in HOAc. Room temperature suffices in the case of aldehydes, whereas reflux temperatures are used for ketones. The other product is Tl(OSO$_2$C$_6$H$_4$CH$_3$). This method is less useful for oxidative cleavage of semicarbazones; long reflux periods are necessary, and acetoxylation is a side reaction.

[1] N. Malaitong and C. Thebtaranonth, *Chem. Letters*, 305 (1980).
[2] R. N. Butler, G. J. Morris, and A. M. O'Donohue, *J. Chem. Res.* (*M*), 808 (1981).

Thallium(I) acetate–Iodine, 5, 654–655; **7,** 359–360; **9,** 458.

Iodolactonization. Unsaturated acids are converted into iodolactones by reaction with TlOAc (1–1.2 equivalent) and iodine in CH$_2$Cl$_2$ or CHCl$_3$ at 20° for an appropriate time.[1] The method is more convenient than an earlier one employing reaction of iodine with an unsaturated thallium(I) carboxylate.[2]

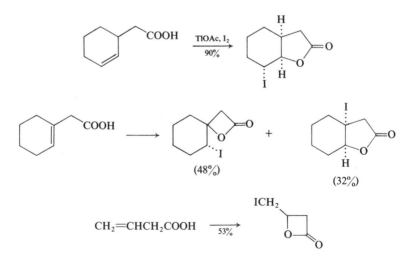

Stereoselective **cis-dihydroxylation.** Details have been published for preparation of either *trans-* or *cis*-1,2-cyclohexenediol by the thallium-based version of the Woodward version of the Prévost reaction (equation I).[3]

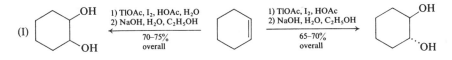

(I)

1) TlOAc, I₂, HOAc, H₂O
2) NaOH, H₂O, C₂H₅OH
70–75% overall

1) TlOAc, I₂, HOAc
2) NaOH, H₂O, C₂H₅OH
65–70% overall

[1] R. C. Cambie, K. S. Ng, P. S. Rutledge, and P. D. Woodgate, *Aust. J. Chem.*, **32**, 2793 (1979).
[2] R. C. Cambie, R. C. Hayward, J. L. Roberts, and P. S. Rutledge, *J.C.S. Perkin I*, 1864 (1974).
[3] R. C. Cambie and P. S. Rutledge, *Org. Syn.*, **59**, 169 (1979).

Thallium(I) ethoxide, 2, 407–411; **4** 501–502; **5**, 656; **6**, 577–578; **7**, 362.

Ether synthesis.[1] A variation of the Williamson reaction involves conversion of an alcohol to the thallium(I) alkoxide followed by alkylation. In practice the method is not useful for simple alcohols because the thallium alkoxide is very sparingly soluble in C_6H_6 or CH_3CN. The method is useful however for substrates containing an additional oxygen function such as OH, OR, COOR, and $CONR_2$ (listed in the order of effectiveness). It is particularly useful for alkylation of chiral hydroxyl groups, since it is free from racemization.

Example:

$$C_2H_5O_2CCHCHCOOC_2H_5$$

OH

OH

1) 2TlOC₂H₅
2) RI
65–95%

$$C_2H_5O_2CCHCHCOOC_2H_5$$

OR

OR

[1] H.-O. Kalinowski, G. Crass, and D. Seebach, *Ber.*, **114**, 477 (1981).

Thallium(III) nitrate, 4, 492–497; **5**, 656–657; **6**, 578–579; **7**, 362–365; **8**, 476–478; **9**, 460–462.

α-Hydroxy nitrate esters; cleavage of ethers.[1] Thallium(III) nitrate cleaves steroidal oxiranes in hexane to *trans*-α-hydroxy nitrate esters.

Example:

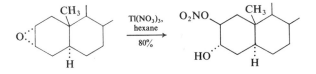

Tl(NO₃)₃, hexane
80%

Thallium(III) nitrate in acetic anhydride cleaves alkyl ethers to the corresponding acetoxy derivatives.

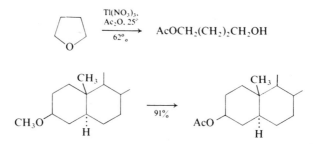

Quinone monoacetals. Wheeler et al.[2] have obtained the highest yields reported to date for oxidation of 1,5-naphthalenediol (**1a**) to juglone (**2a**) by use of TTN in methanol. The yield is improved somewhat by use of TTN supported by Celite for

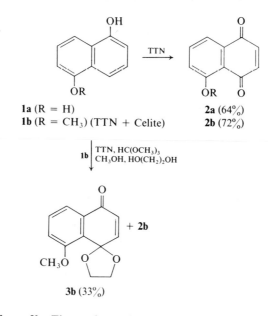

oxidation of **1b** to **2b**. The authors also report the first oxidation of a *para*-unsubstituted phenol to the monoacetal of a 1,4-quinone by use of TTN in trimethyl orthoformate and methanol (**7**, 362) in the presence of ethylene glycol. Thus **1b** can be converted in this way to **3b** in moderate yield.

[1] E. Mincione and F. Lanciano, *Tetrahedron Letters*, **21**, 1149 (1980).
[2] D. J. Crouse, M. M. Wheeler, M. Goemann, P. S. Tobin, S. K. Basu, and D. M. S. Wheeler, *J. Org.*, **46**, 1814 (1981).

Thallium(III) trifluoroacetate, 3, 286–289; **4**, 496–501; **5**, 658–659; **7**, 365; **8**, 478–481; **9**, 462–464.

Biaryls (**8**, 478–479). The details for oxidative dimerization of arenes containing electron-donating groups to biaryls with TTFA have been published. Lead tetraacetate and cobalt(III) fluoride are equally effective reagents.[1]

Use of this coupling for synthesis of a number of isoquinoline alkaloids also has been published.[2] VOF_3 is usually used in this coupling, but this reagent can lead to overoxidation and oxidative demethylation of aryl methyl ether groups.

[1] A. McKillop, A. G. Turrell, D. W. Young, and E. C. Taylor, *Am. Soc.*, **102**, 6504 (1980).
[2] E. C. Taylor, J. G. Andrade, G. J. H. Rall, and A. McKillop, *Am. Soc.*, **102**, 6513 (1980).

Thexylborane (2,3-Dimethyl-2-butylborane), 1, 276; **2**, 148–149; **4**, 175–176; **5**, 232–233; **6**, 207–208.

Stereoselective cyclic hydroboration.[1] The hydroboration of 1,4- and 1,5-dienes with thexylborane generally results in boracycles.[2] If the intial hydroboration produces an asymmetric center, the second intramolecular hydroboration can be effected with remote asymmetric induction. Thus hydroboration of the diene **1** with thexylborane (1.25 equivalent) followed by oxidation results in a 6:1 mixture of the diastereomers **2** and **3**. An even higher induction is obtained in the case of the 1,4-

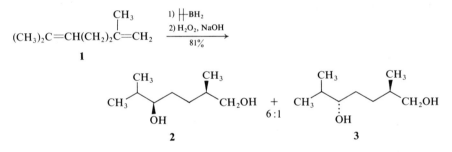

diene **4**. The major product **5** was used for a synthesis of racemic dihydromyoporone (**7**), a potato stress metabolite.

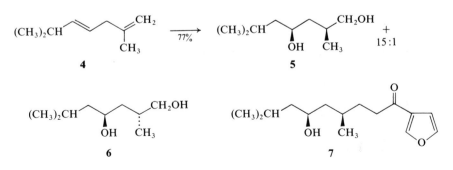

Azacyclanones.[3] The usual syntheses of medium-sized azacyclanones involve Dieckmann or acyloin cyclization conducted under high dilution. An interesting new approach involves hydroboration–cyanidation (**4**, 446–447; **5**, 606–607) of a diunsaturated carbamate such as **1**. Hydroboration of **1** with thexylborane followed by cyanidation under standard conditions gives the cyclic ketone **2** in moderate yield in one step. The product can be reductively cyclized to the indolizidine alkaloid δ-coniceine (**3**).

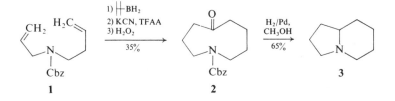

[1] W. C. Still and K. P. Darst, *Am. Soc.*, **102**, 7385 (1980).

[2] H. C. Brown and E.-I. Negishi, *Tetrahedron*, **33**, 2331 (1977).

[3] M. E. Garst and J. N. Bonfiglio, *Tetrahedron Letters*, **22**, 2075 (1981).

Thioanisole–Trifluoroacetic acid.

Deprotection of O-benzyl[1] and N-BOC groups.[2] This combination of a soft nucleophile and a hard electrophile cleaves O-benzyltyrosine without O to C rearrangement, and also cleaves the benzyloxycarbonyl group at 25°. It was used in a synthesis of the pentapeptide Met-enkephalin, Tyr-Gly-Gly-Phe-Met. It was also used for cleavage of BOC groups protecting the N^ε-amino group of lysine in a synthesis of a snake venom mastoparan.[3]

[1] Y. Kiso, K. Ukawa, S. Nakamura, K. Ito, and T. Akita, *Chem. Pharm. Bull. Japan*, **28**, 673 (1980).

[2] Y. Kiso, K. Ukawa, and T. Akita, *J.C.S. Chem. Comm.*, 101 (1980).

[3] H. Yajima, J. Kanaki, M. Kitajima, and S. Funakoshi, *Chem. Pharm. Bull. Japan*, **28**, 1214 (1980).

Thioanisole–Trifluoromethanesulfonic acid.

Deprotection of peptides. In a synthesis of bovine pancreatic ribonuclease, the 33 protective groups required were removed by three treatments with 1 *M* TFMSA–thioanisole at 0° for 60 minutes. The groups were Cbz, *t*-butyl, *p*-methoxybenzyl, and *p*-methoxybenzenesulfonyl.[1] The final step involved air oxidation of the cysteine residues to disulfides mediated by glutathione.[2]

[1] H. Yajima and N. Fujii, *Chem. Pharm. Bull. Japan*, **29**, 600 (1981).

[2] L. G. Chavez, Jr. and H. A. Scheraga, *Biochem.*, **19**, 996 (1980).

Thionyl chloride, 1, 1158–1163; **2**, 412; **3**, 290; **4**, 503–505; **5**, 663–667; **6**, 585; **7**, 366–367; **8**, 481; **9**, 465.

1,2-Benzisothiazole-3(2H)-ones.[1] A novel route to these heterocycles involves cyclization of 2-(methylsulfinyl)benzamides with thionyl chloride (equation I).

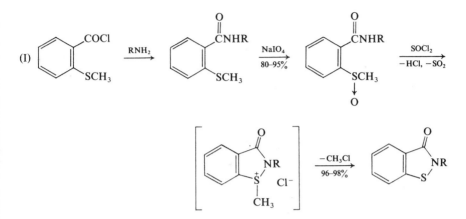

¹ Y. Uchida and S. Kozuka, *J.C.S. Chem. Comm.*, 510 (1981).

Thiophenol, 4, 505; **6**, 585–586; **7**, 367–368; **9**, 465.

Δ⁵,⁷-*Steroids.* The classical method for conversion of a Δ^5-stenyl acetate to the 5,7-diene involves allylic bromination at C_7, followed by dehydrobromination. Unfortunately substantial amounts of the 4,6-diene are formed as well. An expedient is conversion of the major bromide formed (7α) to the 7β-phenyl sulfide (equation I). The sulfide is then oxidized to the (S)- and (R)-sulfoxides, both of which are converted to the 5,7-diene when heated at 70°. In practice it is not

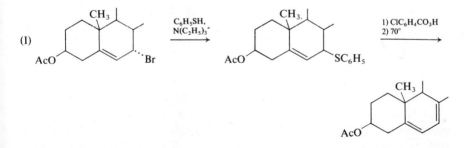

necessary to separate the 7α- and 7β-bromides, since the β-isomer is converted by the subsequent steps into 7α- and 7β-hydroxy derivatives. These are readily separated by a final chromatography. The overall yields from the 5-stenyl acetate are consistently about 50%.[1] Note that the corresponding selenoxide elimination leads to a 1:1 mixture of 4,6- and 5,7-dienes (**8**, 448).

Another route that has been recommended recently is oxidation of the Δ^5-stenyl acetate to the corresponding 7-ketone, conversion to the *p*-tosylhydrazone, and elimination with lithium hydride. Overall yields are in the range 20–50%.[2]

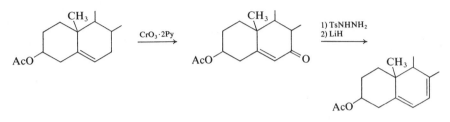

[1] P. N. Confalone, I. D. Kulesha, and M. R. Uskoković, *J. Org.*, **46**, 1030 (1981).
[2] R. M. Moriarty and H. E. Paaren, *J. Org.*, **46**, 970 (1981).

Titanium(III) chloride, 2, 415; **4**, 506–508; **5**, 669–671; **6**, 587–588; **7**, 369; **8**, 482–483; **9**, 467.

Reduction of hydroxamic acids.[1] Buffered $TiCl_3$ can reduce simple hydroxamic acids (equation I). Yields are high when R^1 is an alkyl group; when it is hydrogen, aldehydes are obtained as by-products. The reagent also reduces substituted N-hydroxy-2-azetidinones to β-lactams (equation II).

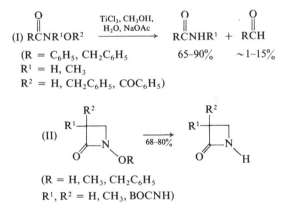

(I) $\overset{O}{\overset{\|}{RCNR^1OR^2}} \xrightarrow[\text{H}_2\text{O, NaOAc}]{\text{TiCl}_3,\ \text{CH}_3\text{OH},} \overset{O}{\overset{\|}{RCNHR^1}} + \overset{O}{\overset{\|}{RCH}}$

(R = C_6H_5, $CH_2C_6H_5$ 65–90%   ~1–15%
R^1 = H, CH_3
R^2 = H, $CH_2C_6H_5$, COC_6H_5)

(II) $\xrightarrow{68–80\%}$

(R = H, CH_3, $CH_2C_6H_5$
R^1, R^2 = H, CH_3, BOCNH)

Reduction of aromatic nitro compounds.[2] Aromatic compounds are reduced to anilines in 90–95% yield by $TiCl_3$ (8 equivalents) and HOAc–H_2O (1:1, v/v) at room temperature (7 minutes). Nitro derivatives of quinoline, isoquinoline, and cinnoline are reduced under the same conditions.

[1] P. G. Mattingly and M. J. Miller, *J. Org.*, **45**, 410 (1980).
[2] M. Somei, K. Kato, and S. Inoue, *Chem. Pharm. Bull.*, **28**, 2515 (1980).

Titanium(III) chloride–Hydrogen peroxide.

Sulfoxides.[1] Sulfides are oxidized to sulfoxides only by $TiCl_3$ and 30% H_2O_2 in aqueous methanol at room temperature within about 5 minutes. Yields are >90%.

[1] Y. Watanabe, T. Numata, and S. Oae, *Synthesis,* 204 (1981).

Titanium(III) chloride–Lithium aluminum hydride, 6, 588–589; **7,** 369–370; **9,** 468.

Reductive coupling. Reductive coupling with low valent Ti, V, and Zr complexes has been reviewed (61 references).[1] $TiCl_3$–$LiAlH_4$ is the most widely used reagent and is the best reagent for coupling of aliphatic carbonyl compounds. $TiCl_3$–K or $TiCl_3$–Li (**7,** 368–369) is somewhat more versatile and is more useful for mixed coupling. $TiCl_3$–$LiAlH_4$ is also the most commonly used reagent for coupling of 1,3-diols and of benzylic and allylic alcohols. $TiCl_3$–CH_3Li (**2,** 415) is also effective for the latter coupling, and is the only known reagent for coupling of allylic alcohols to 1,5-dienes. $TiCl_3$–$LiAlH_4$ and $TiCl_4$–$LiAlH_4$ have both been used for coupling of allylic and benzylic halides, but VCl_3–$LiAlH_4$[2] is apparently more effective and also simpler to use since it is stable to air and moisture.

[1] Y.-H. Lai, *Org. Prep. Proc. Int.,* **12,** 363 (1980).
[2] T. L. Ho and G. A. Olah, *Synthesis,* 170 (1977).

Titanium(IV) chloride, 1, 1169–1171; **2,** 414–415; **3,** 291; **4,** 507–508; **5,** 671–672; **6,** 590–596; **7,** 370–372; **8,** 483–486; **9,** 468–470.

Chiral β-lactams. The synthesis of β-lactams from ketene silyl acetals and Schiff bases (**8,** 484–485) results in asymmetric induction at C_4 in the range 44–78% when optically active Schiff bases are used. An example is shown in equation (I).

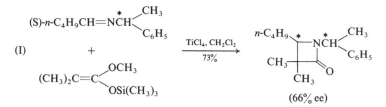

Asymmetric induction is considerably greater when Schiff bases of (S)-α-amino esters are used (equation II).[2]

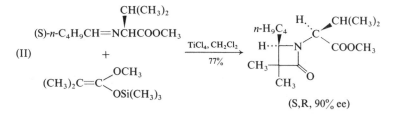

Enantioselective Diels-Alder reaction.[3] The reaction of the chiral acrylate ester **1** with butadiene catalyzed with this Lewis acid followed by hydride reduction gives the alcohol **2** in 70% chemical yield and 86–91% ee. $AlCl_3$ and $SnCl_4$ are inferior in terms of either the chemical or optical yield. The product (**2**) was used for a chiral synthesis of (R)-(−)-sarkomycin (**4**).

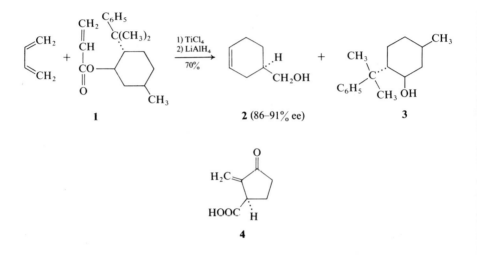

1   2 (86–91% ee)   3

4

1,5-Diketones.[4] A new synthesis of δ-diketones depends on a Michael-type reaction of lithium 1-alkynyltrialkylborates with methyl vinyl ketone in the presence of an excess of $TiCl_4$ (equation I). Other Lewis acids are considerably less effective.

(I) $R_3B + LiC\equiv CR^1 \longrightarrow [R_3\bar{B}C\equiv CR^1]Li^+$ $\xrightarrow{\begin{array}{l}1) CH_2=CHCOCH_3 \\ 2) TiCl_4, CH_2Cl_2\end{array}}$

$$\left[\begin{array}{c} R^1 \quad R \\ BR_2 \\ CH_3 \\ OTiCl_3 \end{array} \right] \xrightarrow[40-80\%]{H_2O_2} RCOCHR^1(CH_2)_2COCH_3$$

1,6-Michael addition. $TiCl_4$ was used to effect addition of the silyl enol ether **1** to **2** to obtain **3**. Base could not be used because of the extreme sensitivity of **2**. The product was used for synthesis of several eudesmane sesquiterpenes containing an α-methylene-γ-butyrolactone unit, such as dl-epiatlantolactone (**4**).[5]

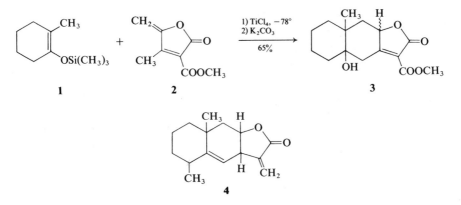

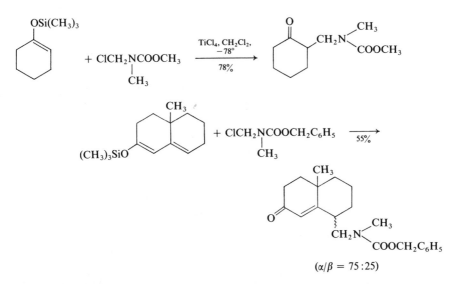

Ureidoalkylation[6] *of silyl enol ethers.* The group —CH$_2$N(CH$_3$)COOR can be introduced regioselectively by reaction of silyl enol ethers with chloromethyl-carbamates under the influence of TiCl$_4$.[7]

Examples:

$(\alpha/\beta = 75:25)$

[1] I. Ojima and S. Inaba, *Tetrahedron Letters*, **21**, 2077 (1980).

[2] I. Ojima and S. Inaba, *Tetrahedron Letters*, **21**, 2081 (1980).

[3] R. K.Boeckman, Jr., P. C. Naegely, and S. D. Arthur, *J. Org.*, **45**, 752 (1980).

[4] S. Hara, K. Kishimura, and A. Suzuki, *Chem. Letters*, 221 (1980).

[5] A. G. Schultz and J. D. Godfrey, *Am. Soc.*, **102**, 2414 (1980).

[6] H. E. Zaugg, *Synthesis*, 49 (1970).

[7] S. Danishefsky, A. Guingant, and M. Prisbylla, *Tetrahedron Letters*, **21**, 2033 (1980).

Titanium(IV) chloride–Sodium borohydride.

Reduction. Carboxylic acids can be reduced satisfactorily to primary alcohols by a slight excess of $TiCl_4$ and $NaBH_4$ in the molar ratio 1:3. A molar ratio of 1:2 is the optimum for reduction of amides, lactams, and sulfoxides.[1] The 1:2 system is also effective for reduction of nitrosamines, R^1R^2N—NO, to secondary amines.[2]

Reduction of alkenes to alcohols.[3] The combination of $TiCl_4$ and $NaBH_4$ in DMF produces a low-valent titanium–borane complex that converts alkenes to alcohols in which the hydroxy group is introduced by an anti-Markovnikoff addition.

Examples:

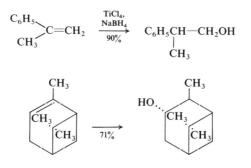

Deoxygenation of N-oxides.[4] $TiCl_4$–$NaBH_4$ (1:2) in DME reduces the N-oxide of pyridine and methylpyridines (picolines) to the corresponding heterocycles in high yield. However the N-oxide of quinolines and isoquinolines is reduced further to dihydro derivatives of the heterocycles. Pyridine, quinoline, and isoquinoline themselves are not reduced by this low-valent titanium species. Reduction of heterocyclic N-oxides with $TiCl_3$ has been reported (6, 588).

[1] S. Kano, Y. Tanaka, E. Sugino, and S. Hibino, *Synthesis*, 695 (1980).
[2] S. Kano, Y. Tanaka, E. Sugino, and S. Hibino, *Synthesis*, 741 (1980).
[3] S. Kano, Y. Tanaka, and S. Hibino, *J.C.S. Chem. Comm.*, 414 (1980).
[4] S. Kano, Y. Tanaka, and S. Hibino, *Heterocycles*, **14**, 39 (1980).

Titanium(IV) isopropoxide, $Ti(O-i-C_3H_7)_4$. Mol. wt. 284.25, m.p. 20°, b.p. 50°/1 mm. Supplier: Alfa.

Epoxy alcohol rearrangement.[1] Titanium alkoxides are weak Lewis acids with no effect on simple epoxides. However, some novel rearrangements of epoxides with an α-hydroxyl group are possible in the presence of $Ti(O-i-Pr)_4$. Thus the epoxide **1** rearranges mainly to **2** and **3** in the presence of this reagent. The corresponding methyl ether of **1** does not rearrange under similar conditions. Of added significance, the *threo*-isomer of **1** is also stable to this weak Lewis acid. The paper also reports cyclization of the acetylenic epoxide **4** to the allene **5** and of **6** to the allylic alcohol **7**.

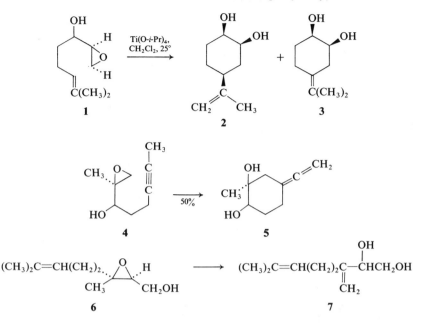

1 D. J. Morgans, Jr., K. B. Sharpless, and S. G. Traynor, *Am. Soc.*, **103**, 462 (1981).

p-Toluenesulfonyl azide, 1, 1178–1179; **2**, 415–417; **3**, 291–292; **4**, 510; **5**, 675; **6**, 597; **9**, 472. Material prepared by reaction of tosyl chloride and sodium azide in aqueous alcohol[1] contains ethyl tosylate as an impurity. Curphey[2] recommends aqueous acetone as the solvent.

1 M. Regitz, J. Hocker, and A. Liedhegener, *Org. Syn. Coll. Vol.*, **5**, 179 (1973).
2 T. J. Curphey, *Org. Prep. Proc. Int.*, **13**, 112 (1981).

(R)-(+)-(p-Tolylsulfinyl)acetic acid (1, R = H). Esters of this acid have been prepared by two routes[1,2] from (S)-(−)-methyl p-toluenesulfinate.[3] The preparation involves inversion of configuration at sulfur.

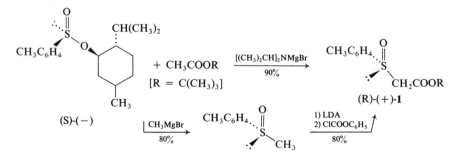

Asymmetric synthesis of β-hydroxy carboxylic esters.[1] Esters of **1** in the presence of *t*-butylmagnesium bromide as base react with aldehydes and ketones to form optically active α-sulfinyl-β-hydroxy carboxylic esters, which are desulfurized by aluminum amalgam in aqueous THF (equation I). Chemical yields are ∼75%; of

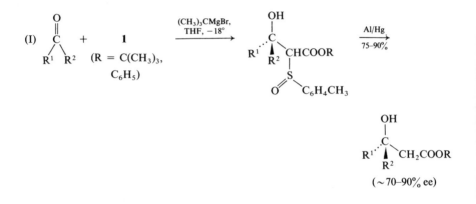

$$(I) \quad \underset{R^1 \quad R^2}{\overset{O}{\overset{\|}{C}}} + \mathbf{1} \quad \xrightarrow[\text{THF, } -18°]{\text{(CH}_3\text{)}_3\text{CMgBr,}} \quad \underset{R^1 \quad R^2}{\overset{OH}{\underset{|}{C}}} \underset{}{\overset{}{\text{CHCOOR}}} \quad \xrightarrow[75-90\%]{\text{Al/Hg}}$$

(R = C(CH₃)₃, C₆H₅)

$$\underset{R^{1''} \overset{|}{\underset{R^2}{C}} CH_2COOR}{\overset{OH}{\overset{|}{C}}}$$

(∼70–90% ee)

more interest, optical yields can be high. The choice of base is important. *t*-Butylmagnesium bromide does not undergo ready addition to carbonyl groups and does not promote retroaldolization; undoubtedly chelation to the bivalent metal is important for asymmetric induction.[1]

This reaction was used to introduce the final two skeletal carbons in a total synthesis of maytansine (**4**).[2] The reaction of the α,β-unsaturated aldehyde (**2**) with **1** (R = C₆H₅) gives the desired 4,5-unsaturated 3-hydroxy ester **3** in ∼80% yield. The ratio of the desired (S)-alcohol to the epimer is 93:7. The resulting amino acid was cyclized to the lactam in ∼80% yield with mesitylenesulfonyl chloride (**8**, 318–319). Epoxidation by the Sharpless procedure (**9**, 78–79) was also highly stereoselective, giving the desired epoxide and the undesired epimer in the ratio >200:1.

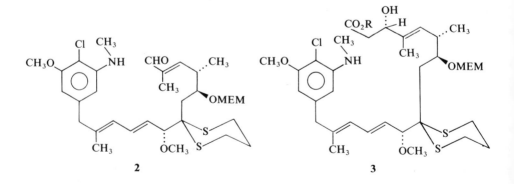

2   3

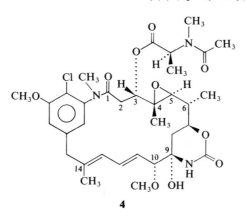

4

Asymmetric Michael additions.[4] In the presence of sodium hydride as base, the *t*-butyl ester of **1** undergoes conjugate addition to α,β-unsaturated esters (but not to ketones) with 12–24% asymmetric induction. The reaction was used to prepare optically active δ-lactones.

Example:

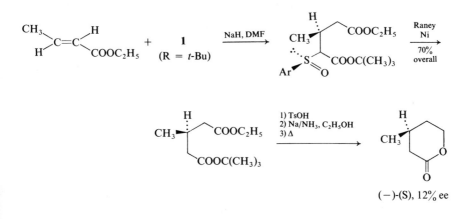

$(-)$-(S), 12% ee

Review.[5] Solladié has reviewed use of nucleophiles containing a chiral sulfoxide group in asymmetric synthesis (140 references).

[1] C. Mioskowski and G. Solladié, *Tetrahedron*, **36**, 227 (1980).
[2] E. J. Corey, L. O. Weigel, A. R. Chamberlin, H. Cho, and D. H. Hua, *Am. Soc.*, **102**, 6613 (1980).
[3] H. Philips, *J. Chem. Soc.*, **127**, 2552 (1925).
[4] F. Matloubi ˋand G. Solladié, *Tetrahedron Letters*, 2141 (1979).
[5] G. Solladié, *Synthesis*, 185 (1981).

(S)-(+)-p-Tolyl p-tolylthiomethyl sulfoxide, 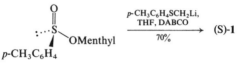   [(S)- 1].

Mol. wt. 292.41, m.p. 78–79°, α_D + 76°.
    Preparation[1]:

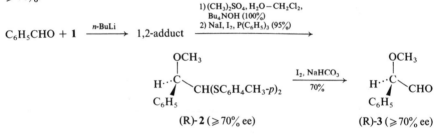

Chiral α-methoxy aldehydes.[2] The anion of **1** undergoes 1,2-addition to benzaldehyde in quantitative yield. The adduct can be methylated under phase-transfer conditions and then reduced[3] to give the dithioacetal **2**, from which the aldehyde **3** is liberated by reaction with I_2 and $NaHCO_3$.[4] The optical yield of **3** is ⩾ 70%.

The sequence can also be used with aliphatic aldehydes; it provides a general route to chiral hydroxy aldehydes, acids, and alcohols.

Conjugate additions.[5] The anion of (S)-**1** undergoes 1,2-addition to α,β-enones in THF at −78°. However 1,4-addition is effected in THF–HMPT. The reaction can be used to obtain optically active aldehydes such as **2**. The adduct is reduced by

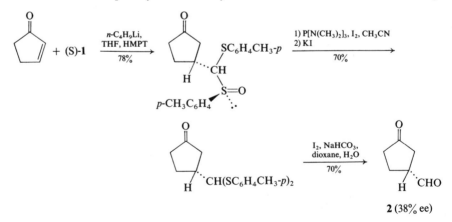

hexamethylphosphorous triamide, iodine, and potassium iodide to the (S,S)-acetal, which is cleaved by iodine in aqueous sodium bicarbonate to the aldehyde.

[1] L. Colombo, C. Gennari, and E. Narisano, *Tetrahedron Letters*, 3861 (1978).
[2] L. Colombo, C. Gennari, C. Scolastico, G. Guanti, and E. Narisano, *J.C.S. Chem. Comm.*, 591 (1979).
[3] G. A. Olah, B. G. B. Gupta, and S. C. Narang, *Synthesis*, 137 (1978).
[4] G. A. Russel and L. A. Ochrymowycz, *J. Org.*, **34**, 3618 (1969).
[5] L. Colombo, C. Gennari, G. Resnati, and C. Scolastico, *Synthesis*, 74 (1981).

Tosylmethyl isocyanide, 4, 514–516; **5**, 684–685; **6**, 600; **7**, 377; **8**, 493–494.

Reductive cyanation (**5**, 684; **6**, 600).[1] The original conditions for conversion of ketones into nitriles give low yields when applied to aldehydes. Satisfactory results are obtained, however, if the initial reaction with TosMIC is conducted at −50° in DME before addition of methanol and reflux. Yields of ~50–70% are then possible.

Cyclobutanones.[2] A method for preparation of cyclobutanones is formulated in equation (I) for preparation of the parent ketone.

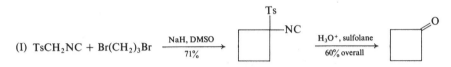

(I) $TsCH_2NC + Br(CH_2)_3Br$ $\xrightarrow[71\%]{NaH, DMSO}$ ... $\xrightarrow[60\% \text{ overall}]{H_3O^+, \text{ sulfolane}}$ ...

[1] A. M. van Leusen and P. G. Oomkes, *Syn. Comm.*, **10**, 399 (1980).
[2] D. van Leusen and A. M. van Leusen, *Synthesis*, 325 (1980).

1,2,4-Triazine, 1, 1188; **2**, 423.

Pyridine annelation.[1] 1,2,4-Triazine can function as a reactive azadiene in Diels-Alder reactions with enamines, particularly pyrrolidine enamines, to form substituted pyridines.

Example:

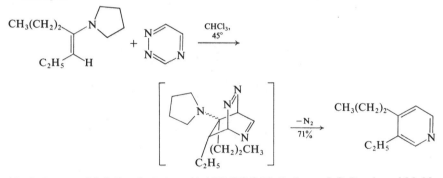

[1] D. L. Boger and J. S. Panek, *J. Org.*, **46**, 2179 (1981); D. L. Boger, J. S. Panek, and M. M. Meier, *J. Org.*, **47**, 895 (1982).

1,3,5-Triazine, 2, 423.

Isoflavones.[1] 1,3,5-Triazine can be used as a formylating reagent in a synthesis of isoflavones from *o*-hydroxydeoxybenzoins. The yields are about the same as those obtained with methanesulfonyl chloride and DMF (**7**, 423–424).

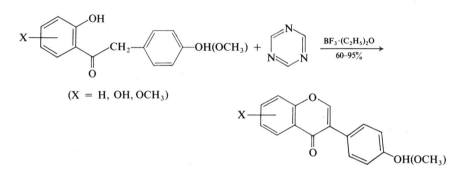

(X = H, OH, OCH₃)

[1] H. C. Jha, F. Zilliken, and E. Breitmaier, *Angew. Chem. Int. Ed.*, **20**, 102 (1980).

Tri-*n*-butylborane, [CH₃(CH₂)₃]₃B. Mol. wt. 182.16. Supplier Aldrich.

erythro-1,3-Diols. β-Hydroxy ketones can be converted stereoselectively into *erythro* (or *meso*)-1,3-diols by treatment with tri-*n*-butylborane to form a chelate (**a**), which is not isolated but rather is reduced directly with sodium borohydride.[1]

Examples:

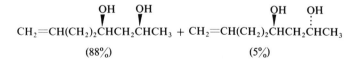

[1] K. Narasaka and H. C. Pai, *Chem. Letters*, 1415 (1980).

Tri-*n*-butylcrotyltin (1). (Z)- or (E)-Crotyl chloride reacts with tri-*n*-butyltinlithium to form (Z)- or (E)- **1** with retention of the geometry of the double bond[1]:

$$(n\text{-Bu})_3\text{SnLi} + \text{ClCH}_2\text{CH}=\text{CHCH}_3 \xrightarrow[\sim 70\%]{\text{THF}, -30°} \text{CH}_3\text{CH}=\text{CHCH}_2\text{Sn}(n\text{-Bu})_3$$
$$\mathbf{1}$$

erythro-*Selective addition to aldehydes.* [2] In the presence of BF_3 etherate **1** adds to aldehydes stereoselectively to give *erythro-β*-methylhomoallyl alcohols; the stereo-selectivity is independent of the geometry of the crotyl group and of the nature of R group of the aldehyde (equation I). This stereoselectivity has been observed before

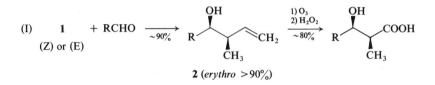

(I)   **1**   + RCHO
(Z) or (E)

2 (*erythro* >90%)

with (Z)-metal enolates and has been attributed to chelation involving the bivalent metal. In the present case the *erythro*-selectivity is attributed to steric factors, since isomerization of (E)-**1** to (Z)-**1** was shown not to be involved.

[1] E. Mattarasso-Tchiroukhine and P. Cadiot, *J. Organometal. Chem.*, **121**, 155, 159 (1976).
[2] Y. Yamamoto, H. Yatagai, Y. Naruta, and K. Maruyama, *Am. Soc.*, **102**, 7107 (1980).

Tri-*n*-butylcyanotin, $(n\text{-C}_4\text{H}_9)_3\text{SnCN}$ **(1).** Mol. wt. 318.05. The reagent is prepared by reaction in CH_3CN with tri-*n*-butylchlorotin with KCN–18-crown-6. The chemical is hygroscopic but stable to storage in a desiccator.

Acyl nitriles, RCOCN.[1] These products are obtained in high yield by reaction of **1** with acid chlorides.

[1] M. Tanaka, *Tetrahedron Letters*, **21**, 2959 (1980).

Tri-*n*-butyltin hydride, 1, 1192–1193; **2**, 424; **3**, 294; **4**, 518–520; **6**, 604; **7**, 379–380; **8**, 497–498; **9**, 379–380.

Reduction of acyl chlorides. [1] The known reduction by tributyltin hydride of acyl chlorides to aldehydes (**1**, 1193) and esters[2] when conducted in the presence of $Pd[P(C_6H_5)_3]_4$ is completely selective in favor of the former products and is much more rapid. Yields of aldehydes are ~75–95%, even on a preparative scale.

Deoxygenation of allylic alcohols. [3] A method for conversion of allylic alcohols to 1-alkenes is outlined in equation (I). The first step is an allylic rearrangement of an O-allyl xanthate to **2**. The second step is an allyl transfer from sulfur to tin with tri-*n*-butyltin hydride to give the allylic stannane (**3**). The last step, destannylation, is a well-known route to terminal alkenes.[4]

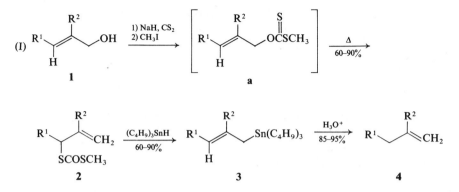

RCOOH → RH. Another method for degradation of carboxylic acids to the noralkane involves conversion to the phenyl carboselenoate **1**, which is then reduced with tri-*n*-butyltin hydride (AIBN initiation) (equation I). Reduction can result in

$$(I) \quad ROCl + C_6H_5SeH \longrightarrow RCOSeC_6H_5 \xrightarrow[\text{AIBN}, \Delta]{\text{Bu}_3\text{SnH},} RH\,[+\,RCHO]$$

1

the noralkane or the corresponding aldehyde. Formation of the aldehyde is favored at temperatures of 80° or by irradiation at 25°. Decarbonylation to alkanes is favored by use of temperatures of 144–164°. It is the only reaction observed with esters of tertiary carboxylic acids, and in this case can be effected even at 80°. Esters of α,β-unsaturated acids are converted only into aldehydes, regardless of temperature. By use of the proper conditions, yields of 80–95% of the noralkanes can be obtained. The method compares favorably with other methods for reductive decarboxylation.[5]

The stannane reduction is also applicable to selenocarbonates (**2**) of primary and secondary alcohols (equation II) and can result in the noralkane, the formate, or the

$$(II) \quad R^1R^2CHOH \xrightarrow[\text{2) } C_6H_5SeH]{\text{1) COCl}_2} R^1R^2CHO\overset{\overset{\displaystyle O}{\|}}{C}SeC_6H_5 \xrightarrow[\text{AIBN}, \Delta]{\text{Bu}_3\text{SnH},}$$

2

$$R^1R^2CH_2\,[+\,R^1R^2CHOCHO + R^1R^2CHOH]$$

starting alcohol. Again optimal yields of the alkane are obtained at 144–164°; they range from 65 to 90%. The method fails with selenocarbonates of phenols.

Hydrodeamination.[6] α-Amino acid esters can be deaminated by conversion to isonitrile esters followed by reduction with this hydride (equation I).

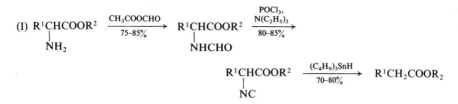

(I) $R^1CHCOOR^2$ $\underset{75-85\%}{\xrightarrow{CH_3COOCHO}}$ $R^1CHCOOR^2$ $\underset{80-85\%}{\overset{POCl_3,}{\underset{}{\xrightarrow{N(C_2H_5)_3}}}}$
 |
 NH_2 $NHCHO$

 $R^1CHCOOR^2$ $\underset{70-80\%}{\xrightarrow{(C_4H_9)_3SnH}}$ $R^1CH_2COOR_2$
 |
 NC

Decarboxylation.[7] Esters of carboxylic acids are converted to the corresponding nor-hydrocarbons by reaction with *n*-Bu$_3$SnH in the presence of AIBN as radical initiator. The reaction involves fragmentation to $RCO_2\cdot$ followed by loss of CO_2 to give $R\cdot$, which is reduced to RH. Yields are in the range 60–90%.

–NO$_2$ → –H.[8] Tertiary nitro groups can be replaced by hydrogen by treatment with Bu$_3$SnH (AIBN initiation). The reduction is also applicable to secondary nitro compounds substituted by electron-attracting groups. The nitro group is reduced preferentially in the presence of keto, ester, chloro, or sulfur groups.

Examples:

$$C_6H_5CH_2CHNO_2 \xrightarrow{78\%} C_6H_5CH_2CH_2COC_6H_5$$
$$|$$
$$COC_2H_5$$

$$\underset{CH_3}{\overset{CH_3}{\underset{|}{\overset{|}{C_6H_5C-NO_2}}}} \xrightarrow[92\%]{\underset{AIBN, C_6H_6}{n\text{-}Bu_3SnH,}} \underset{CH_3}{\overset{CH_3}{\underset{|}{\overset{|}{C_6H_5CH}}}}$$

Desulfurization of 1,3-dithiolanes.[9] Selective hydrogenolysis of dithioketals and dithioacetals is possible with 4 equivalent of this hydride in the presence of AIBN. Yields are 75–95%. The other products are ethane and $(Bu_3Sn)_2S$.

[1] F. Guibe, P. Four, and H. Riviere, *J.C.S. Chem. Comm.*, 432 (1980).

[2] E. J. Kupchik and R. J. Keisel, *J. Org.*, **31**, 456 (1966).

[3] Y. Ueno, H. Sano, and M. Okawara, *Tetrahedron Letters*, **21**, 1767 (1980).

[4] J. A. Verdone, J. A. Mangravite, M. Scarpa, and H. G. Kuivila, *Am. Soc.*, **97**, 843 (1975).

[5] J. Pfenninger, C. Heuberger, and W. Graf, *Helv.*, **63**, 2328 (1980).

[6] D. H. R. Barton, G. Bringmann, W. B. Motherwell, *Synthesis*, 68 (1980).

[7] D. H. R. Barton, H. A. Dowlatshahi, W. B. Motherwell, and D. Villemin, *J.CS. Chem. Comm.*, 732 (1980).

[8] N. Ono, H. Miyake, R. Tamura, and A. Kaji, *Tetrahedron Letters*, **22**, 1705 (1981).

[9] C. G. Gutierrez, R. A. Stringham, T. Nitasaka, and K. G. Glasscock, *J. Org.*, **45**, 3393 (1980).

Tri-*n*-butyltinlithium, 8, 495–497.

Coupling of aldehydes to alkenes.[1] The stannylcarbinols **1** obtained by reaction of aldehydes with Still's reagent can be converted into α-stannylalkyl halides (**2**) by

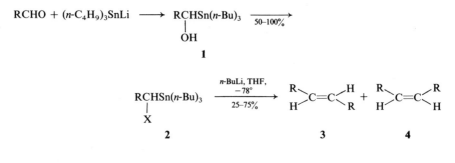

several methods, such as use of $P(C_6H_5)_3$–CX_4. When treated with n-butyllithium, **2** is converted into a mixture of **3** and **4**, in which **3** usually predominates. The process provides a method for coupling of aldehydes to alkenes.

It is also possible to effect cross-coupling of two different aldehydes (equation I).

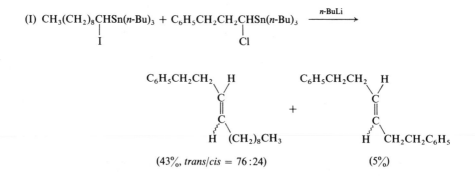

If **2** bears an α-methyl substituent as in **5**, treatment with n-butyllithium results in elimination to a terminal alkene (**6**).

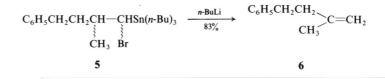

[1] Y. Torisawa, M. Shibasaki, and S. Ikegami, *Tetrahedron Letters*, **22**, 2397 (1981).

Trichloroethylene 9, 479–480. *Caution:* Suspected carcinogen.

Dichlorovinylation. This reaction can be used to prepare α- and γ-acetylenic ketones (equation I).[1]

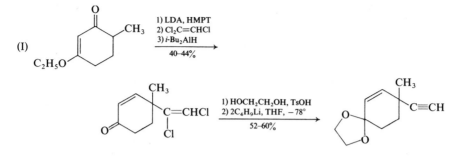

¹ A. S. Kende and P. Fludzinski, *Org. Syn.*, submitted (1980).

Triethylaluminum, 1, 1197–1198; 2, 427; 3, 299; 4, 526; 5, 688–689.

Allylic aluminum ate complexes. Carbanions of allylic sulfides and ethers generally react with electrophiles at both the α- and the γ-position. If the carbanion is converted into an ate complex with triethylaluminum, electrophiles react predominately at the α-position. After the reaction is completed the trialkylaluminum is destroyed with aqueous methanol. Boron ate complexes are not so generally useful for control of regioselectivity.¹

Example:

$$(CH_3)_2CHO\overset{\alpha}{C}H_2\overset{\gamma}{C}H=CH_2 \xrightarrow[2) Al(C_2H_5)_3]{1) sec-BuLi}$$

$$\left[(CH_3)_2CHOCH=CHCH_2\bar{A}l(C_2H_5)_3 \right] Li^+ \xrightarrow[81\%]{C_6H_5CHO} CH_2=CHCHCHC_6H_5 \begin{matrix} OH \\ | \\ | \\ OCH(CH_3)_2 \end{matrix}$$

(>99% selectivity)

In the absence of triethylaluminum, the reaction leads to the products of α- and γ-substitution in the ratio 28:72. Unfortunately reaction of these ate complexes with alkyl halides appears to be limited to benzylic and allylic types, but alkylation at the α-position is highly favored.

¹ Y. Yamamoto, H. Yatagai, and K. Maruyama, *J. Org.*, **45**, 195 (1980).

Triethylammonium hydrofluoride, $(C_2H_5)_3\overset{+}{N}HF^-$. Preparation (5, 632).¹

*Methyl γ-oxocarboxylates.*² Silyl enol ethers react with methyl diazoacetate in the presence of a copper salt to form siloxy-substituted cyclopropanes (1) in useful yields. Cleavage of the siloxy group with fluoride ion (4, 477–478) affords γ-oxo esters (2) in good yields.

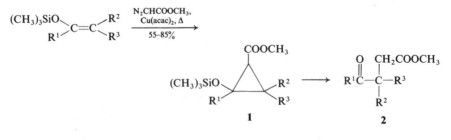

¹ S. Hünig and G. Wehner, *Synthesis*, 180 (1975).
² H.-U. Reissig and E. Hirsch, *Angew. Chem. Int. Ed.*, **19**, 813 (1980).

Triethylborane, 9, 482.

   Stereoselective aldol condensation. 2-Butenyllithium (**1**) reacts with aldehydes to form the *threo*- and *erythro*-β-methyl alcohol in equal amounts. However, if a trialkylborane is present, *threo*-products predominate. Presumably an allylboronate complex (**a**) is involved.¹ An example is formulated in equation (I). The products are converted into β-hydroxy ketones (**4**) by a Wacker-type oxidation.²

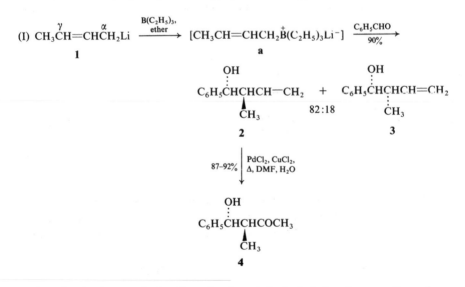

   A regioselective synthesis of *erythro*-β-methyl alcohols involves reaction of an aldehyde with the (Z)-2-butenyltin compound **5** (equation II).³

[1] Y. Yamamoto, H. Yatagai, and K. Maruyama, *J.C.S. Chem. Comm.*, 1072 (1980).
[2] J. Tsuji, I. Shimizu, and K. Yamamoto, *Tetrahedron Letters*, 2975 (1976).
[3] H. Yatagai, Y. Yamamoto, and K. Maruyama, *Am. Soc.*, **102**, 4548 (1980).

Triethyl orthoacetate, 3, 300–302; 4, 527; 6, 607–610.

Chiral allenes.[1] A key step in a recent synthesis of the optically active pheromone of the male boll weevil (**3**) used the orthoester Claisen rearrangement for transfer of the chirality of the acetylenic alcohol **1** to the allene **2**. The final product (**3**) shows a higher optical rotation than the naturally occurring material, which may

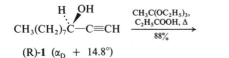

(R)-**1** (α_D + 14.8°)

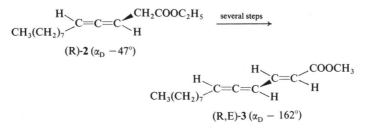

(R)-**2** (α_D −47°)

(R,E)-**3** (α_D − 162°)

not be homogeneous. Another approach to **3** has been reported by Pirkle and Boeder (**8**, 356–357).

[1] K. Mori, T. Nukada, and T. Ebata, *Tetrahedron*, **37**, 1343 (1981).

Triethyloxonium tetrafluoroborate, 1, 1210–1212; 2, 430–431; 3, 303; 4, 527–529; 6, 691–693; 7, 386–387; 8, 500–501; 9, 482–483.

Thionolactones.[1] Lactones can be converted into thionolactones by O-alkylation with Meerwein's salt followed by reaction with sodium hydrosulfide in acetonitrile. Hydroxy thionoesters are usually formed as by-products.

Example:

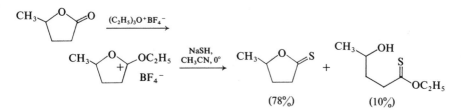

(78%)   (10%)

[1] M. K. Kaloustian and F. Khouri, *Tetrahedron Letters*, **22**, 413 (1981).

Triethylsilane–Boron trifluoride, 7, 387–388; 8, 501.

Lactones → ethers. γ- and δ-Lactones can be converted into tetrahydrofuranes and tetrahydropyranes, respectively, by a two-step procedure. The first step is the well-known reduction of lactones to lactols with DIBAH (1, 261; 2, 140). The second step is deoxygenation of the alcohol with triethylsilane and BF₃ etherate.[1] This reaction is compatible with several other functional groups, even including hydroxyl groups.

Example:

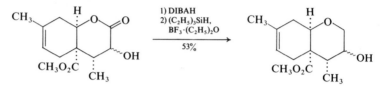

[1] G. A. Kraus, K. A. Frazier, B. D. Roth, M. J. Taschner, and K. Neuenschwander, *J. Org.*, **46**, 2417 (1981).

Trifluoroacetic acid, 1, 1219–1221; 2, 433–434; 3, 305–308; 4, 530–532; 5, 695–700; 6, 613–615; 7, 388–389; 8, 503; 9, 483.

Benzoisocoumarins.[1] A regiospecific synthesis of this system (2) involves the acid-catalyzed C—C coupling of 1 with substituted phenols. The products can be elaborated to naphthoisocoumarins (4).

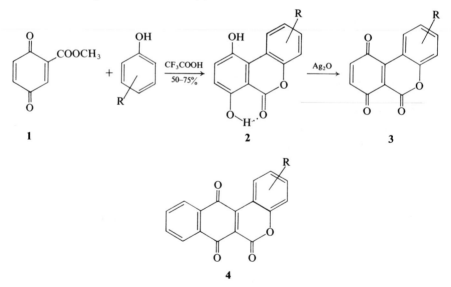

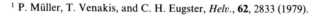

[1] P. Müller, T. Venakis, and C. H. Eugster, *Helv.*, **62**, 2833 (1979).

Trifluoroacetyl chloride, CF_3COCl. The reagent can be generated *in situ* from lithium chloride and trifluoroacetic anhydride.

Olefin inversion (*cf.* **7**, 338).[1] Trifluoroacetyl chloride reacts with 1,2-dialkyl epoxides in DMF stereospecifically by *trans* opening to give *vic*-chlorohydrin trifluoroacetates. These products are reduced stereospecifically by NaI to alkenes with *syn*-elimination to give inverted alkenes. Reductions with zinc are less selective. Inversion of olefins is also possible by addition of NCS in CF_3COOH (actual reagent is trifluoroacetyl hypochlorite) followed by reduction with NaI.

The method is not suitable for trisubstituted alkenes or for a highly strained alkene such as *trans*-cyclooctene.

[1] P. E. Sonnet, *J. Org.*, **45**, 154 (1980).

Trifluoroacetyl hypoiodite (1). The reagent is generated *in situ* from bis(trifluoroacetoxy)phenyliodide and iodine:

$$C_6H_5I(OCOCF_3)_2 + I_2 \rightleftharpoons C_6H_5I + 2CF_3COOI$$

1

Aryl iodides.[1] The reagent iodinates aromatic compounds at 25° in reasonable yields (50–85%).

[1] E. B. Merkushev, N. D. Simakhina, and G. M. Koveshnikova, *Synthesis*, 480 (1980).

Trifluoromethanesulfonic anhydride, 4, 533–534; **5**, 702–705; **6**, 618–620; **7**, 390.

Epoxides from hydroperoxides.[1] The biosynthesis of "slow reacting substances" (SRS) from arachidonic acid is considered to involve the hydroperoxide **1** [(S)-5-HPETE] and then the oxide **2** (leukotriene A). The latter substance is considered to be the precursor to the family of SRSs including leukotriene C-1 (LTC-1, **3**).[2] Since the epoxide (**2**) is exceedingly labile, nonacidic and mild conditions are required for conversion of **1** into **2**. This reaction was achieved in fair yield by use of $(CF_3SO_2)_2O$ at −110° with the highly hindered 1,2,2,6,6-pentamethylpiperidine[3] as a proton sponge. Even so, the corresponding dienone formed by a 1,2-elimination of H_2O was a major by-product.

The oxide (**2**) was converted into the methyl ester of **3** by reaction with glutathione.

[1] E. J. Corey, A. E. Barton, and D. A. Clark, *Am. Soc.*, **102**, 4278 (1980).
[2] E. J. Corey, J. O. Albright, A. E. Barton, and S. Hashimoto, *Am. Soc.*, **102**, 1435 (1980); the authors obtained this hydroperoxide as a single enantiomer by use of a lipoxygenase derived from potato tubers.
[3] H. Z. Sommer, H. I. Lipp, and L. L. Jackson, *J. Org.*, **36**, 824 (1971).

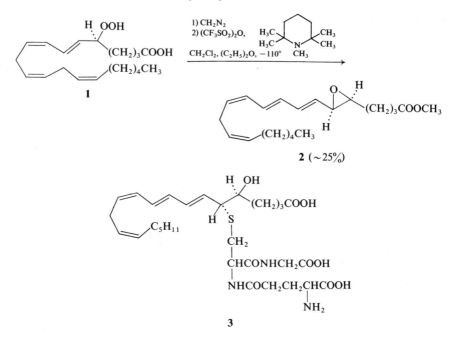

2 (~25%)

3

Trifluoromethanesulfonic–carboxylic anhydrides, 4, 533–534.

Acylation of arenes. These mixed anhydrides have been used for acylation of arenes. However these reagents are sensitive to moisture and tend to disproportionate. One solution is to generate the reagents *in situ*. Thus acyl trifluoroacetates can be generated in the presence of the arene from a carboxylic acid, TFAA, and 85% phosphoric acid (as catalyst) in CH_3CN. Ketones of the type

$$\overset{O}{\overset{\|}{ArCR}}$$

$ArCR$ can be prepared in this way in about 50–80% yield. In the absence of H_3PO_4 longer reaction times are required. The most satisfactory yields result when the ratio of H_3PO_4 to TFAA is $1:3$.[1]

Another solution employs 2-acyloxypyridines in conjunction with TFA. Yields are generally > 70%.[2] However, electron-attracting groups in the acyloxypyridine or aroyloxypyridine lower the yield.

[1] C. Galli, *Synthesis*, 303 (1979).
[2] T. Keumi, R. Tanigushi, and H. Kitajima, *Synthesis*, 139 (1980).

Trifluoromethyl hypofluorite (Fluoroxytrifluoromethane), 2, 200.

α-Fluoro carbonyl compounds. Trimethylsilyl enol ethers of ketones, aldehydes, esters, and amides are converted to α-fluoro carbonyl compounds by reaction with CF_3OF at −70° in a solvent such as CCl_3F.[1]

Examples:

$$C_6H_5CH_2COOC_2H_5 \xrightarrow[\text{ClSi(CH}_3)_3]{\text{LDA,}} C_6H_5CH=\overset{\overset{\displaystyle OSi(CH_3)_3}{|}}{C}OC_2H_5 \xrightarrow[\substack{77\% \\ \text{overall}}]{\substack{CF_3OF, \\ CCl_3F, -70°}}$$

$$C_6H_5CHFCOOC_2H_5 + COF_2 + FSi(CH_3)_3$$

$$(CH_3)_2CHCH_2C_6H_4CH_2COOH \longrightarrow (CH_3)_2CHCH_2C_6H_4CH=C[OSi(CH_3)_3]_2 \longrightarrow$$

$$(CH_3)_2CHCH_2C_6H_4CHFCOOSi(CH_3)_3 \xrightarrow[\substack{80\% \\ \text{overall}}]{H_2O} (CH_3)_2CHCH_2C_6H_4CHFCOOH$$

[1] W. J. Middleton and E. M. Bingham, *Am. Soc.*, **102**, 4845 (1980).

Trifluoroperacetic acid, 1, 821–827; **8**, 505.

Baeyer-Villiger oxidation (**1**, 823–824). Baeyer-Villiger oxidation of very hindered methyl ketones can be effected with trifluoroperacetic acid [(CF$_3$CO)$_2$O–H$_2$O$_2$] in the presence of 1 equivalent of Na$_2$HPO$_4$. Use of more than 1 equivalent of Na$_2$HPO$_4$ results in a slower reaction and a lower yield.[1]

Baeyer-Villiger oxidation of the 5α-6-keto steroid **1** with trifluoroperacetic acid is 1000 times faster than oxidation with *m*-chloroperbenzoic acid and also is more regioselective. This oxidation was used in the last step in a synthesis of brassinolide (**2**), a natural steroid that promotes plant growth.[2]

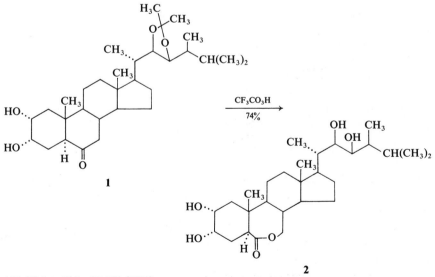

[1] H. Welter, *Helv.*, **64**, 761 (1981).
[2] S. Fung and G. B. Siddall, *Am. Soc.*, **102**, 6580 (1980).

Triisopropoxymethyltitanium (1). Mol. wt. 240.19, b.p. 35–40°/0.018 mm, stable yellow liquid, soluble in common aprotic solvents. Preparation (equation I).[1]

$$(I) \quad [(CH_3)_2CHO]_4Ti \xrightarrow[85\%]{CH_3COCl} [(CH_3)_2CHO]_3TiCl \xrightarrow[69\%]{CH_3Li} [(CH_3)_2CHO]_3TiCH_3$$

$$\mathbf{1}$$

Methylation of aldehydes and ketones.[2] Unlike Grignard and alkyllithium reagents, this reagent does not react with esters, S-thiolates, nitriles, and epoxides, but does react with aldehydes even at -70 to $-20°$ and ketones at -25 to $-80°$ at reasonable rates. Reaction of **1** with 4-*t*-butylcyclohexanone results in the corresponding adducts in the ratio *cis/trans* = $\sim 85:15$.

A series of alkyl- and aryl-substituted titanium compounds of type **1** have been prepared[3]; they exhibit the same reactivity for carbonyl compounds, but vary in stability although all are more stable than the lithium counterparts.

Enantioselective methylation with a chiral reagent of this type has been reported (equation I).

$$CH_3Ti(OCH_2CHC_2H_5)_3 + C_6H_5CHO \longrightarrow$$
$$\underset{CH_3}{|}$$
$$(S)\text{-}(-)$$

$$\overset{H\cdots}{\underset{C_6H_5}{\diagdown}}C\overset{OH}{\diagup}CH_3$$
$$(8\% \text{ ee})$$

[1] M. D. Rausch and N. B. Gordon, *J. Organometal. Chem.*, **74**, 85 (1974).
[2] B. Weidmann and D. Seebach, *Helv.*, **63**, 2451 (1980).
[3] B. Weidmann, L. Widler, A. G. Olivero, C. D. Maycock, and D. Seebach, *Helv.*, **64**, 357 (1981).

2,4,6-Triisopropylbenzenesulfonyl azide, 5, 706.

Cyclic α-diazo ketones.[1] Diazo transfer can be conducted under phase-transfer conditions (tetra-*n*-butylammonium bromide or 18-crown-6) wth this arylsulfonyl azide. Other arylsulfonyl azides (mesityl, tosyl) are unsatisfactory. The method may not offer significant advantages with simple ketones, but is the most satisfactory route to hindered diazo ketones. Isolated yields in five cases were 48–84%.

[1] L. Lombardi and L. N. Mander, *Synthesis*, 368 (1980).

2,4,6-Triisopropylbenzenesulfonylhydrazine, 4, 535; 5, 706–707; 7, 392; 9, 486–488.

McFadyen-Stevens aldehyde synthesis.[1] Use of this hydrazine in place of *p*-toluenesulfonylhydrazine for the McFadyen-Stevens synthesis of aromatic aldehydes has the advantage that decomposition of the intermediate acyl derivatives proceeds under mild conditions (K_2CO_3 in refluxing CH_3OH); however yields are unsatisfactory unless hydrazine is added.

Lithium vinylcuprates.[2] Treatment of trisylhydrazones with 2 equivalents of *t*-butyllithium to generate the vinyllithium reagent (**9**, 486–487) followed by addition of phenylthiocopper results in mixed alkenylcuprates, $(R^1CH=CR^2)(C_6H_5S)CuLi$. These cuprates undergo conjugate addition to α,β-enones in fair to good yield. Toluenesulfonylhydrazones are unsatisfactory precursors to these cuprates.

Example:

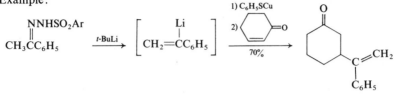

[1] C. C. Dudman, P. Grice, and C. B. Reese, *Tetrahedron Letters*, **21**, 4645 (1980).
[2] A. S. Kende and L. N. Jungheim, *Tetrahedron Letters*, **21**, 3849 (1980).

Trimethylaluminum–Dichlorobis(cyclopentadienyl)zirconium.

Homoallylic alcohols.[1] An efficient synthesis of homoallylic alcohols of type **1** involves carbometallation of 1-alkynes (**8**, 506) followed by reaction of the alkenylaluminate (**a**) with an epoxide.[2]

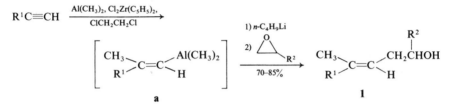

[1] M. Kobayashi, L. F. Valente, E. Negishi, and W. Patterson, *Synthesis*, 1034 (1980).
[2] E. Negishi, S. Baba, and A. O. King, *J.C.S. Chem. Comm.* 17 (1976).

Trimethylamine N-oxide, 1, 1230–1231; **2**, 434; **3**, 309–310; **6**, 624–625; **7**, 392; **8**, 507; **9**, 488–489.

Oxidation of organoboranes (**6**, 624; **8**, 507). A new synthesis of acylsilanes (**2**) involves oxidation of the hydroboration products of alkynylsilanes (**1**).[1] Oxidation with the commercially available dihydrate of trimethylamine N-oxide is sluggish; however, anhydrous reagent[2] is satisfactory for this purpose (equation I).

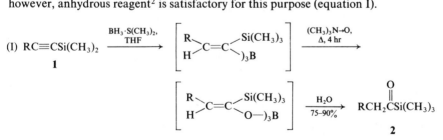

[1] J. A. Miller and G. Zweifel, *Synthesis*, 288 (181).
[2] Anhydrous reagent is obtained by azeotropic distillation of the dihydrate with toluene.

2,2,6-Trimethyl-1,3-dioxolenone, 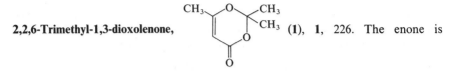 (1), 1, 226. The enone is

obtained in 90% yield by acid-catalyzed reaction of diketene and acetone.

Cyclohexenones.[2] This enone undergoes photochemical [2 + 2]cycloaddition to alkenes. The products, obtained in high yield, can be converted into cyclohexenones in two steps. An example is formulated in equation (I). The regioselectivity of the cycloaddition depends on both steric and electronic factors.[1]

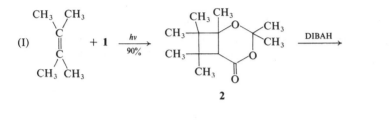

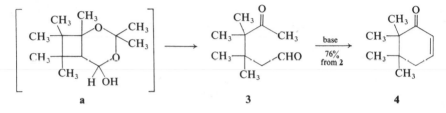

For a related synthesis of cyclohexenones *see* 2,2-dimethyl-3(2*H*)-furanone, this volume.

[1] S. W. Baldwin and J. M. Wilkinson, *Am. Soc.*, **102**, 3634 (1980).

Trimethyl β-(methoxy)orthopropionate, $CH_3OCH_2CH_2C(OCH_3)_3$ (1). Mol. wt. 164.20, b.p. 89–90°/20 mm. The orthoester is obtained in 72% yield from the corresponding nitrile.[1]

Claisen orthoester rearrangement.[2] This reagent can be used to prepare methyl α-substituted acrylates from allylic alcohols. 2,4,6-Trimethylbenzoic acid is used as the acid catalyst.[1]

Example:

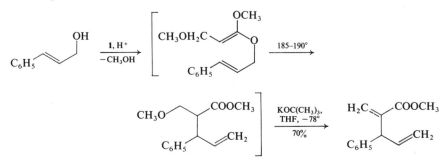

[1] S. Raucher, J. E. Macdonald, and R. F. Lawrence, *Tetrahedron Letters*, **21**, 4335 (1980).
[2] S. J. Rhoads and N. R. Raulins, *Org. React.*, **22**, 1 (1975).

Trimethyl orthoformate, 3, 313; **4**, 540; **5**, 714; **8**, 507–508.

Cycloalkenones.[1] Attempted preparation of the methyl enol ether of **1** with trimethyl orthoformate and perchloric acid results in formation of **2** in 41% yield.

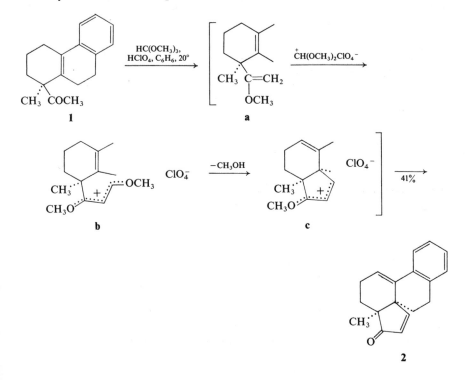

The additional carbon atom is believed to arise from dimethoxycarbonium perchlorate. Use of triethyl orthoformate results in a more complex structure.
Another example of this formylation–cyclization is the conversion of **3** to **4**.

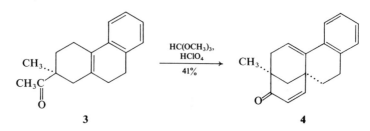

3		**4**

[1] U. R. Ghatak, B. Sanyal, S. Ghosh, M. Sarkar, M. S. Raju, and E. Wenkert, *J. Org.*, **45**, 1081 (1980).

Trimethyloxonium tetrafluoroborate, 1, 1232; **2**, 438; **3**, 314–315; **4**, 541; **5**, 716.

Dethioacetalization.[1] Thioacetals can be cleaved to the ketones in high yield by treatment with trimethyloxonium tetrafluoroborate (6 equivalents) followed by addition of water (*cf.* **4**, 528–529).

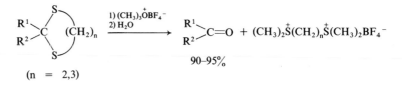

$$90\text{–}95\%$$

(n = 2,3)

[1] I. Stahl, *Synthesis*, 135 (1981).

Trimethyl(phenylthio)silane (1). Mol. wt. 182.35, b.p. 72–74°/3 mm.
Preparation[1]:

$$C_6H_5SH + HN[Si(CH_3)_3]_2 \xrightarrow[93\%]{\text{imidazole}} C_6H_5SSi(CH_3)_3$$

1

Cleavage of methyl and benzyl ethers.[2] These ethers are cleaved by the combination of this silane and zinc iodide. The rate of dealkylation is enhanced by the presence of $(n\text{-}C_4H_9)_4NI$. Trimethyl(methylthio)silane serves the same purpose. CH_2Cl_2 and $ClCH_2CH_2Cl$ are satisfactory solvents. The cleavage probably does not involve *in situ* formation of iodotrimethylsilane.

Examples:

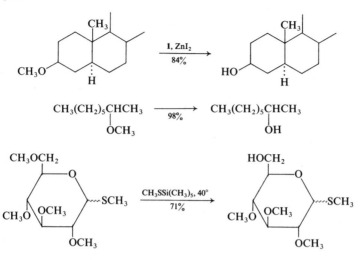

$$CH_3(CH_2)_5\underset{OCH_3}{CHCH_3} \xrightarrow{98\%} CH_3(CH_2)_5\underset{OH}{CHCH_3}$$

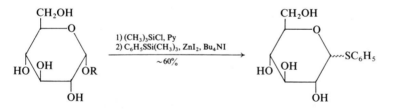

1-Thioglycosides.[3] Glycosides are converted into an anomeric mixture of 1-phenylthioglycosides by treatment with trimethylphenylthiosilane (or methylthiosilane), zinc iodide, and tetra-*n*-butyl ammonium iodide in 1,2-dichloroethane at 60°.[1] Under these conditions some cleavage of O-methyl ethers of a primary alcohol can occur. This reaction is useful for hydrolysis of glycosides since 1-thioglycosides are more rapidly hydrolyzed than glycosides themselves.[4]

Example:

[1] R. S. Glass, *J. Organometal. Chem.*, **61**, 8 (1973); I. Ojima, M. Nihonyangi, and T. Nagai, *Organometal. Chem.* **50**, C26 (1973).

[2] S. Hanessian and Y. Guindon, *Tetrahedron Letters*, **21**, 2305 (1980).

[3] S. Hanessian and Y. Guindon, *Carbohydrate Res.*, **86**, C3 (1980).

[4] S.-H. L. Chu and L. Anderson, *Carbohydrate Res.*, **50**, 227 (1976).

Trimethylsilylacetonitrile, $(CH_3)_3SiCH_2CN$ (**1**). Mol. wt. 113.24, b.p. 65–70°/20 mm. The reagent is obtained in 80% yield by reaction of $BrCH_2CN$, $ClSi(CH_3)_3$, and zinc (2–3 equivalents) in C_6H_6–THF (1:1). The yield is 60% if $ClCH_2CN$ is substituted for $BrCH_2CN$.[1]

Cleavage of epoxides.[1] The anion (2) of 1 reacts with carbonyl compounds to give α,β-unsaturated nitriles and $LiOSi(CH_3)_3$, as expected.[2] The reaction with epoxides is unexpected: the products are γ-trimethylsilyloxynitriles, which are useful precursors to γ-lactones (equation I). The formulation indicates that a 1,4-migration of the trimethylsilyl group is involved.[2]

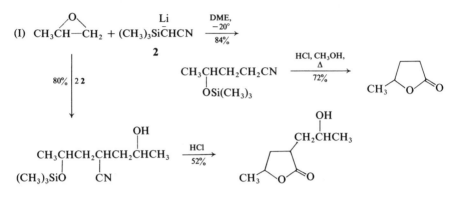

[1] I. Matsuda, S. Murata, and Y. Ishii, *J.C.S. Perkin I*, 26 (1979).
[2] *See also* I. Ojima, M. Kumagai, and Y. Nagai, *Tetrahedron Letters*, 4005 (1974).

Trimethylsilylallene (1). Mol. wt. 112.25.
  Preparation:

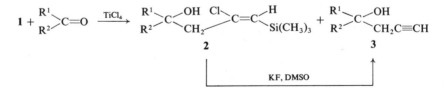

Homopropargylic alcohols.[1] In the presence of $TiCl_4$ this allene reacts with carbonyl compounds to form vinyl chlorides (2) and homopropargylic alcohols (3).

Treatment of the mixture with KF in DMSO converts 2 into 3. The reaction is also observed with 1-substituted trimethylsilylallenes.

Examples:

$$C_6H_5CH_2COCH_3 + 1 \xrightarrow[72\%]{TiCl_4}$$

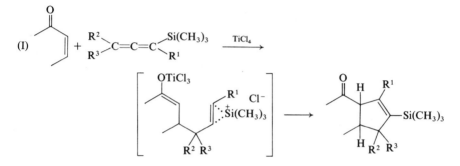

(*Trimethylsilyl)cyclopentenes.*[2] In the presence of $TiCl_4$, trimethylsilylallenes react with cyclic and acyclic α,β- enones to form annelated (trimethylsilyl)cyclopentenes (equation I). The reaction is most efficient with 1-substituted (trimethyl-

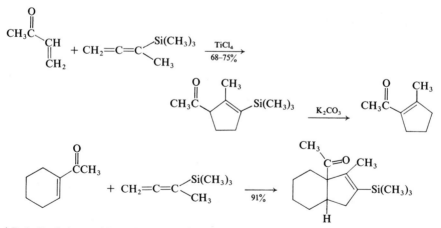

silyl)allenes. It fails with β,β-disubstituted α,β-enones. In all cases, *cis*-fused adducts are formed.

Examples:

[1] R. L. Danheiser and D. J. Carini, *J. Org.*, **45**, 3925 (1980).
[2] R. L. Danheiser, D. J. Carini, and A. Basak, *Am. Soc.*, **103**, 1604 (1981).

[(Trimethylsilyl)allyl]lithium, 8, 273–274.

γ-Butyrolactones.[1] The complete paper on preparation of γ-lactones from alde-hydes and ketones with this anion has been published.[1] The synthesis (equation I) has been examined in eight cases.

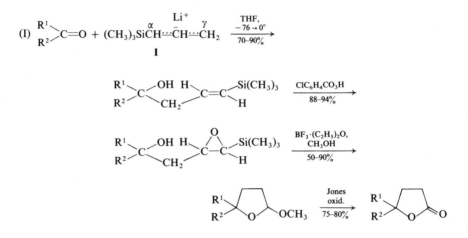

In the case of a 17-keto steroid the reaction proceeds in higher yield when Li^+ is replaced with ^+ZnCl. The Sharpless procedure can be used for the epoxidation step.

[1] E. Ehlinger and P. Magnus, *Am. Soc.*, **102**, 5004 (1980).

α-Trimethylsilylcrotyl-9-borabicyclo[3.3.1]nonane (1).

Preparation[1]:

$$CH_3CH=CHCH_2Br \xrightarrow[2)\ 2/3\ CH_3O-9-BBN]{1)\ 2/3\ Al}$$

$$CH_3CH=CHCH_2B\bigcirc \xrightarrow[40\%]{\substack{LiTMP, \\ ClSi(CH_3)_3}} CH_3CH=CHCHB\bigcirc$$
$$\underset{Si(CH_3)_3}{|}$$

1

Stereoregulated condensation with aldehydes.[2] In the presence of a base (pyridine, *n*- or *sec*-butyllithium, to form an ate complex) **1** reacts with aldehydes to form predominately adduct with the structure **2**. Thus stereochemical control of four consecutive carbons in one step is achieved. The products are useful for further elaboration.

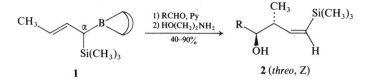

1   **2** (*threo, Z*)

[1] H. Yatagai, Y. Yamamoto, and K. Maruyama, *Am. Soc.*, **102**, 4548 (1980).
[2] Y. Yamamoto, H. Yatagai, and K. Maruyama, *Am. Soc.*, **103**, 3229 (1981).

Trimethylsilyldiazomethane, $(CH_3)_3SiCHN_2$. Mol. wt. 114.23, thermally stable, greenish-yellow, b.p. 96°/775 mm. This material is obtained by treatment of N-nitroso-N-(trimethylsilylmethyl)urea, $(CH_3)_3SiCH_2N(NO)CONH_2$, with KOH.[1]

Arndt-Eistert reaction.[2] This reagent is recommended for use in place of diazomethane in the Arndt-Eistert synthesis (equation I).

$$(I) \quad RCOCl \xrightarrow[\substack{60-80\%}]{\substack{1)\,(CH_3)_3SiCHN_2,\,N(C_2H_5)_3,\,0° \\ 2)\,C_6H_5CH_2OH,\,180-185°}} RCH_2COOCH_2C_6H_5$$

Homologaton of ketones.[3] The reagent is generally superior to diazomethane for homologation of ketones, particularly alicyclic ones. The reaction with fluorenone, followed by hydrolysis, gives 9-phenanthrol in 80% yield.

[1] D. Seyferth, H. Menzel, A. W. Dow, and T. C. Flood, *J. Organometal. Chem.*, **44**, 279 (1972).
[2] T. Aoyama and T. Shioriri, *Tetrahedron Letters*, **21**, 4461 (1980).
[3] N. Hashimoto, T. Aoyama, and T. Shioriri, *Tetrahedron Letters*, **21**, 4619 (1980).

β-(Trimethylsilyl)ethoxymethyl chloride, $ClCH_2OCH_2CH_2Si(CH_3)_3$ (**1**). Mol. wt. 166.73, b.p. 57–59°/8 mm. The reagent is prepared by reaction of β-(trimethylsilyl)ethanol with paraformaldehyde in the presence of HCl gas at 0° (87% yield).

Protection of hydroxyl groups.[1] Alcohols are converted into "SEM" ethers in high yield on reaction with **1** and diisopropylethylamine. Deprotection is effected by treatment with $n\text{-Bu}_4NF$ in THF or HMPT at 45°.

[1] B. H. Lipshutz and J. J. Pegram, *Tetrahedron Letters*, **21**, 3343 (1980).

2-(Trimethylsilyl)ethyl chloroformate, $(CH_3)_3SiCH_2CH_2OCOCl$ (**1**). Mol. wt. 180.71, b.p. 43°/4 mm, decomposes on standing.
Preparation.[1]

Protection of hydroxyl groups.[2] The reagent in the presence of pyridine converts alcohols into the β-(trimethylsilyl)ethoxycarbonyl derivative (**2**). The group is stable to 80% acetic acid at 20°, but is susceptible to alkaline hydrolysis. Of greater interest, it is rapidly cleaved by $n\text{-Bu}_4NF$ at 20° (equation I) or by $ZnCl_2$ or $ZnBr_2$ in CH_3NO_2 or, more slowly, in CH_2Cl_2. It is stable to KCl–18-crown-6.

$$(CH_3)_3SiCH_2CH_2O\overset{\overset{\displaystyle O}{\|}}{C}OR \xrightarrow[90-95\%]{F^-} FSi(CH_3)_3 + H_2C{=}CH_2 + CO_2 + ROH$$

2

The β-(trimethylsilyl)ethoxycarbonyl group has been recommended for protection of amines (**8**, 470–471).

[1] V. P. Kozyukov, V. D. Skeludyakov, and V. F. Mironov, *Zh. Obsch. Kim.*, **38**, 1179 (1968).
[2] C. Giveli, N. Balgobin, S. Josephson, and J. B. Chattopadhyaya, *Tetrahedron Letters*, **22**, 969 (1981).

2-Trimethylsilylmethyl-1,3-butadiene, 9, 493–494.

Diels-Alder reactions.[1] This diene (**1**) undergoes stereospecific Diels-Alder reactions with symmetrical dienophiles (equations I and II). The reaction of **1** with

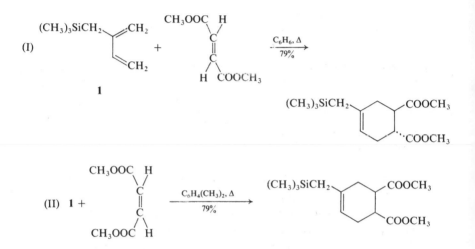

unsymmetrical dienophiles is more regioselective than that of isoprene itself (equation III). 2-Trimethylstannylmethyl-1,3-butadiene reacts with higher regioselectivity than **1**.

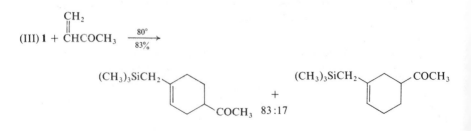

The products of these reactions are functionalized allylsilanes and are useful for further transformations, such as the reaction formulated in equation (IV).

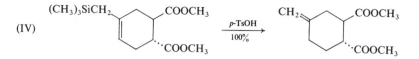

(IV)

[1] A. Hosomi, M. Saito, and H. Sakurai, *Tetrahedron Letters*, **21**, 355 (1980).

Trimethylsilylmethyllithium (1), 6, 635–636; 9, 495–496. The reagent can be generated *in situ* by transmetalation of (trimethylsilylmethyl)tributyltin (2) with *n*-butyllithium at 0° in THF–hexane (equation I).[1]

(I) $(CH_3)_3SiCH_2Cl + (n\text{-}C_4H_9)_3SnLi \xrightarrow{>90\%}$

$$(CH_3)_3SiCH_2Sn(n\text{-}C_4H_9)_3 \xrightarrow{n\text{-}C_4H_9Li} (CH_3)_3SiCH_2Li + (n\text{-}C_4H_9)_4Sn$$

$$\textbf{2} \qquad\qquad\qquad\qquad\qquad \textbf{1}$$

The reagent is useful for methylenation of aldehydes and ketones.[2]
Examples:

$$CH_3(CH_2)_6CHO + 2 \xrightarrow{n\text{-}BuLi} \left[CH_3(CH_2)_6\overset{\overset{\displaystyle OH}{|}}{C}HCH_2Si(CH_3)_3 \right] \xrightarrow[61\%]{H_2SO_4,\ THF,\ 65°}$$

$$CH_3(CH_2)_6CH{=}CH_2$$

$$(C_6H_5)_2CO \xrightarrow{78\%} (C_6H_5)_2C{=}CH_2$$

[1] D. E. Seitz and A. Zapata, *Tetrahedron Letters*, **21**, 3451 (1980).
[2] D. J. Peterson, *J. Org.*, **33**, 780 (1968).

Trimethylsilylmethylmagnesium chloride, $(CH_3)_3SiCH_2MgCl$ (1). Mol. wt. 146.99. The Grignard reagent is prepared in the usual way by reaction of magnesium with chloromethyl(trimethyl)silane.[1]

α-Trimethylsilylallenes.[2] In the presence of CuBr–LiBr, this Grignard reagent reacts with a propargylic tosylate or acetate to furnish an α-trimethylsilylallene

$$(I)\quad \overset{\overset{\displaystyle OR}{|}}{\underset{R^2}{\overset{R^1}{C}}}{-}C{\equiv}C{-}R^3 + (CH_3)_3SiCH_2MgCl \xrightarrow[50\text{-}85\%]{\substack{CuBr_2Li,\\ THF}} \underset{R^2}{\overset{R^1}{C}}{=}C{=}\underset{CH_2Si(CH_3)_3}{\overset{R^3}{C}}$$

$$(R = Ac, Ts)$$

(equation I). This allene synthesis is an extension of a previous synthesis of α-trimethylsilylallenes (equation II).[3]

$$\text{(II) } (CH_3)_3Si-C\equiv C-\underset{\underset{R^2}{|}}{\overset{\overset{OR}{|}}{C}}-R^1 + R^3MgBr \xrightarrow[\sim 80-95\%]{CuBr_2Li} \underset{R^3}{\overset{(CH_3)_3Si}{>}}C=C=C\underset{R^2}{\overset{R^1}{<}}$$

(R = Ms, Ts)

[1] C. R. Hauser and C. R. Hance, *Am. Soc.*, **74**, 5091 (1952).
[2] M. Montury, B. Psaume, and J. Goré, *Tetrahedron Letters*, **21**, 163 (1980).
[3] H. Westmijze and P. Vermeer, *Synthesis*, 390 (1979).

Trimethylsilylmethyl trifluoromethanesulfonate, $CF_3SO_2OCH_2Si(CH_3)_3$ **(1).** Mol. wt. 236.29, b.p. 156–158°. The reagent is prepared by reaction of trimethylsilylmethanol with the complex of triflic anhydride and pyridine (76% yield). It is stable to storage (refrigerator). Since it is a potent alkylating reagent, it should be handled with care.

Methylides. The reagent reacts with sulfides, phosphines, amines, and imines to form α-trimethylsilylmethylsulfonium, -phosphonium, -ammonium, and -immonium triflates at 20°. These salts are desilylated by F⁻ at the same temperature to give an intermediate showing ylide reactivity. For this purpose cesium fluoride is superior to KF–18-crown-6 or tetra-*n*-butylammonium fluoride. The sulfur ylides generated in this way undergo immediate rearrangement to alkenes.

Example:

$$C_6H_5CH=CHCH_2SCH_2COOC_2H_5 \xrightarrow{\mathbf{1}} C_6H_5CH=CHCH_2\overset{\overset{-OTf}{+}}{\underset{\underset{CH_2Si(CH_3)_3}{|}}{S}}CH_2COOC_2H_5 \xrightarrow{CsF}$$

$$C_6H_5CH=CHCH_2\overset{\overset{+}{}}{\underset{\underset{-CH_2}{|}}{S}}CH_2COOC_2H_5 \xrightarrow[81\%]{} \underset{CH_2SCH_2COOC_2H_5}{\overset{C_6H_5CHCH=CH_2}{|}}$$

Application of this two-step sequence to phosphines is a useful alternative to Wittig methylenation (equation I).[1]

$$\text{(I) } (C_6H_5)_3P \xrightarrow{\mathbf{1}} (C_6H_5)_3\overset{+}{P}CH_2Si(CH_3)_3 {}^-OTf \xrightarrow{CsF}$$

Ammonium ylides generated in this way are useful as a route to alkenes. Of more interest, nonstabilized immonium ylides can be prepared *in situ* from imines and trapped with a dipolarophile. This sequence provides a route to pyrrolines by reaction of azomethine ylides with acetylenedicarboxylic acid esters.

Example:

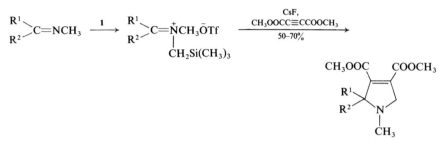

This CsF desilylation of α-trimethylsilylimmonium salts provides the key step in a recent synthesis of the pyrrolizidine retronecine (**2**), as outlined in equation (II).[2]

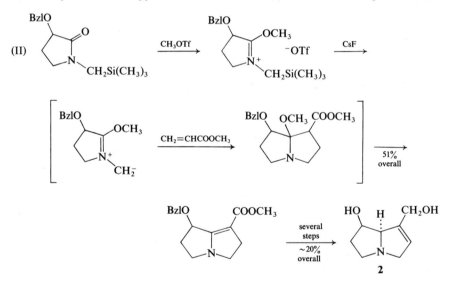

Although the yield in the 1,3-dipolar addition is rather low, the overall yield of the alkaloid is about 20%.

Propargyltrimethylsilanes. The reaction of **1** with the lithium salt of terminal alkynes provides an improved route to propargyltrimethylsilanes[3] (equation I).

$$(I) \quad CH_3(CH_2)_5C{\equiv}CH \xrightarrow[\substack{2)\,1 \\ \cdot \quad 78\%}]{\substack{1)\,n\text{-BuLi, ether,} \\ HMPT}} CH_3(CH_2)_5C{\equiv}CCH_2Si(CH_3)_3$$

The propargylsilane group can participate in intramolecular cyclization reactions to give a five-membered ring with an exocyclic allene group. An example is the cyclization of **2** to a mixture of **3** and **4** when treated with acetic acid.[4]

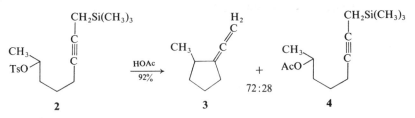

A particularly interesting cyclization utilizes a propargylsilane group as a terminator for a biomimetic synthesis of a tetracyclic hydrocarbon (**6**) with a potential cortical side chain.[5]

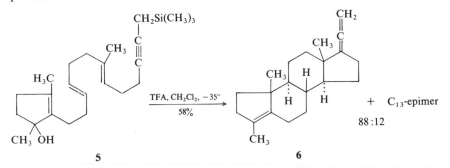

Propargyltrimethylsilanes are converted into 3-alkyl-3-halo-1,2-dienes by iodine or bromine. Reaction with acetyl chloride and $AlCl_3$ results in a 3-acetyl-3-alkyl-1,2-diene, but in low yield.[6]

[1] E. Vedejs and G. R. Martinez, *Am. Soc.*, **101**, 6453 (1979).

[2] E. Vedejs and G. R. Martinez, *Am. Soc.*, **102**, 7993 (1980).

[3] S. Ambasht, S. K. Chiu, and P. E. Peterson, *Synthesis*, 318 (1980); S. K. Chiu and P. E. Peterson, *Tetrahedron Letters*, **21**, 4047 (1980).

[4] A. D. Despo, S. K. Chiu, T. Flood, and P. E. Peterson, *Am. Soc.*, **102**, 5120 (1980).

[5] R. Schmid, P. L. Heusmann, and W. S. Johnson, *Am. Soc.* **102**, 5122 (1980).

[6] T. Flood and P. E. Peterson, *J. Org.*, **45**, 5006 (1980).

1-Trimethylsilyl(pentadienyl)lithium (1). Mol. wt. 146.24.

Preparation[1]:

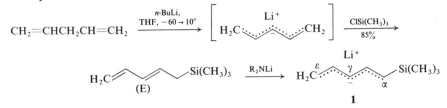

(Z)-2,4-Pentadienyltrimethylsilane is obtained by reaction of pentadienyl potassium with chlorotrimethylsilane in THF.[2]

Electrophilic substitution; hexahydrophenanthrenes.[1] Electrophilic attack could occur at three sites in **1**, but no products of α-substitution have been reported. Benzyl bromide reacts exclusively by γ-attack. Aldehydes and ketones react usually by ε- and/or γ-attack, but with variable regioselectivity. ε-Attack is useful because products can be obtained that undergo an intramolecular Diels-Alder reaction. An example is the synthesis of the two isomeric hexahydrophenanthrenes (**3**, equation I).

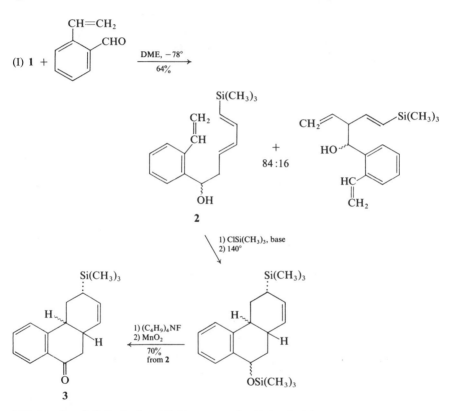

[1] W. Oppolzer, S. S. Burford, and F. Marazza, *Helv.*, **63**, 555 (1980).
[2] H. Yasuda, M. Yamauchi, and A. Nakamura, *J. Organometal. Chem.*, 3202, C1 (1980).

1-Trimethylsilyl polyphosphate (PPSE).[1] The reagent is prepared from P_2O_5 and $[(CH_3)_3Si]_2O$. It is a colorless, volatile liquid, soluble in the usual organic solvents. It is comparable to polyphosphate ester for the Beckmann rearrangement of oximes to amides (**3**, 230–231), but it is prepared more easily.

[1] T. Imamoto, H. Yokoyama, and M. Yokoyama, *Tetrahedron Letters*, **22**, 1803 (1981).

Trimethylsilyl trifluoromethanesulfonate, 8, 514–515; **9**, 497–498. Supplier: Fluka.

Hydrolysis of t-butyl esters.[1] The reaction can be conducted under nonacidic conditions in two steps: conversion to the trimethylsilyl ester with trimethylsilyl triflate and triethylamine followed by treatment with water (equation I).

$$(I)\ RCOOC(CH_3)_3 \xrightarrow[\substack{80-90\%}]{\substack{(CH_3)_3SiOTf, \\ N(C_2H_5)_3}} RCOOSi(CH_3)_3 \xrightarrow[\substack{80-95\% \\ overall}]{H_2O} RCOOH$$

Stereoselective aldol-type condensation.[2] Enol silyl ethers do not undergo aldol condensation with aldehydes or ketones in the presence of this triflate, but the reaction occurs at −78° (4–12 hours) with the corresponding acetals or ketals (and certain orthoesters). Moreover the *erythro*-aldol is formed with high stereoselectivity. Examples:

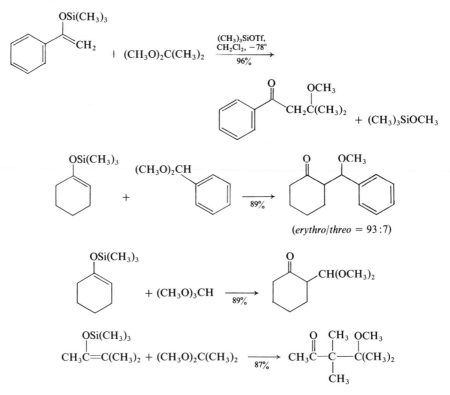

(*erythro/threo* = 93:7)

α-Alkoxyalkyl-α,β-enones.[3] The aldol-type condensation of silyl enol ethers with acetals coupled with the conjugate addition of phenyl trimethylsilyl selenide (**9**, 373) to α,β-enones provides a one-pot procedure for α-alkoxyalkylation of α,β-enones, as illustrated for the preparation of 2-diethoxymethyl-2-cyclohexenone (equation I).

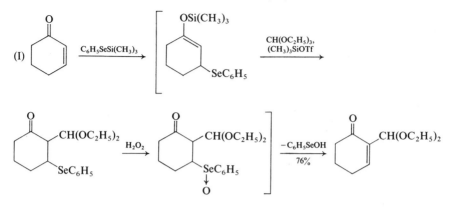

The products are useful substrates for conjugate addition of organocopper reagents (see this volume).

Acetalization.[4] Aldehydes and ketones are converted into acetals and ketals, respectively, by reaction with alkoxytrimethylsilanes in the presence of trimethylsilyl triflate as catalyst.

Examples:

Homoallyl ethers.[5] Trimethylsilyl triflate catalyzes a reaction between dimethyl ketals and allyltrimethylsilane to form homoallyl ethers. Allylation is not possible with the parent ketones.

Examples:

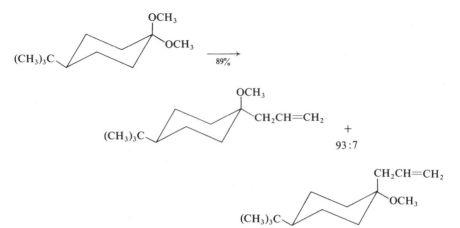

1,3-Bis(trimethylsilyloxy)-1,3-dienes.[6] The reaction of 1,3-dicarbonyl compounds with 2 equivalents of trimethylsilyl triflate and triethylamine results in 1,3-bis(trimethylsilyloxy)-1,3-dienes in about 85–95% yield (crude).

Examples:

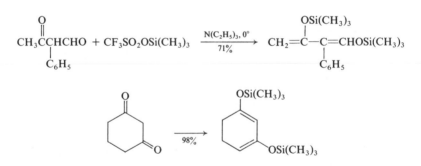

α-Alkoxymethyl ketones.[7] The condensation of enol silyl ethers and dialkoxymethanes to give α-alkoxymethyl ketones can be effected by use of catalytic amounts of trimethylsilyl triflate and a hindered base such as 2,6-di-*t*-butylpyridine.

Examples:

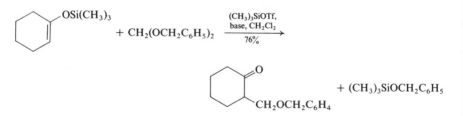

$$\underset{C_6H_5C=CH_2}{\overset{OSi(CH_3)_3}{|}} + CH_2(OCH_2C_6H_5)_2 \xrightarrow[92\%]{} \underset{C_6H_5CCH_2CH_2OCH_2C_6H_5}{\overset{O}{\overset{\|}{}}}$$

Simplified nucleoside synthesis.[8] The known synthesis of nucleosides from silylated heterocycles and a protected sugar derivative in the presence of $(CH_3)_3SiClO_4$ or $(CH_3)_3SiOTf$ **(6**, 639–640) has been adapted to a one-pot synthesis based on *in situ* silylation and Lewis acid catalysis. The reagent **(1)** is prepared *in situ* (equation I) and is added to the free base and acylated sugar; then triflic acid, potassium nonaflate, or $SnCl_4$ is added as catalyst. The last Lewis acid is the most active and allows condensation to proceed at 24°. Acetonitrile is the most useful solvent. The method is generally applicable and yields are about the same as those obtained in the two-step procedure.

(I) $3CF_3SO_3H + (CH_3)_3SiNHSi(CH_3)_3 + ClSi(CH_3)_3 \xrightarrow[-NH_4Cl]{} 3(CH_3)_3SiOSO_3CF_3$

                                                             1

[1] J. Borgulya and K. Bernauer, *Synthesis*, 545 (1980).
[2] S. Murata, M. Suzuki, and R. Noyori, *Am. Soc.*, **102**, 3248 (1980).
[3] M. Suzuki, T. Kawagishi, and R. Noyori, *Tetrahedron Letters*, **22**, 1809 (1981).
[4] T. Tsunoda, M. Suzuki, and R. Noyori, *Tetrahedron Letters*, **21**, 1357 (1980).
[5] T. Tsunoda, M. Suzuki, and R. Noyori, *Tetrahedron Letters*, **21**, 71 (1980).
[6] K. Krägeloh and G. Simchen, *Synthesis*, 30 (1981).
[7] S. Murata, M. Suzuki, and R. Noyori, *Tetrahedron Letters*, **21**, 2527 (1980).
[8] H. Vorbrüggen and B. Bennua, *Ber.*, **114**, 1279 (1981).

Trimethylsilylvinylketene (1). Mol. wt. 140.26, yellow-green liquid, stable for 1–2 weeks at 0°.
    Preparation:

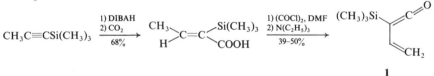

Diels-Alder reactions.[1] This ketene does not undergo [2 + 2]cycloadditions, but does undergo regiospecific Diels-Alder reactions with moderately reactive dienophiles.
    Examples:

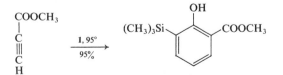

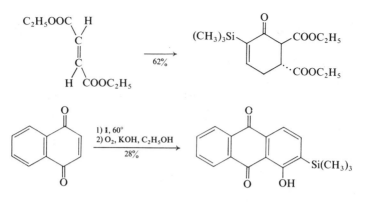

[1] R. L. Danheiser and H. Sard, *J. Org.*, **45**, 4810 (1980).

α-Trimethylsilylvinyllithium, $H_2C=C\overset{Si(CH_3)_3}{\underset{Li}{}}$ **(1).** This vinyllithium is prepared[1] by reaction of α-bromovinyltrimethylsilane[2] with *n*-butyllithium.

Ethylidenation.[1] The ketone (2) undergoes Wittig reaction with ethylidene-triphenylphosphorane in yields of only about 40%, with recovery of 2. The difficulty may be enolization promoted by the basic ylide. Use of the Grignard reagent from α-chloroethyltrimethylsilane results only in reduction of the carbonyl group. The problem is solved by use of 1, which reacts with 2 to give the internal ketal 3. Hydro-

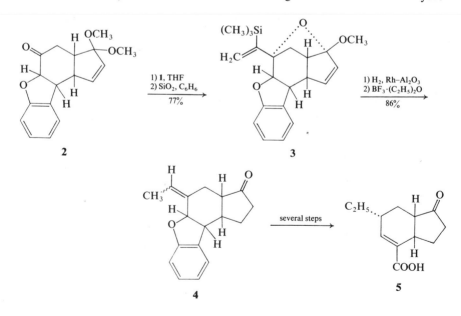

genation of **3** results in a tetrahydro derivative that can be desilylated by BF_3 etherate to a mixture of olefins (**4**). This product was converted into the target (**5**) of this investigation in several steps. This compound (**5**) is known as coronafacic acid, produced by a phtyopathogenic bacterium (*Pseudomonas coronafacience*).

[1] M. E. Jung and J. P. Hudspeth, *Am. Soc.*, **102**, 2463 (1980).
[2] R. K. Boeckman, *Am. Soc.*, **96**, 6179 (1974); G. Stork and J. Singh, *Am. Soc.*, **96**, 6181 (1974).

(E)-β-(Trimethylsilyl)vinyllithium (1).

Preparation[1]:

$(CH_3)_3SiC{\equiv}CH \xrightarrow[98\%]{(n\text{-Bu})_3SnH}$ (CH₃)₃Si∖ $C{=}C$ ∕H ∖Sn(n-Bu)₃ over H $\xrightarrow{n\text{-BuLi}}$ (CH₃)₃Si∖ $C{=}C$ ∕H ∖Li over H

1

Spiroannelation.[2] A new route to spiro[4.5]decadienones via this vinylsilane is formulated in scheme (I).

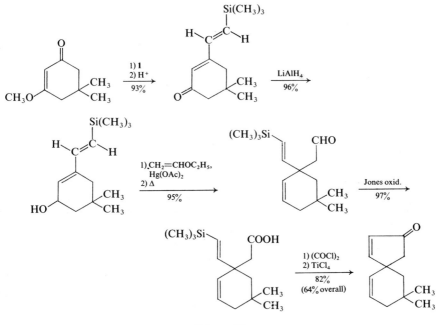

Scheme (I)

[1] R. F. Cunico and F. J. Clayton, *J. Og.*, **41**, 1480 (1976).
[2] S. D. Burke, C. W. Murtiashaw, M. S. Dike, S. M. S. Strickland, and J. O. Saunders, *J. Org.*, **46**, 2400 (1981).

Trimethyltrityloxysilane, $(C_6H_5)_3COSi(CH_3)_3$ **(1).** Mol. wt. 332.53, m.p. 35–36°. The silane is prepared from triphenylmethoxide and chlorotrimethylsilane (THF, 80% yield)[1]

O-Tritylation.[2] A trimethylsilylated alcohol, phenol, or carboxylic acid can be converted into the corresponding trityl derivative by reaction with **1** catalyzed by trimethylsilyl triflate (equation I). *t*-Butoxytrimethylsilane can be used, but it is less effective.

$$\text{(I)} \quad ROSi(CH_3)_3 + \mathbf{1} \xrightarrow[\text{75–95\%}]{(CH_3)_3SiOTf} ROC(C_6H_5)_3 + [(CH_3)_3Si]_2O$$

[1] K. Uhle and U. Werner, *Z. Chem.*, **13**, 224 (1973).
[2] S. Murata and R. Noyori, *Tetrahedron Letters*, **22**, 2107 (1981).

Trimethylvinylsilane, $CH_2\!=\!CHSi(CH_3)_3$ **(1),** 9, 498–499.

Aryl vinyl sulfides and sulfoxides.[1] Arylsulfenyl chlorides add to trimethylvinyl-silane in CH_2Cl_2 at $-78 \to 20°$ to form 2-chloro-1-(trimethylsilyl)ethyl aryl sulfides in 90–95% yield. These adducts are converted into aryl vinyl sulfides in high yield by $KF\cdot2H_2O$–DMSO at 70–100°. The sulfides, as expected, can be oxidized to the corresponding sulfoxides by *m*-chloroperbenzoic acid in CH_2Cl_2 at 20°.

Example:

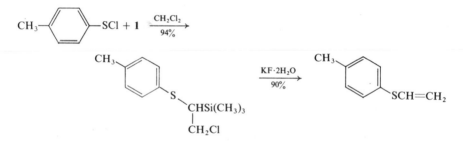

Attempted elimination of the elements of hydrogen chloride from the adducts is generally not successful, but rather results in intractable tars. Thus **1** can function as an equivalent of ethylene.

Bicyclopentenones.[1] Cyclic α,β-unsaturated acid chlorides in the presence of $SnCl_4$ react with **1** in CH_2Cl_2 at $-70 \to 20°$ to form bicyclopentenones in yields of $\sim 50\%$.

Example:

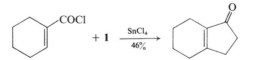

[1] F. Cooke, R. Moerck, J. Schwindeman, and P. Magnus, *J. Org.*, **45**, 1046 (1980).

Trioxo(*t*-butylimido)osmium(VIII), 6, 641–642.

Oxyamination.[1] The ratio of amino alcohol to diol formed by reaction of alkenes with the reagent is considerably improved by the presence of tertiary alkyl bridgehead amines. Of these ligands, quinuclidine (**1**, 976; **4**, 417) is the most efficient. In this case DME is used in place of pyridine as solvent.

[1] S. G. Hentges and K. B. Sharpless, *J. Org.*, **45**, 2257 (1980).

Triphenylarsonium ethylide (1). Mol. wt. 334.30, decomposes above $-20°$. Since triphenylarsine is moderately toxic, the reagent should be handled in a hood.

Preparation:

$$(C_6H_5)_3As + (C_2H_5)_3\overset{+}{O}BF_4{}^-\longrightarrow$$

$$(C_6H_5)_3\overset{+}{A}sCH_2CH_3BF_4{}^- \xrightarrow[\text{THF, HMPT, } -40°]{\text{KN[Si(CH}_3)_3]_2,} (C_6H_5)_3As{=}CHCH_3$$

$$\mathbf{1}$$

trans-Epoxides.[1] This unstable ylide (**1**), when generated as formulated above, reacts with an aliphatic aldehyde at $-78°$ to give a *trans*-epoxide with almost complete stereoselectivity. The stereochemical selectivity is markedly dependent on the base and also on the counterion of the arsonium salt. Optimum selectivity for the *trans*-epoxide is obtained with conditions similar to those that induce *cis*-olefination in Wittig reactions.[2] Stereoselection is not so high with aromatic aldehydes. The reagent also reacts with ketones to form trisubstituted epoxides.

Examples:

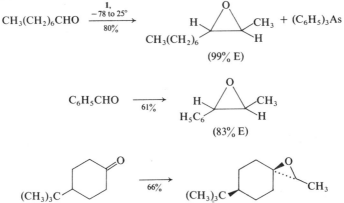

Stereoselectivity of $\sim 50{:}1$ is observed with most aldehydes and with various arsonium ylides. Although the method used to prepare **1** is not general, the paper describes general routes to primary and to α-branched ylides.

[1] W. C. Still and V. T. Novack, *Am. Soc.*, **103**, 1283 (1981).
[2] C. Sreekumar, K. D. Darst, and W. C. Still, *J. Org.*, **45**, 4260 (1980).

Triphenylbismuth carbonate, $(C_6H_5)_3BiCO_3$, **9,** 501.

Arylation. Although some oxidations with triphenylbismuth carbonate occur with quantitative conversion to $(C_6H_5)_3Bi$, some proceed with Bi–C cleavage. Although attempts to trap aryl radicals have failed, this cleavage can be used in a synthesis for transfer of an aryl group from Bi to N.[1]

Examples:

$$C_6H_5NHOH \xrightarrow{(C_6H_5)_3BiCO_3} C_6H_5NO + (C_6H_5)_2N\!-\!O^{\boldsymbol{\cdot}}$$
$$\qquad\qquad\qquad\qquad (22\%) \qquad\qquad (64\%)$$

$$CH_3COCH_2COOC_2H_5 \xrightarrow[59\%]{} CH_3COCHCOOC_2H_5$$
$$\qquad\qquad\qquad\qquad\qquad\qquad\qquad\overset{|}{C_6H_5}$$

[1] D. H. R. Barton, D. J. Lester, W. B. Motherwell, and M. T. B. Papoula, *J.C.S. Chem. Comm.*, 246 (1980).

Triphenylmethylphosphonium permanganate, $(C_6H_5)_3\overset{+}{P}CH_3MnO_4{}^-$. Mol. wt. 396.26, violet powder, stable in dark for several weeks, explodes above 70°. The material is prepared (86% yield) by reaction of triphenylmethylphosphonium bromide with $KMnO_4$ in water.[1]

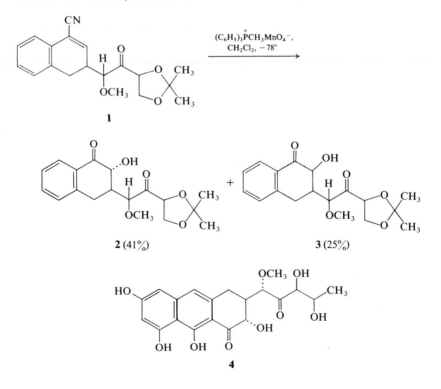

cis-*Dihydroxylation.* The reagent converts alkenes into *cis*-diols at $-70°$ (1–12 hours). CH_2Cl_2 is a suitable solvent. Yields are 40–80%, and are generally higher than those obtained with $KMnO_4$ in H_2O or with $KMnO_4$ under phase-transfer or crown ether catalysis.[1]

The method was used for oxidation of an unsaturated nitrile (1) to an aromatic acyloin (2) in a projected synthesis of a model for olivin (4), an aglycone of olivomycin antibiotics.[2]

[1] W. Reischl and E. Zbiral, *Tetrahedron*, **35**, 1109 (1979).
[2] R. W. Franck and T. V. John, *J. Org.*, **45**, 1170 (1980).

Triphenylphosphine–Carbon tetrachloride, 1, 1247; **2**, 445; **3**, 320; **4**, 551–552; **5**, 727; **6**, 644–645; **7**, 404; **8**, 516; **9**, 503.

β-Lactams.[1] A biomimetic synthesis of *β*-lactams from chiral amino acids such as L-serine has been developed by Mattingly and co-workers. The protected amino acid (1) is first converted into the O-alkyl or O-acyl hydroxamate (2), which undergoes cyclization to derivatives of 1-hydroxy-2-azetidinones on treatment with triphenylphosphine–carbon tetrachloride. This cyclization is also possible with triphenylphosphine–diethyl azodicarboxylate.[2] The final step involves reduction of the N—OH group with $TiCl_3$.[3] The advantage of this method over that of Wasserman (**9**, 428), which involves cyclization of *β*-haloamides, is that a strong base such as NaH is not required.

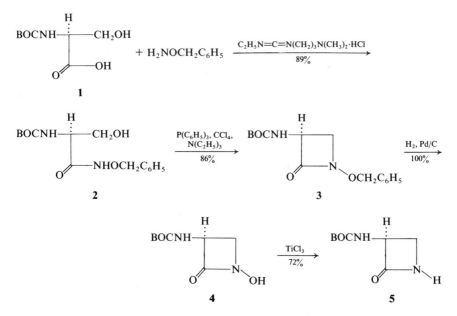

Arylamides of *α*-phthaloylamino-*β*-hydroxy acids (6) also cyclize to *β*-lactams (7) when treated with $P(C_6H_5)_3$–DEAD.[4] The triphenylphosphine can be replaced by hexamethylphosphorus triamide, but not by trialkylphosphines. Use of either CbO or BOC derivatives results in cyclization to aziridines.

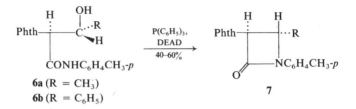

6a (R = CH$_3$)

6b (R = C$_6$H$_5$)

[1] M. J. Miller, P. G. Mattingly, M. A. Morrison, and J. F. Kerwin, Jr., *Am. Soc.*, **102**, 7026 (1980); *Org. Syn.*, submitted (1980).

[2] P. G. Mattingly, J. F. Kerwin, Jr., and M. J. Miller, *Am. Soc.*, **101**, 3983 (1979).

[3] P. G. Mattingly and M. J. Miller, *J. Org.*, **45**, 410 (1980).

[4] A. K. Bose, D. P. Sahu, and M. S. Manhas, *J. Org.*, **46**, 1229 (1981).

Triphenylphosphine–Dibromodifluoromethane.

gem-*Difluoroalkenes.*[1] The ylide difluoromethylenetriphenylphosphorane can be generated *in situ* from CBr$_2$F$_2$ and P(C$_6$H$_5$)$_3$ and can be used for Wittig reactions with aldehydes. Addition of zinc dust (1 equivalent) markedly improves the yields. The ylide prepared from CBr$_2$F$_2$ and P[N(CH$_3$)$_2$]$_3$ is somewhat more reactive and reacts with both aldehydes and unactivated ketones. The resulting 1,1-difluoroalkenes are reduced selectively to monofluoroalkenes with sodium bis(2-methoxyethoxy)aluminum hydride.

The reagent was used to prepare the α-(difluoromethylene)-γ-lactone (**1**). Grignard and alkyllithium reagents react with **1** with replacement of one or both fluorines.[2]

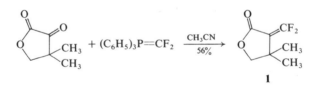

1

[1] S. Hayashi, T. Nakai, N. Ishikawa, D. I. Burton, D. G. Naae, and H. S. Keshing, *Chem. Letters*, 983 (1979).

[2] M. Suda, *Tetrahedron Letters*, **22**, 1421 (1981).

Triphenylphosphine–Diethyl azodicarboxylate (DEAD), 1, 245–247; 4, 553–555; 5 727–728; 6, 645; 7, 404–406; 8, 517; 9, 504–506.

Stereochemical inversion of a δ-lactone.[1] The δ-lactone **2**, obtained by standard lactonization of **1**, was inverted to **4** by hydrolysis to the hydroxy acid **3** followed by lactonization with triphenylphosphine and diethyl azodicarboxylate. This method evidently involves activation of the hydroxyl group followed by an intramolecular S$_N$2 attack by the carboxyl group. Two other reagents can be used for this purpose:

$P(C_6H_5)_3$ combined with $O=C(COOC_2H_5)_2$ and $O=P(C_6H_5)_3$ combined with $(CF_3SO_2)_2O$.

A similar inversion of a large-ring lactone has been reported.[2]

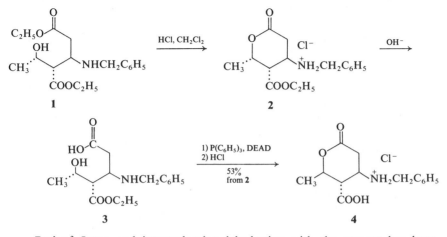

1   **2**

3   **4**

Review.[3] Inter- and intramolecular dehydration with the reagent has been reviewed with particular emphasis on applications to synthesis and transformations of natural products (100 references).

[1] D. G. Melillo, T. Liu, K. Ryan, M. Sletzinger, and I. Shinkai, *Tetrahedron Letters,* **22**, 913 (1981).
[2] B. Seuring and D. Seebach, *Ann.,* 2044 (1978).
[3] O. Mitsunobu, *Synthesis,* 1 (1981).

Triphenylphosphine–2,2′-Dipyridyl disulfide, 5, 285–286; 6, 246–247; 7, 141–142.

β-Lactams.[1] *β*-Amino acids are cyclized to *β*-lactams in good to excellent yield by this reagent when acetonitrile is used as solvent. Even moist acetonitrile can be used for water-soluble *β*-amino acids. High dilution and higher temperatures (reflux) improve yields.

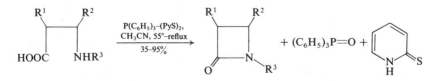

[1] S. Kobayashi, T. Iimori, T. Izawa, and M. Ohno, *Am. Soc.,* **103**, 2406 (1981).

Triphenylphosphine–Hexachloroacetone, 8, 516.

Allylic chlorides (8, 516). The use of these two reagents for conversion of allylic alcohols to the largely unrearranged allylic chloride[1] has been improved by use of only 0.5–1 equivalent of hexachloroacetone and of sulfolane as solvent. With these improvements reasonable yields of the unrearranged allylic chloride can be obtained from tertiary allylic alcohols (equation I). The method is applicable to saturated alcohols.[2]

$$(I) \quad \underset{\underset{OH}{|}}{H_2C=CHC(CH_3)_2} \xrightarrow[\text{CCl}_3\text{COCCl}_3]{P(C_6H_5)_3,} \underset{\underset{Cl}{|}}{H_2C=CHC(CH_3)_2} + ClCH_2CH=C(CH_3)_2$$

$$\qquad\qquad\qquad\qquad\qquad\qquad\qquad\quad (37\%) \qquad\qquad (22\%)$$

[1] R. M. Magid, O. S. Fruchey, W. L. Johnson, and T. G. Allen, *J. Org.*, **44**, 359 (1979).
[2] R. M. Magid, B. G. Talley, and S. K. Souther, *J. Org.*, **46**, 824 (1981).

Triphenylphosphine–Iodine.

Alkyl iodides. Although this mixture reduces arsenesulfonic acids, it converts aliphatic sulfonic acids,[1] sulfinic acids, thiols, sulfonates, and disulfides to alkyl iodides in 70–100% yield.[2]

[1] K. Fujimoto, H. Togo, and S. Oae, *Tetrahedron Letters*, **21**, 4921 (1980).
[2] S. Oae and H. Togo, *Synthesis*, 371 (1981).

Triphenylphosphine–Iodoform–Imidazole.

Alkenes from vic-diols (*cf.* **9**, 507–508). Hexapyranoside *vic*-diols are converted into alkenes by reaction with this combination of reagents.[1] The method is applicable to both *cis*- and *trans*-diols, but highest yields of alkenes are usually obtained from diequatorial *trans*-diols. The reaction is believed to involve formation of a *vic*-diiodide derivative (inversion), which then undergoes reductive elimination with imidazole.

[1] M. Bessodes, E. Abushanab, and R. P. Panzica, *J.C.S. Chem. Comm.*, 26 (1981).

Triphenylphosphine hydroiodide–Triphenylphosphine diiodide.

Deoxygenation of epoxides.[1] Epoxides are converted to alkenes with retention of configuration by treatment with this combination in hexane at $0 \rightarrow 25°$. The reaction involves *anti*-opening of the epoxide to an iodohydrin, which undergoes *trans*-elimination to an alkene.

[1] P. E. Sonnet, *Synthesis*, 828 (1980).

(Triphenylphosphorylidene)ketene, $(C_6H_5)_3P=C=C=O$ (1). Mol. wt. 302.29, m.p. 172–173.5°.

Preparation[1]:

$$(C_6H_5)_3P=C=P(C_6H_5)_3 + CO_2 \xrightarrow[96\%]{} \underset{\underset{O^{\cdots}\overset{|}{C}{\cdots}O}{}}{\overset{\overset{+}{(C_6H_5)_3P}\overset{}{C}P(C_6H_5)_3}{}} \xrightarrow{\Delta} 1$$

α,β-Unsaturated γ-lactones. The reagent reacts with benzoin to form **2** in 58% yield.[2] It can also be used for preparation of cardenolides.[3]

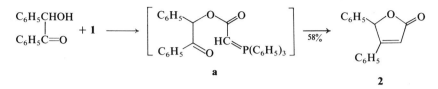

a

2

[1] H. J. Bestmann, *Angew. Chem. Int. Ed.*, **16**, 349 (1977).
[2] H. J. Bestmann, *Angew. Chem. Int. Ed.* **15**, 115 (1976).
[3] K. Nickisch, W. Klose, and F. Bohlmann, *Ber.*, **113**, 2038 (1980).

Triphenyltin chloride, $(C_6H_5)_3SnCl$. Mol. wt. 385.46, m.p. 106°.

erythro-Selective aldol condensation.[1] Triphenyltin enolates, prepared by reaction of lithium enolates with triphenyltin chloride, react with aldehydes to form mainly *eythro*-products, regardless of the geometry of the enolate. Assistance with a Lewis acid is not necessary.

Example:

$$CH_3CH=C\underset{C_2H_5}{\overset{OSn(C_6H_5)_3}{<}} + C_6H_5CHO \xrightarrow{80\%} C_6H_5\overset{OH}{\underset{CH_3}{\overset{|}{C}H}}CH\overset{O}{\overset{||}{C}}C_2H_5$$

(E/Z = 86 : 14)   (*erythro/threo* = 82 : 18)
(E/Z = 8 : 92)   (*erythro/threo* = 74 : 26)

The corresponding tri-*n*-butyltin and trimethyltin enolates react with benzaldehyde to form about equal amounts of *erythro*- and *threo*-aldols.

The same laboratory has reported that both (E)- and (Z)-crotyltributyltin undergo *erythro*-selective coupling to aldehydes (equation I). A Lewis acid is required.[2]

$$\text{(I)} \quad CH_3CH=CHCH_2Sn(Bu)_3 + C_6H_5CHO \xrightarrow{BF_3}{90\%} C_6H_5\overset{OH}{\underset{CH_3}{\overset{|}{C}H}}CHCH=CH_2$$

(E) or (Z)

(*erythro/threo* = 95–98 : 5–2)

[1] Y. Yamamoto, H. Yatagai, and K. Maruyama, *J.C.S. Chem. Comm.*, 162 (1981).
[2] Y. Yamamoto, H. Yatagai, Y. Naruta, and K. Maruyama, *Am. Soc.*, **102**, 7107 (1980).

Triphenyltin hydride, 1, 1250–1251; **3,** 448; **4,** 559; **5,** 734; **6,** 649; **8,** 521–522.

Reduction of selenides (**8,** 521–522). More than 20 examples of use of this hydride for reduction of selenides and selenoacetals have been published. The paper also includes similar reduction of tellurides.[1]

The paper reports a selective reduction of an epoxide in the presence of a ketone group (equation I).

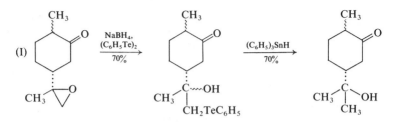

¹ D. L. J. Clive, G. J. Chittattu, V. Farina, W. A. Kiel, S. M. Menchen, C. G. Russell, A. Singh, C. K. Wong, and N. J. Curtis, *Am. Soc.*, **102**, 4438 (1980).

Tris(*p*-bromophenyl)aminium hexachlorostibnate, $(p\text{-}BrC_6H_4)_3\overset{+}{N}\cdot SbCl_6^-$ (**1**). Mol. wt. 816.48. Preparation.¹

*Diels-Alder catalysis.*² This cation radical enhances the reactivity of a neutral or electron-rich *cis*-1,3-diene in Diels-Alder reactions. Thus 1,3-cyclohexadiene undergoes Diels-Alder dimerization only at temperatures around 200°. The presence of 5–10 mole % of this salt effects dimerization even at −78°, with the usual *endo*/ *exo* selectivity (5:1). It also permits facile condensation of 1,3-cyclohexadiene with a hindered dienophile such as 2,5-dimethyl-2,4-hexadiene (equation I); the dimer of the former diene is a minor product (20% yield).

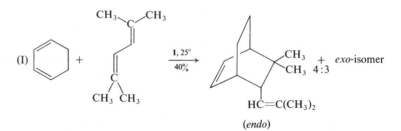

¹ F. A. Bell, A. Ledwith, and D. C. Sherrington, *J. Chem. Soc. C*, 2719 (1969).
² D. J. Bellville, D. D. Wirth, and N. L. Bauld, *Am. Soc.*, **103**, 718 (1981).

Tris(diethylamino)sulfonium   difluorotrimethylsiliconate, $[(C_2H_5)_2N]_3\overset{+}{S}(CH_3)_3\overset{-}{Si}F_2$ (**1**). Mol. wt. 359.64. Preparation.¹

Super anions. Unlike quaternary ammonium fluorides, **1** can be obtained anhydrous. It reacts with enol silyl ethers to form an unsolvated tris(diethylamino)-sulfonium (TAS) enolate in which there is negligible interaction between the ions.² The "naked" enolate undergoes C-alkylation at −78 to −30°.³

Examples:

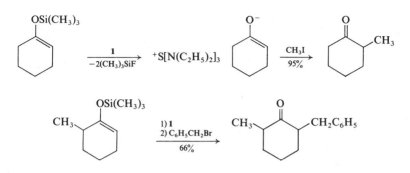

Aldol condensation.[4] The aldol condensation of silyl enol ethers with an aldehyde in the presence of **1** (0.1–5 equivalents) results mainly or even exclusively in *erythro*-adducts (equations I and II) regardless of the stereochemistry of the enolate.

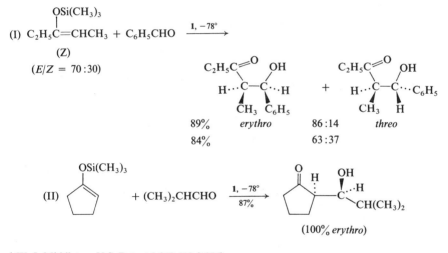

[1] W. J. Middleton, U.S. Patent 3,940,402 (1976).
[2] R. Noyori, I. Nishida, J. Sakata, and M. Nishizawa, *Am. Soc.*, **102**, 1223 (1980).
[3] R. Noyori, I. Nishida, and J. Sakata, *Tetrahedron Letters*, **21**, 2085 (1980).
[4] R. Noyori, I. Nishida, and J. Sakata, *Am. Soc.*, **103**, 2106 (1981).

Tris(methylthio)methyllithium, 7, 412.

Ring expansion of cyclic ketones.[1] A new method for expansion of cyclic ketones to the expanded α-dithioketals involves addition of $(CH_3S)_3CLi$ followed by treatment of the adduct with tetrakis(acetonitrile)copper(I) perchlorate or tetrafluoroborate.[2]

Example:

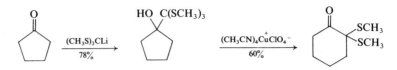

[1] S. Knapp, A. F. Trope, and R. M. Ornaf, *Tetrahedron Letters*, **21**, 4301 (1980).
[2] M. Kubota and D. L. Johnson, *J. Inorg. Nucl. Chem.*, **29**, 769 (1967).

Tris(phenylseleno)borane, 9, 511.

Cleavage of epoxides.[1] Tris(phenylseleno)borane and tris(methylseleno)borane[2] react rapidly with terminal and α,β-disubstituted epoxides to give β-hydroxy selenides, or in some cases alkenes. In the latter case, there are marked differences between the *cis-* and *trans-*epoxides; the latter react very slowly to give mainly *trans-*alkenes.

Examples:

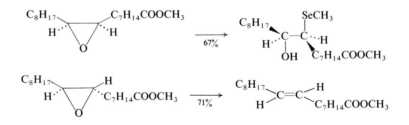

Trisubstituted epoxides react to give allylic alcohols in which the hydroxyl group is linked to the less substituted carbon.

Example:

$$C_6H_{13}CH-C \overset{C_2H_5}{\underset{C_2H_5}{\diagdown}} \xrightarrow[76\%]{B(SeC_6H_5)_3} C_6H_{13}\underset{\underset{OH}{|}}{CH}-\underset{\underset{C_2H_5}{|}}{C}=CHCH_3$$

[1] A. Cravador and A. Krief, *Tetrahedron Letters*, **22**, 2491 (1981).
[2] Preparation: W. Siebert, W. Ruf, and R. Full, *Z. Naturforsch.*, **30b**, 642 (1975).

Tris(tetra-*n*-butylammonium) hydrogen pyrophosphate, $[(n\text{-}C_4H_9)_4N]_3HP_2O_7$ **(1)**. Mol. wt. 417.42. The salt is prepared by treatment of $Na_2H_2P_2O_7$ (Stauffer) with an acidic ion-exchange resin and then with tetra-*n*-butylammonium hydroxide.

Allylic pyrophosphates.[1] Primary allylic bromides (prepared by reaction of the alcohol with PBr_3) can be converted into the corresponding pyrophosphate esters by reaction with this reagent in 46–54% overall yield (equation I).

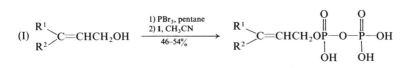

[1] V. M. Dixit, F. M. Laskovics, W. I. Noall, and C. D. Poulter, *J. Org.*, **46**, 1969 (1981).

Trityl tetrafluoroborate, 1, 1256–1258; **2,** 454; **4,** 565–567; **6,** 657; **7,** 414–415; **8,** 524–525; **9,** 512.

Alkene synthesis.[1] A regio- and stereoselective alkene synthesis is formulated in equations (I)–(III). The first step involves formation of the alkyliron complex **1**. Treatment of **1** with trityl tetrafluoroborate abstracts a β-hydrogen to give a cationic iron complex **2**. Liberation of the free alkene is effected by NaI in acetone. This sequence is useful because 1- and 2-alkyliron complexes (**1**) are converted into 1-alkenes exclusively; 3-alkyliron complexes are converted exclusively into the less stable (Z)-2-alkenes. The paper includes some 20 examples of alkenes prepared in this way.

$$(I)\ [Fe(CO)_2Cp]_2 \xrightarrow[\text{THF}]{\text{Na/Hg,}} [Fe(CO)_2Cp]^-Na^+ \xrightarrow{RX} [Fe(CO)_2Cp]R + NaX$$

$$\underset{\mathbf{1}}{}$$

$$(II)\ CpFe(CO)_2CHR^1CH_2R^2 + (C_6H_5)_3C^+BF_4^- \longrightarrow$$

$$\underset{\mathbf{1}}{}$$

$$CpFe(CO)_2\,(alkene)BF_4 + (C_6H_5)_3CH$$

$$\underset{\mathbf{2}}{}$$

$$(III)\ \mathbf{2} \xrightarrow{NaI,\ (CH_3)_2C=O} alkene$$

$$\underset{\mathbf{3}}{}$$

[1] D. E. Laycock, J. Hartgerink, and M. C. Baird, *J. Org.*, **45**, 291 (1980).

Tungsten(VI) chloride–Tetramethyltin, $WCl_6\text{–}Sn(CH_3)_4$.

Long-chain alkyl iodides.[1] Even-numbered alkyl compounds of chain lengths up to 42 carbon atoms can be prepared by metathesis of 1-alkenes with WCl_6–$Sn(CH_3)_4$, originally used for metathesis of unsaturated fatty esters to alkenes and unsaturated diesters.[2] Hydrozirconation of the internal olefin obtained in this way gives the terminal zirconium compound, which can then be converted to various alkyl derivatives (**6**, 175–177).

Example:

$$C_{11}H_{23}CH{=}CH_2 \xrightarrow[50\%]{\substack{WCl_6,\ Sn(CH_3)_4, \\ CH_3COOC_2H_5}} C_{11}H_{23}CH{=}CHC_{11}H_{23} \xrightarrow{Cp_2ZrHCl}$$

$$[C_{22}H_{45}CH_2CH_2ZrCp_2Cl] \xrightarrow[75\%]{I_2} C_{22}H_{45}CH_2CH_2I$$

[1] T. Gibson and L. Tulich, *J. Org.*, **46**, 1821 (1981).

[2] P. B. Van Dam, M. C. Mittlemeijer, and C. Boelhouwer, *J.C.S. Chem. Comm.*, 1221 (1972).

V

Vanadium(II) chloride. 7, 418.

Reductive cleavage of oximes.[1] Vanadium(II) chloride in THF is a convenient reagent for deoximation (75–90% yield).

[1] G. A. Olah, M. Arvanaghi, and G. K. Surya, *Synthesis*, 220 (1980).

Vanadyl acetylacetonate–Azobisisobutyronitrile, $VO(C_5H_7O_2)_2$. Mol. wt. 265.16. Supplier: Alfa.

α,β-Epoxy alcohols. $VO(acac)_2$–AIBN (1:10) is an effective catalyst for oxygenation of cyclohexene to give mainly the *cis-α,β*-epoxy alcohol (equation I). The system is comparable to $CpV(CO)_4$ (**5,** 174). Other cyclic olefins react in the same way except for cyclooctene, which is converted only into the epoxide.

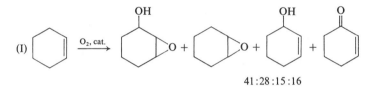

(I)

41:28:15:16

[1] K. Kaneda, K. Jitsukawa, T. Itoh, and S. Teranishi, *J. Org.*, **45,** 3005 (1980).

Vilsmeier reagent, 1, 284–298; **2,** 159; **3,** 116; **4,** 186; **5,** 251; **6,** 220; **7,** 422–424; **8,** 529–530; **9,** 514–515.

Nitriles from aldoximes.[1] The Vilsmeier reagent reacts with ketoximes to form the O-formate, but converts aldoximes into nitriles in excellent yields.

Examples:

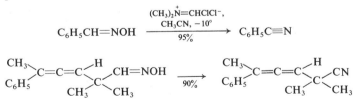

[1] J.-P. Dulcere, *Tetrahedron Letters*, **22,** 1599 (1981).

X

(S)-2-(2,6-Xylidinomethyl)pyrrolidine, 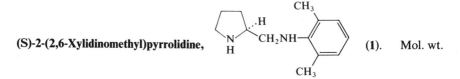 (1). Mol. wt.

218.29, b.p. 116–123°/0.7 mm, α_D + 19.1°. The reagent is prepared from (S)-proline and 2,6-dimethylaniline.

Asymmetric reduction of ketones.[1] Lithium aluminium hydride in conjunction with this chiral ligand reduces prochiral aromatic ketones to (S)-secondary alcohols in 90–95% optical yields. Optical yields are lower (10–40% ee) in the case of alkyl aryl ketones. It is superior to (S)-2-(anilinomethyl)pyrrolidine for this reduction. Evidently the two methyl groups enhance the enantioselectivity.

[1] M. Asami and T. Mukaiyama, *Heterocycles*, **12**, 499 (1979).

Z

Zinc, 1, 1276–1284; **2**, 459–462; **3**, 334–337; **4**, 574–577; **5**, 753–756; **6**, 672–675; **7**, 426–428; **8**, 532.

Protection of phenols.[1] Zinc, previously activated by treatment with HCl, is an effective catalyst for deacylation of aryl acetates in methanol in high yield. Aliphatic acetates are not affected.

[1] A. G. González, Z. D. Jorge, H. L. Dorta, and F. R. Luis, *Tetrahedron Letters*, **22**, 335 (1981).

Zinc–Copper couple, 1, 1292–1293; **5**, 758–760; **7**, 428–429; **8**, 533–534.

Amino lactones.[1] The Reformatsky reagent prepared from ethyl 2-bromopropionate with Zn–Cu (**8**, 533) adds to azirines to form two isomeric products, which are converted to 4-amino lactones when treated with aqueous hydrochloric acid or HF–Py (**5**, 538, **6**, 473).

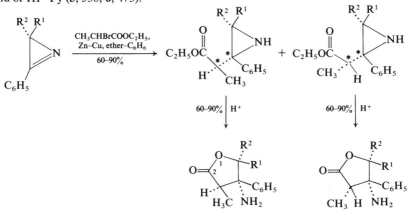

Reduction of alkynes to alkenes. The zinc–copper couple of Smith and Simmons (**1**, 1292) reduces internal alkynes to (Z)-alkenes exclusively in yields generally > 95%.[2] Terminal alkynes are reduced to 1-alkenes.[3]

Example:

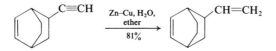

[1] G. Alvernhe, S. Lacombe, A. Laurent, and B. Marquet, *J. Chem. Res.* (*M*), 858 (1980).
[2] B. L. Sondengam, G. Charles, and T. M. Akam, *Tetrahedron Letters*, **21**, 1069 (1980).
[3] M. G. Veliev, M. M. Guseinov, and S. A. Mamedov, *Synthesis*, 400 (1981).

Zinc–Copper–Isopropyl iodide.

1,3-Cycloalkanediones.[1] In the presence of a zinc reagent prepared from Zn–Cu and this alkyl iodide, chloromethyl methyl ether reacts with 1,2-bis(trimethyl-silyloxy)cycloalkenes (1) to form 2-hydroxy-2-methoxymethylcycloalkanones (2) (*cf.* **8**, 415). Ring-enlarged 1,3-cycloalkanediones (3) are formed from (2) by treatment with acid.

Examples:

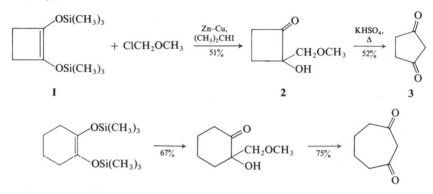

1   **2**   **3**

[1] I. Nishiguchi and T. Hirashima, T. Shono, and M. Sasaki, *Chem. Letters,* 551 (1981).

Zinc–Silver couple, 4, 436; 5, 760–761; 9, 519–520.

β,γ-Unsaturated ketones.[1] The reaction of allyl halides with nitriles in the presence of zinc to give allyl ketones has been reported, but yields were low. Use of zinc–silver couple has been found to improve yields to 70–80% in reactions with allyl bromide. The reaction usually proceeds better in benzene or THF than in ether. The reaction involves attack at the more substituted position.

Examples:

$$CH_3CN + CH_3CH{=}CHCH_2Br \xrightarrow[78\%]{Zn{-}Ag, C_6H_6} CH_3\overset{\overset{\displaystyle O}{\|}}{C}\underset{\underset{\displaystyle CH_3}{|}}{C}HCH{=}CH_2$$

$$C_2H_5CN + (CH_3)_2C{=}CHCH_2Br \xrightarrow[71\%]{} C_2H_5\overset{\overset{\displaystyle O}{\|}}{C}{-}\overset{\overset{\displaystyle CH_3}{|}}{\underset{\underset{\displaystyle CH_3}{|}}{C}}{-}CH{=}CH_2$$

[1] G. Rousseau and J. M. Conia, *Tetrahedron Letters,* **22**, 649 (1981).

Zinc borohydride, 3, 337–338; 5, 761–762.

Stereoselective reduction of β-keto esters. A few years ago Canceill and Jacques[1] noted that methyl 2-benzoylpropionate (1) is reduced by $LiAlH_4$ mainly to the

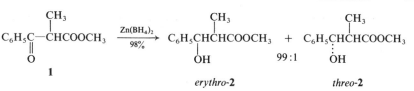

erythro-**2**   threo-**2**

erythro-alcohol (**2**) and by KBH$_4$ mainly to the *threo*-alcohol. Reduction with Zn(BH$_4$)$_2$ is even more selective than with LiAlH$_4$. Thus reduction of **1** with this hydride gives a 99:1 mixture of the *erythro/threo* hydroxy esters. The superiority of Zn(BH$_4$)$_2$ over LiAlH$_4$ is attributed to the greater ability of Zn^{2+} to coordinate with the carbonyl oxygen. This stereoselectivity is fairly general for β-keto esters.[2]

[1] J. Canceill, J.-J. Basselier, and J. Jacques, *Bull. Soc.*, 1024 (1967); 2180 (1970).
[2] T. Nakata and T. Oishi, *Tetrahedron Letters*, **21**, 1641 (1980).

Zinc bromide, 2, 463–464; **8**, 535; **9**, 520–522.

Detritylation.[1,2] Zinc bromide is effective for removal of 5′-trityl ethers of deoxynucleosides with only a slight effect on 3′-trityl ethers (*cf.* cleavage of MEM ethers, 7, 228). The reaction may involve chelation of ZnBr$_2$ with the ether oxygen and the deoxyribose ring oxygen.

[1] V. Kohli, H. Blöcker, and H. Köster, *Tetrahedron Letters*, **21**, 2683 (1980).
[2] M. D. Mattenci and M. H. Caruthers, *Tetrahedron Letters*, **21**, 3243 (1980).

Zinc chloride, 1, 1289–1292; **2**, 464; **3**, 338; **5**, 763–764; **6**, 676; **7**, 430; **8**, 536–537; **9**, 522–523.

Allenyl cations. In the presence of a zinc chloride–ether complex allenyl cations (**a**) can be generated from a propargyl halide such as (**1**). The cation can undergo either [3 + 2]- or [2 + 2]cycloaddition with an alkene. When the R group is aryl, [2 + 2]cycloaddition is the major pathway (**2**); when R is CH$_3$ or C$_2$H$_5$, [3 + 2]-cycloaddition becomes the major reaction (**3**) (equation I).[1]

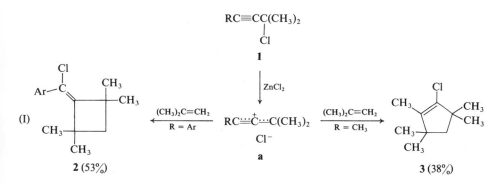

The reaction of **a** with cyclopentadiene also can follow two paths: [3 + 4]- or [2 + 4]cycloaddition. The former is seen when R is alkyl, the latter when R is aryl (equation II).[2]

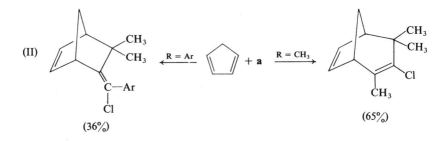

(II)

(36%)

(65%)

o-Allylphenols.[3] The reaction of potassium phenolates with allylic halides in xylene catalyzed by zinc chloride results in exclusive allylation at the *ortho*-position. γ-Substituted primary allylic halides react at the α-carbon, and α-substituted secondary allylic halides react at the γ-carbon.

Example:

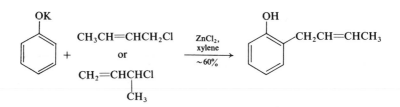

[1] H. Mayr, B. Seitz, and I. K. Halberstat-Kausch, *J. Org.*, **46**, 1041 (1981).
[2] H. Mayr and I. K. Halberstadt, *Angew. Chem. Int. Ed.*, **19**, 814 (1980).
[3] F. Bigi, G. Casiraghi, G. Casnati, and G. Sartori, *Synthesis*, 310 (1981).

Zinc iodide, 1, 1294.

Cleavage of epoxides with alkylthiotrialkylsilanes.[1] In the presence of ZnI_2 these thiosilanes ($R'SSiR_3$) convert epoxides into β-trimethylsilyloxy thioethers, which can be hydrolyzed to β-hydroxy thioethers. The regioselectivity can be reversed by use of *n*-butyllithium as the catalyst.

Examples:

$$
\underset{C_6H_5}{\triangle}^O + C_6H_5SSi(CH_3)_3 \xrightarrow[\text{2) HOAc}]{\text{1) Cat.}} \underset{65\%}{C_6H_5\overset{SC_6H_5}{CH}-CH_2OH} + \underset{10\%}{C_6H_5\overset{OH}{CHCH_2SC_6H_5}}
$$

ZnI₂   65%   10%
n-BuLi   18%   59%

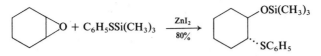

[1] Y. Guidon, R. N. Young, and R. Frenette, *Syn. Comm.*, **11**, 391 (1981).

INDEX OF REAGENTS
ACCORDING TO TYPES

CARBONYLATION: Dicarbonylbis(triphenylphosphine)nickel. Palladium(II) acetate. Palladium(II) chloride.

CHLORINATION: Iodobenzene dichloride.

CLAISEN REARRANGEMENT: Benzylamine. Bis(benzonitrile)dichloropalladium(II). Trimethyl β-(methoxy)orthopropionate.

CLEAVAGE OF:

ACETALS: Sodium hydride–Dimethyl sulfoxide.

ALKENYL SULFONES: Potassium–Graphite.

AMIDES: Di-*t*-butyl dicarbonate.

BENZYL ETHERS: Palladium catalysts.

t-BUTYLDIMETHYLSILYL ETHERS: Lithium tetrafluoroborate.

CYCLOBUTANES: Iodotrimethylsilane.

CYCLOPROPANES: Sodium benzeneselenoate.

CYCLOPROPYL KETONES: Iodotrimethylsilane.

ENOL ESTERS: Hydroxylamine.

ENOL ETHERS: Chlorotrimethylsilane–Sodium iodide. Dimethyl sulfoxide.

EPOXIDES: Alumina. *t*-Butyldimethyliodosilane. *n*-Butyllithium–Magnesium bromide. Cyclohexylisopropylaminomagnesium bromide. Dialkylaluminum amides. Iodotrimethylsilane. Lithium 1-α,α′-dimethyldibenzylamide. Nafion-H. Organoaluminum compounds. Pyridinium chloride. Raney nickel. Trifluoroacetyl chloride. Trimethylsilylacetonitrile. Tris(phenylseleno)borane. Zinc iodide.

ETHERS: Boron tribromide. Ceric ammonium nitrate. N-Chlorosuccinimide. Sodium N-methylanelide–Hexamethylphosphoric triamide. Thallium(III) nitrate. Trimethyl(phenylthio)silane.

ETHYLENE THIOKETALS: Mercury(II) oxide–Tetrafluoroboric acid. Periodic acid.

FURANES: Pyridinium chlorochromate.

GLYCOLS: N-Iodosuccinimide.

LACTONES: Sodium benzeneselenoate.

OXIMES: Hydrogen peroxide. Pyridinium chlorochromate–Hydrogen peroxide. Sodium dithionite. Vanadium(II) chloride.

STYRENES: Boron trifluoride–Ethanedithiol.

SULFONES: Potassium–Graphite.

THIOKETALS: Periodic acid. Pyridinium bromide perbromide. Trimethyloxonium tetrafluoroborate.

TRITYL ETHERS: Zinc bromide.

CONJUGATE ADDITION: Hexamethylphosphoric triamide.

COPE REARRANGEMENT: Bis(benzonitrile)dichloropalladium(II).

CYANATION: Diethyl phosphorocyanidate.

[1.3] CYCLOADDITION: Diazoacetaldehyde dimethyl acetal.

[2+2] CYCLOADDITION: Alumina chloride. 1,2-Bis(trimethylsilyloxy)cyclobutene. 2,2-Dimethyl-3(2*H*)-furanone. Ethylaluminum dichloride. Ketene diethyl acetal. Methyl cyclobutenecarboxylate. Trimethyl-1,3-dioxolenone. Zinc chloride.

[3+2] CYCLOADDITION: Phenyl isocyanate.

[3+4] CYCLOADDITION: Zinc chloride.

[2+2+2] CYCLOADDITIONS: Dicarbonylcyclopentadienylcobalt.

CYCLOALKYLATION: (Z)-1,4-Dichloro-2-butene.

CYCLOAROMATIZATION: (E)-1-Methoxy-1,3-bis(trimethylsilyl)-1,3-butadiene.

CYCLODEHYDRATION: Phosphorus(V) oxide. Polyphosphoric acid.

CYCLOPROPANATION: Dimethylsulfonium ethoxycarbonylmethylide. Molybdenum carbonyl. Rhodium(II) carboxylates. Tetra-μ^3-carbonyldodecacarbonylhexarhodium.

[4+2] CYCLOREVERSION: Tetra-*n*-butylammonium fluoride.

DEACETOXYLATION: Sodium α-(N,N-dimethylamino)naphthalenide.

DEACETYLATION: Boron trifluoride–Ethanedithiol.

DEALKYLATION, ESTERS: Aluminum chloride–Ethanediol.

DEAMINATION: Tri-*n*-butyltin hydride.

DEBENZYLATION: Boron trifluoride–Dimethyl sulfide.

DECARBOXYLATION: Iodosobenzene.

DECYANATION: Oxygen. Potassium–Alumina.

DEETHOXYCARBONYLATION: Magnesium chloride.

DEHALOGENATION: Chlorotrimethylsilane–Sodium iodide. Chromium(II) perchlorate–Butanethiol. Hypochlorous acid. Lithium triethylborohydride. Sodium iodide. Sodium methylselenolate.

DEHYDRATION: Copper(II) sulfate. Triphenylphosphine–Diethyl diazodicarboxylate.

DEHYDROBROMINATION: Silver fluoride–Pyridine.

DEHYDROCYANATION: Potassium *t*-butoxide.

DEHYDROGENATION: Dichlorodicyanobenzoquinone. Platinum catalysts. Potassium nitrosodisulfonate. Sulfuryl chloride.

DEHYDROHALOGENATION: Crown ethers.

DEMERCURATION: Sodium trithiocarbonate.

DEMETHYLATION: Lithium tri-*sec*-butylborohydride.

DEOXYGENATION OF:

ALCOHOLS: Chlorotrimethylsilane–Sodium iodide. Iron carbonyl. Phenyl chlorothionocarbonate. Potassium–18-Crown-6. Tri-*n*-butyltin hydride.

CARBONYL COMPOUNDS: Bis(benzoyloxy)borane. Bis(triphenylphosphine)copper(I) tetrahydroborate. Catecholborane.

ENONES: Catecholborane.

EPOXIDES: Diphosphorus tetraiodide. Lithium. 3-Methyl-2-selenoxo-1,3-benzothiazole. Phosphorus triiodide. Lithium. Potassium iodide–Zinc–Phosphorus(V) oxide. Samarium(II) iodide. Sodium O,O-diethyl phosphorotelluroate. Triphenylphosphine hydroiodide–Triphenylphosphine diiodide.

ETHERS: Chlorotrimethylsilane–Sodium iodide.

NITRONES: Hexachlorodisilane.

N-OXIDES: Diphosphorus tetraiodide. Hexachlorodisilane. Molybdenum(V) chloride–Zinc. Titanium(IV) chloride–Sodium borohydride.

PHENOLS: 5-Chloro-1-phenyltetrazole.

SULFINES: Phosphorus pentasulfide.

SULFONIC ACIDS: Iodotriphenylphosphonium iodide.

SULFOXIDES: Chlorotrimethylsilane–Zinc. Cyanuric fluoride. Phosphorus triiodide.

DESILYLATION: Potassium fluoride.

DESULFURATION: Molybdenum carbonyl. Sodium triethylborohydride–Iron(II) chloride. Tri-*n*-Butyltin hydride.

DETHIOKETALIZATION: Pyridinium bromide perbromide. Trimethyloxonium tetrafluoroborate.

vic-DIAMINATION: Cyclopentadienylnitrosocobalt dimer.

DIELS-ALDER CATALYSTS: Boron trifluoride. Boron trifluoride etherate. Ethylaluminum dichloride. Silica gel. Stannic chloride. Titanium(IV) chloride. Tris(*p*-bromophenyl)aminium hexachlorostibnate.

DIELS-ALDER REACTIONS: Bis(methylthio)-1,3-ketadiene. 1,3-Bis(trimethylsilyloxy)-1,3-butadiene. 2,5-Bis(trimethylsilyloxy)furanes. 5,5-Dimethoxy-1,2,3,4-tetrachlorocyclopentadiene. Diperoxo-oxohexamethylphosphoramidomolybdenum(VI). Ethynyl *p*-tolyl sulfone. Furane. (E)-1-Methoxy-1,3-

bis(trimethylsilyloxy)-1,3-butadiene.
1-Methoxy-1,3-butadiene. 1-Methyl-2-
trimethylsilyloxy-1,3-butadiene. Silver(I)
oxide. 2,3,5,6-Tetra-methylidene-7-
oxanorborane. 1,2,4-Triazine. 2-Tri-
methylsilyl-1,3-butadiene. Trimethylsilylvinylketene.
DIENOPHILES: Acrolein. Ethynyl p-tolyl
sulfone. 3-Methylsulfonyl-2,5-dihydro-
furane. 3-Nitro-2-cyclohexenone. Phenyl
vinyl sulfone.
DIHYDROXYLATION: Osmium tetroxide–
Dihydroquinine acetate. Osmium tetroxide–Diphenyl selenoxide. Osmium
tetroxide–Trimethylamine N-oxide–
Pyridine. Thallium(I) acetate–Iodine.
Triphenylmethylphosphonium permanganate.

ENE REACTIONS: Diethyl oxomalonate.
Ethylaluminum dichloride.
EPOXIDATION, ASYMMETRIC: (−)-
Benzylquininium chloride. t-Butyl
hydroperoxide–Vanadyl acetoacetate.
Hydrogen peroxide-1,1,3,3-Tetrachloro-
acetone. (S)-(2-Hydroxy-N,N-dimethyl-
propanamide-O,O′)oxodiperoxymolyb-
denum(VI).
ESCHENMOSER FRAGMENTATION:
N-Bromosuccinimide. Mesitylenesulfonyl
hydrazine.
ESTERIFICATION: Alumina. Dimethyl-
aminopyridine.
ETHYLIDENATION: Dicarbonylcyclo-
pentadienyliron ethyl(methylphenyl)sul-
fonium fluorosulfonate.

FISCHER INDOLE SYNTHESIS: Phosphorus trichloride. Polyphosphoric acid.
FLUORINATION: Acetyl hypofluorite.
Cesium fluoroxysulfate. (Diethylamino)-
sulfur trifluoride.
FORMYLATION: 2-(N-Methyl-N-formyl)-
aminopyridine. 1,3,5-Triazine. Trimethyl orthoformate.
FRIEDEL-CRAFTS REACTION: Copper(I)

trifluoromethanesulfonate. Iodotri-
methylsilane.
FRIEDLANDER QUINOLINE SYN-
THESIS: Cobalt(II) phthalocyanine.

GLYCOL CLEAVAGE: N-Iodosuccinimide.

HYDROALUMINATION: Dichlorobis-
(cyclopentadienyl)titanium. Dichloro-
bis(cyclopentadienyl)zirconium. Lithium
aluminum hydride–Dichlorobis(cyclopentadienyl)zirconium. Borane-Di-
methyl sulfide.
HYDROBORATION: Bis[3-methyl-2-
butyl]borane. 9-Borabicyclo[3.3.1]-
nonane. Borane-1,4-oxathiane. Thexylborane.
HYDROGENATION CATALYSTS: Car-
bonylchlorobis(triphenylphosphine)-
iridium. (1,5-Cyclooctadiene)(pyridine)-
(triphenylphosphine)iridium(I) hexa-
fluorophate. Palladium–Graphite.
Raney nickel. Ruthenium–Silica. Sodium
hydride–Nickel acetate–t-Amyl oxide.
HYDROMAGNESATION: Dichlorobis-
(cyclopentadienyl)titanium.
HYDROXYLATION: Benzeneseleninic
anhydride. Diperoxo-oxohexamethyl-
phosphoramidomolybdenum(VI).
Hydrogen peroxide. Iodosobenzene.
Oxygen. Ozone–Silica gel. Perchloryl
fluoride.

INVERSION, ALCOHOLS: Potassium
nitrite. Silver nitrate.
INVERSION, ALKENES: Tellurium chlo-
ride. Trifluoroacetyl chloride.
INVERSION, LACTONES: Triphenyl-
phosphine–Diethyl azodicarboxyl-
ate.
IODINATION: Iodine.
IODOLACTONIZATION: Thallium(I)
acetate–Iodine.
ISOMERIZATION OF ALLYLIC GROUPS:
Iron carbonyl. Palladium catalysts.

OXIDATIVE LACTONIZATION: Lead tetraacetate.

OXIDATIVE PHENOL COUPLING: Ferric chloride–Silica.

OXYAMINATION: Mercury(II) oxide–Tetrafluoroboric acid. Trioxo(*t*-butylimido)osmium(VIII).

OXY-COPE REARRANGEMENT: Grignard reagents. Potassium hydride.

OXYSELENATION: Copper(II) chloride.

PERMETHYLATION: Phase-transfer catalysts. Potassium methylsulfinylmethylide.

PHENYLSELENOETHERIFICATION: Benzeneselenenyl halides.

PICTET-SPENGER CYCLIZATION: Dimethyl(methylene)ammonium salts.

PROTECTION OF:

ALCOHOLS: Allyltrimethylsilane. (1,5-Cyclooctadiene)bis(methyldiphenylphosphine)iridium hexafluorophosphate. Guaiacylmethyl chloride. Levulinic acid. β-(Trimethylsilyl)ethoxymethyl chloride. 2-(Trimethylsilyl)ethyl chloroformate. Trimethyltrityloxysilane.

AMINES: Chlorotris(triphenylphosphine)rhodium(I). 1,4-Dichloro-1,1,4,4-tetramethyldisilethylene.

AMINO ACIDS: 2-*t*-Butoxycarbonyloxyimino-2-phenylacetonitrile. Palladium catalysts. Thioanisole–Trifluoroacetic acid.

CARBOXYLIC ACIDS: Allyltrimethylsilane. Bistrimethylsilyl ether. Chlorotrimethylsilane. Hexamethyldisiloxane.

PHENOLS: Zinc.

THIOLS: *o*-Nitrobenzenesulfenyl chloride.

REARRANGEMENT OF EPOXY ALCOHOLS: Titanium(IV) isopropoxide.

REDUCTION, REAGENTS:

AMMONIA–BORANE: μ-Bis(cyanotrihydroborato)tetrakis(triphenylphos-phine)-dicopper. Bis(triphenylphosphine)copper(I) tetrahydroborate. Borane–Dimethyl-sulfide. Chromium(III) chloride–Lithium aluminum hydride. Cobalt(II) phthalocyanine. Diisobutylaluminum hydride. 2,2′-Dihydroxy-1,1′-binaphthyl–Lithium aluminum hydride. Hydriodic acid. Hydrogen telluride. Iodotriphenylphosphonium iodide. Lithium–Ammonia. Lithium aluminum hydride. Lithium aluminum hydride–Copper(I) iodide. Lithium aluminum hydride–N-Methylephedrine. Lithium *t*-butylbis(2-methylpropyl)aluminate. Lithium tri-*t*-butoxyaluminum hydride. Lithium tri-*sec*-butylborohydride. Lithium triethylborohydride. Magnesium–Methanol. Molybdenum(V) chloride–Zinc. Phosphorus triiodide. β-(3)-α-Pinanyl-9-borabicyclo-[3.3.1]nonane. Potassium graphite. Raney nickel. Samarium(II) iodide. Sodium–Ammonia. Sodium bis(2-methoxy-ethoxy)aluminum hydride. Sodium borohydride. Sodium borohydride–Cadmium chloride. Sodium borohydride–Cerium(III) chloride. Sodium borohydride–Cobalt(II) chloride. Sodium borohydride–Pyridine. Sodium cyanoborohydride. Sodium dithionite. Sodium hydrogen telluride. Sodium iodide–Hydrochloric acid. Sodium tris(3,5-dimethylphenoxy)borohydride. Tetra-*n*-Butylammonium borohydride. Titanium(III) chloride. Titanium(IV) chloride–Sodium borohydride. Tri-*n*-Butyltin hydride. Triphenyltin hydride. Zinc–Copper couple. Zinc borohydride.

REDUCTIVE ALKYLATION: Lithium–Ammonia.

REDUCTIVE AMINATION: Sodium cyanoborohydride.

REDUCTIVE COUPLING: Titanium(III) chloride–Lithium aluminum hydride.

REDUCTIVE CYANATION: Tosylmethyl isocyanide.

peroxide–Nickel bromide. [(Trimethyl-silyl)allyl] lithium.

2-CARBAPENAMS: Rhodium(II) car-boxylates. Sodium dicarbonyl-cyclo-pentadienylferrate.

CARBODIIMIDE: Bromotriphenyl-phosphonium bromide.

CARBOXYLIC ACIDS, ARYL: Palladi-um(II) acetate.

α-CHLORO CARBOXYLIC ACIDS: Chlorine.

CHLOROHYDRINS: Chloromine-T. Hypochlorous acid. Pyridinium hydro-chloride.

α-CHLORO KETONES: Copper(II) chloride.

4-CHROMANONES: 4-Dimethylamino-pyridine.

CINNAMYL ALCOHOLS: Selenium dioxide.

CYANOHYDRINS: Acetyl cyanide.

β-CYANO KETONES: Cyanotrimethyl-silane.

CYCLIC CARBONATES: Molybden-um(V) chloride–Triphenylphosphine.

CYCLOALKENONES: Trimethyl ortho-formate.

CYCLOBUTANONES: Tosylmethyl isocyanide.

CYCLOBUTENES: Ketene diethyl acetal. α-Lithiomethylselenocyclobutene.

CYCLOHEXANONES: Disodium tetra-carbonylferrate.

2-CYCLOHEXENOLS: Organocuprates.

CYCLOHEXENONES: 2,2-Dimethyl-3(2H)-furanone. Methanesulfonic acid. 2,2,6-Trimethyl-1,3-dioxolenone.

CYCLOPENTADIENONES: Dicarbonyl-cyclopentadienylcobalt.

3-CYCLOPENTANOLS: 2-(Chloro-ethoxy)carbene.

CYCLOPENTANONES: Bis[1,2-bis(di-phenylphosphino)ethane] palladium. Chlorotris(triphenylphosphine)rhodi-um(I). Dichloroketene. Disodium tetracarbonyl ferrate.

CYCLOPENTENONES: Boron trifluoride etherate. Copper(I) trifluoromethyl-sulfonate. 5,5-Dimethoxy-1,2,3,4-tetra-chlorocyclopentadiene. 1-(Dimethyl-amino)-3-pentenotrile. Methanesulfonic acid. Phosphoric acid–Formic acid. Tetrakis(triphenylphosphine)palladium.

CYCLOPROPANES: Iron carbonyl. Molybdenum carbonyl. Tetra-μ^3-car-bonyldodecacarbonylhexarhodium.

CYCLOPROPYL ALDEHYDES: Diazo-acetaldehyde dimethyl acetal.

DECALINS, trans: Methyl cyclobutene-carboxylate.

DEHYDROPEPTIDES: 2,3-Dichloro-5,6-dicyano-1,4-benzoquinone.

2,3-DIALKYL-1,3-CYCLOHEXA-DIENES: Butane-1,4-bis(triphenyl-phosphonium) dibromide.

1,2-DIAMINES: Cyclopentadienylnitrosyl-cobalt dimes.

DIARYLALKYNES: Alumina.

DIARYL ETHERS: Tetraphenylbismuth trifluoroacetate.

α-DIAZO KETONES: 2,4,6-Triisopropyl-benzenesulfonyl azide.

1,3-DIENES: Copper(I) bromide–Di-methyl sulfide. Grignard reagents. Iron carbonyl. α-Lithiomethylselenocyclo-butane. Organocopper reagents. Palla-dium(II) acetate. Tetrakis(triphenyl-phosphine)palladium.

1,4-DIENES: Copper(I) bromide–Di-methyl sulfide. Tetrakis(triphenyl-phosphine)palladium.

1,5-DIENES: Benzylchlorobis(triphenyl-phosphine)palladium(II). Copper(I) bromide–Dimethyl sulfide. Palladium(II) chloride. Tetrakis(triphenyl-phosphine)-palladium. Titanium(IV)chloride–Lithium aluminum hydride.

2,4-DIENONES: n-Butyllithium.

gem-DIFLUOROALKENES: Triphenyl-phosphine–Dibromodifluoromethane.

1,2-DIKETONES: 2-Lithio-1,3-dithianes. Ozone.

1,3-DIKETONES: Disodium tetrachloro-palladate-*t*-Butyl hydroperoxide. 1-Lithio-2,4,6-trimethylbenzene. Tetra-kis(triphenylphosphine)palladium. Zinc–Copper–Isopropyl iodide.

1,4-DIKETONES: 1-Benzyl-4-(2-hydroxy-ethyl)-5-methyl-1,3-thiazolium chloride. Copper(II) trifluoromethanesulfonate.

1,5-DIKETONES: Allyltrimethylsilanes. Titanium(IV) chloride.

1,3-DIOLS: Tri-*n*-butylborane.

DIOSPHENOLS: Sodium methoxide–Dimethyl sulfoxide.

DISACCHARIDES: Iodonium di-*sym*-collidine perchlorate.

DISULFIDES: Potassium iodide–Boron triiodide.

ENDOPEROXIDES: Hydrogen per-oxide.

ENOL CARBONATES AND CARBA-MATES: Benzyltrimethylammonium fluoride.

ENOL ETHERS: Dimethyl diazomethyl-phosphonate. Grignard reagents.

1,3-ENYNES: N,N-Methylphenylamino-(tri-*n*-butylphosphonium)iodide.

1,5-ENYNES: Di-*μ*-carbonylhexacarbonyl-dicobalt. Tetrakis(triphenyl-phosphine)-palladium.

EPOXIDES: Dimethylsulfoxonium methylide. Hydrogen peroxide. Hydro-gen peroxide–Triethyl orthoacetate. Permonophosphoric acid. Silica gel. Trifluoromethanesulfonic anhydride. Triphenylarsonium ethylide.

α,β-EPOXY ALCOHOLS: Vanadyl acetyl-acetonate–Azobisisobutyronitrile.

α,β-EPOXY ALDEHYDES: L-Proline.

ESTERS: N,N-Bis(2-oxo-3-oxazolidinyl)-phosphordiamidic chloride. Copper(II) acetate. Di-*n*-butyltin oxide.

ETHERS: *m*-Chloroperbenzoic acid. Thallium(I) ethoxide. Triethylsilane–Boron trifluoride.

ETHYL 2,4-DIENOATES: Ethyl 2-phenylsulfinylacetate.

α-FLUORO CARBOXYLIC ACIDS: Trifluoromethyl hypofluorite.

FULVENES: Crown ethers.

FURANES: Dimethylformamide dimethyl acetal. Lithiomethyl isocyanide. 1-Nitro-1-(phenylthio)propene. Organo-aluminum compounds.

FURANONES: Phase-transfer catalysts.

α-GLUCOSIDES: Silver imidazolate.

GLYCIDIC ESTERS: Dimethylsulfonium ethoxy carbonylmethylide.

GLYCOSIDES: Bromotrimethylsilane–Cobalt(II) bromide. Iodonium di-*sym*-collidine perchlorate. Mercury(II) nitrate. Silver imidazolate. Stannous chloride–Silver perchlorate.

HOMOALLYLIC ALCOHOLS: Allyltin difluoroiodide. Dichlorobis(cyclo-penta-dienyl)titanium. Ethylaluminum di-chloride. Stannous fluoride. Trimethyl-aluminum–Dichlorobis(cyclopenta-dienyl)zirconium.

HOMOALLYLIC ETHERS: Iodotri-methylsilane. Trimethylsilyltrifluoro-methanesulfonate.

HOMOPROPARGYLIC ALCOHOLS: Trimethylsilylallene.

α-HYDROXY AMIDES: Cyanotrimethyl-silane.

α-HYDROXY AMINES: Iodonium di-*sym*-collidine perchlorate.

γ-HYDROXYBUTENOLIDES: 2,5-Bis-(trimethylsilyloxy)furanes.

β-HYDROXY CARBOXYLIC ACIDS: Dihydropyrane.

α-HYDROXY CARBOXYLIC ESTERS: *m*-Chloroperbenzoic acid.

β-HYDROXY CARBOXYLIC ESTERS: (R)-(+)-*p*-Tolsulfinylacetic acid.

2-HYDROXYCYCLOPENTENEONES: 1-Benzyl-4-(2-hydroxyethyl)-5-methyl-1,3-thiazolium chloride.

3-HYDROXYHOMOPHTHALATES: 1,3-Bis(trimethylsilyloxy)-1-methoxy-1,3-butadiene.

α-HYDROXY KETONES: Diethyl 1-phenyl-1-trimethylsilyloxymethane-phosphonate. Iodosobenzene. Osmium tetroxide–N-Methylmorpholine N-oxide.

β-HYDROXY KETONES: Triethylborane.

2-HYDROXY-1,4-QUINONES: Silver(II) oxide.

β-HYDROXY SELENIDES: Benzeneselenenyl halides.

IMIDAZOLES: Potassium cyanide.

INDANES: Phosphorus(V) oxide.

INDOLES: n-Butyllithium. Phosphorus trichloride.

α-IODO KETONES: Iodine–Copper(II) acetate. Iodotrimethylsilane.

β-IODO KETONES: Iodotrimethylsilane.

α-IODO-α,β-UNSATURATED ALDEHYDES: Pyridinium dichromate.

IRIDOIDS: Diphenyl disulfide.

ISOCYANATES: Azidotrimethylsilane.

ISOFLAVONES: 1,3,5-Triazine.

ISOQUINOLINES: Thallium(III) trifluoroacetate.

KETENE S, S-ACETALS: 1,3-Dithiolan-2-yltriphenylphosphonium tetrafluoroborate.

α-KETO ACIDS: Bis(ethylthio)acetic acid.

γ-KETO ALDEHYDES: Allyltrimethylsilane. (S)-(+)-p-Tolyl-p-tolylthiomethyl sulfoxide.

β-KETO ESTERS: Disodium tetrachloropallate–t-Butyl hydroperoxide. Ethyl malonate. Meldrum's acid.

γ-KETO ESTERS: Palladium(II) chloride. Triethylammonium hydrofluoride.

KETONES: Bis(phenylthio)methane. Chlorothexylborane. Grignard reagents. Nickel chloride–Zinc. o-Nitrophenyl selenocyanate. Organomagnesium(II) iodides. Phenylthiophenyl(trimethylsilyl)methane.

KETONES, DIARYL: Bis(cyclooctadiene)nickel.

KETONES, MACROCYCLIC: 1,3-Bis(dimethylphosphono)-2-propanone.

γ-KETO-α,β-UNSATURATED ESTERS: Chromic anhydride.

LACTAMS: Di-n-butyltin oxide. Hydroxylamine-O-sulfonic acid. Iodine azide. Sodium cyanoborohydride.

β-LACTAMS: Cyanuric chloride. Grignard reagents. Ion-exchange resins. Lithium phenylethynolate. Sodium dicarbonylcyclopentadienylferrate. Titanium(III) chloride. Titanium(IV) chloride. Triphenylphosphine–Carbon tetrachloride. Triphenylphosphine–Diethyl azodicarboxylate. Triphenylphosphine–2,2'-Dipyridyl disulfide.

LACTONES: Bis(3-dimethylaminopropyl)phenylphosphine. Copper(I) trifluoromethanesulfonate. Di-n-butyltin oxide. Dichlorobis(triphenyl-phosphine)-palladium. Ruthenium tetroxide.

α-LACTONES: Iodosobenzene.

γ-LACTONES: Trimethylsilylacetonitrile. (Trimethylsilyl)allyllithium.

δ-LACTONES: 1-Methoxy-1,3-butadiene. (R)-(+)-p-Tolylsulfinylacetic acid.

MACROLIDES: Copper(II) acetate–Iron(II) sulfate. Copper(I) trifluoromethylsulfonate. Cyanuric chloride. Di-n-butyltin oxide. N-Methyl-N-phenylaminoethynyllithium. Oxygen, singlet.

α-METHOXY ALDEHYDES, ChiraL: (S)-(+)-p-Tolyl p-tolylthiomethyl sulfoxide.

METHOXYCYCLOPROPANES: Methoxyallene.

2-METHYL-1-ALKENES: Organocuprates.

α-METHYLAMINO ACIDS: (3S,6S)-(+)-2,5-Dimethoxy-3,6-dimethyl-3,6-dihydropyrazine.

METHYL ARYLACETATES: Lead tetraacetate.

METHYL 2,4-DIENOATES: Methyl (allylthio)acetate.

α-METHYLENE-γ-BUTYROLACTONES:
Bis(dimethylamino)methoxymethane.
Boron trifluoride etherate. Bromo-
methyl methyl ether.

METHYLENE CYCLOALKANES: 9-
Borabicyclo[3.3.1]nonane.

α-METHYLENELACTONES: Dicarbonyl-
bis(triphenylphosphine)nickel.

α-METHYLENE-δ-LACTONES: Sodium
selenophenolate.

METHYL KETONES: Allyltrimethylsilane. Bis(acetonitrilo)chloronitro-
palladium(II). Dicarbonylbis(triphenylphosphine)nickel. Dichloro-dicyano-
benzoquinone. Hydrogen peroxide–
Palladium acetate. Meldrum's acid.
Palladium t-butyl peroxide trifluoro-
acctate. Palladium(II) chloride.

α-METHYLSELENO KETONES,
ESTERS: Selenium.

NAPHTHALENES: N,N-Dimethylforma-
mide dimethyl acetal.

NAPHTHOQUINONES: Iron carbonyl.

NITRILES: S,S-Diphenylsulfilimine.
Potassium cyanide. Potassium(III)
iodide. Tosylmethyl isocyanide.
Vilsmeier reagent.

1-NITROALKENES: Benzene selenenyl
halides. Tetranitromethane.

NITROCYCLOALKANES: Hypochlorous
acid.

NUCLEOSIDES: Trimethylsilyl trifluoromethanesulfonate.

OXAZOLES: Copper(I) oxide. Lithiomethyl isocyanide.

PEPTIDES: Bis(o-nitrophenyl)phenyl-
phosphonate. Dicyclohexylcarbodiimide. Diphenyl N-succinimidyl phosphate. N-Methyl-N-phenylbenzohydrazonyl bromide. n-Propylphosphonic
anhydride. Trifluoromethanesulfonic
acid–Thioanisole.

PHENOLS: Methyl vinyl ketone.

PHENYL SELENIDES: Lithium benzeneselenolate.

α-PHENYLTHIOALDEHYDES: Methoxy-

phenylthiomethyllithium.

PIPERIDINES: Benzeneselenenyl halides.

POLYNUCLEODIDES: Chloro-N,N-
dimethylaminomethoxyphosphine.

PYRIDINES: 1,2,4-Triazine.

PYRROLES: Tetrakis(triphenylphosphine)palladium.

PYRROLIDINES: Benzeneselenenyl
halides. Sodium cyanoborohydride.

PYRROLIZIDINES: Carboethoxycyclopropyltriphenylphosphonium tetrafluoroborate. Trimethylmethyl trifluoromethanesulfonate.

PYRYLIUM SALTS: Sulfoacetic acid.

o-QUINODIMETHANES: Cesium fluoride. 1,3-Dihydrobenzo[c]thiophene
2,2-dioxide.

p-QUINODIMETHANES: Tetra-n-butylammonium fluoride.

QUINOLINES: Cobalt(II) phthocyanine.

QUINONE MONOACETALS: Thalli-
um(III) nitrate.

o-QUINONES: Selenium dioxide.

SELENIDES: N-Phenylselenophthalimide. Sodium pyridylselenate.

SILYL ENOL ETHERS: Chlorotrimethylsilane–Sodium iodide. Organocuprates.

SPIROLACTONES: Allyltrimethylsilylzinc chloride. Grignard reagents.

STEROID SIDE CHAIN: 9-Borabicyclo-
[3.3.1]nonane. Ethylaluminum dichloride.

SULFILIMINES: Dimethyl sulfoxide.

SULFONES: Hydrogen peroxide–
Tungstic acid.

SULFONES, DIARYL: Sulfuric acid–
Trifluoroacetic anhydride.

SULFOXIMINES: Dimethyl sulfoxide.

TETRAHYDROFURANES: Ethylene
glycol–p-Toluenesulfonic acid.

TETRAHYDROTHIOPHENES: Potassium ethanethiolate.

TETRALINS: Phosphorus(V) oxide.

TETRALONES: Dimethylsulfonium
ethoxycarbonylmethyldie. Stannic
chloride.

TETRAZENES: Benzeneseleninic acid.

THIOAMIDES: Diphenylphosphinodithioic acid.

1-THIOGLYCOSIDES: Trimethyl(phenylthio)silane.

THIONOLACTONES: Triethyloxonium tetrafluoroborate.

TRITYL ETHERS: 4-Dimethylaminopyridine.

α,β-UNSATURATED ALDEHYDES: Benzeneselenenyl halides. Benzenesulfenyl chloride. Benzo-1,3-dithiolium tetrafluoroborate. Ethyl vinyl ether. Grignard reagents.

α,β-UNSATURATED ESTERS: O-Methyl C,O-bis(trimethylsilyl)ketene acetal. Organocopper reagents.

γ,δ-UNSATURATED-β-KETO ACIDS: Ethyl 4-diethoxyphosphinoyl-3-oxobutanoate.

α,β-UNSATURATED KETONES: n-Butyllithium. Cesium fluoride. m-Chloroperbenzoic acid. Lithium diisopropylamide. Nickel carbonyl. Propargyl bromide.

β,γ-UNSATURATED KETONES: Bis-(1,5-cyclooctadiene)nickel(0). Dimethyl-(phenylthio)aluminum. α-Phenylglycine methyl ester. Phenylselenoacetaldehyde. Zinc–Silver couple.

α,β-UNSATURATED γ-LACTONES: (Triphenylphosphorylidene)ketene.

α,β-UNSATURATED NITRILES: Cyanotrimethylsilane. Phenyl cyanate. Trimethylsilylacetonitrile.

VINYL CHLORIDES: Grignard reagents.

VINYL ETHERS: μ-Chlorobis(η-2,4-cyclopentadien-1-yl)(dimethylaluminum)-μ-methylenetitanium.

VINYL IODIDES: Iodo-monochloride.

VINYL SILANES: Chlorotrimethyl-silane. Dimethylphenylsilyllithium.

VINYL SULFIDES: Iodotrimethylsilane.

VINYL SULFONES: Phenyl p-tolueneselenosulfonate.

VINYL TRIFLATES: Di-t-butyl-4-methylpyridine.

o-XYLYLENES: Cesium fluoride. Tetra-n-butylammonium fluoride.

TELLUROXIDE ELIMINATION: Chloramine-T.

THALLATION–CARBONYLATION: Palladium(II) chloride.

THIONATION: 2,4-Bis(4-methoxyphenyl)-1,3,2,4-dithiadiphosphetane-2,4-disulfide.

TRANSSILYLATION: t-Butyldimethylchlorosilane.

ULLMANN REACTION: trans-Bromo-o-tolylbis(triethylphosphine)nickel.

UREIDOALKYLATION: Titanium(IV) chloride.

WACKER OXIDATION: Allyltrimethylsilane. Bis(acetonitrile)chloronitropalladium(II). Palladium(II) chloride. Triethylborane.

WILLGERODT-KINDLER REACTION: Lead tetraacetate.

WILLIAMSON REACTION: Thallium(I) ethoxide.

WITTIG-HORNER REACTION: Alkyl diphenyl phosphonates. 1,3-Bis(dimethyl-phosphono)-2-propanone. Crown ethers. Dimethyl diazomethylphosphonate. Tetra-n-hexylammonium bromide.

WITTIG REACTIONS: Methylenetriphenylphosphorane. Potassium hexamethyldisilazide.

AUTHOR INDEX

Shishiyama, Y., 114
Shoda, S., 6, 374
Shono, T., 310, 460
Siddall, G. B., 421
Siebert, W., 454
Sieburth, S. M., 328
Sikowski, J. A., 95
Sillion, B., 85
Silveira, A., Jr., 243
Simakhina, N. D., 419
Simchen, G., 1, 96
Simon, R. M., 146
Sims, R. J., 112
Sinay, P., 21
Sinclair, J. A., 127
Singh, A., 21, 452
Singh, J., 443
Singh, K., 26
Singh, R., 26
Singh, R. P., 346
Singleton, D. M., 140
Sipio, W. J., 21
Skeludyakov, V. D., 432
Skulnick, H. I., 182
Slade, J. S., 289
Sletzinger, M., 252, 449
Smadja, W., 343
Smart, J. C., 38
Smith, A. B., III, 55, 56, 265, 289, 343
  356
Smith, D. J. H., 62
Smith, J. C., 50
Smith, K. M., 52
Smith, L. W., 330
Snider, B. B., 11, 180, 181, 187,
  216
Snieckus, V., 75
Sohn, J. E., 133
Solladié, G., 407
Somei, M., 400
Sommer, H. Z., 419
Sondengam, B. L., 459
Sonnet, P. E., 332, 419, 450
Sonoda, N., 206, 276, 312
Sonogashira, K., 117
Sorenson, C. M., 5

Sorgi, K. L., 66
Sorrell, T. N., 175, 378
Souchi, T., 36
South, M. S., 223
Souther, S. K., 450
Sozzani, P., 324
Spindell, D. K., 11, 180
Springer, J. P., 330
Spry, D. O.,
Sreckumar, C., 327, 445
Srogl, J., 188
Stahl, I., 426
Stahle, M., 186
Stahnke, M., 216
Stammer, C. H., 136
Stang, P. J., 123, 289
Stanovnik, B., 274
Stark, C. I., 204
Stathom, A. A., 126
Stavber, S., 85
Steckhan, E., 101
Steglich, W., 151, 308
Steinbach, R., 138, 271
Steinbeck, K., 365
Steliou, K., 125
Stephan, W., 276
Stephenson, L. M., 144
Stephenson, T. A., 303
Sternberg, E. D., 127
Stetter, H., 27
Stevens, R. V., 365
Stewart, P., 25
Stewart, R. F., 63
Stibor, I., 188
Still, I. W. J., 320
Still, W. C., 194, 327, 398, 445
Stille, J. K., 26, 133
Stojanac, Z., 55
Stoll, M., 150
Stoodley, R. J., 260
Stork, G., 236, 443
Stott, P. E., 112
Stout, E. J., 39
Stowell, J. C., 201, 221
Strickand, S. M. S., 443
Stringham, R. A., 413

SUBJECT INDEX